AF775495

Springer
Tokyo
Berlin
Heidelberg
New York
Barcelona
Hong Kong
London
Milan
Paris
Singapore

Y.H. Mori, K. Ohnishi (Eds.)

Energy and Environment

Technological Challenges for the Future

With 178 Figures, Including 38 in Color

Springer

Yasuhiko H. Mori
Professor
Department of Mechanical Engineering
Keio University
3-14-1 Hiyoshi, Kohoku-ku, Yokohama 223-8522, Japan

Kouhei Ohnishi
Professor
Department of System Design Engineering
Keio University
3-14-1 Hiyoshi, Kohoku-ku, Yokohama 223-8522, Japan

ISBN 978-4-431-68327-8

Library of Congress Cataloging-in-Publication Data

Energy and environment : technological challenges for the future / Y.H. Mori, K. Ohnishi (eds.).
p. cm.
Includes bibliographical references and index.
ISBN 978-4-431-68327-8 ISBN 978-4-431-68325-4 (eBook)
DOI 10.1007/978-4-431-68325-4
1. Energy development—Environmental aspects—Congresses. 2. Energy development—Technological innovations—Congresses. I. Mori, Y.H. (Yasuhiko H.), 1947– II. Ohnishi, K. (Kouhei), 1952–
TD195.E49 E5112 2000
333.79′14—dc21 00-061237

Printed on acid-free paper

© Springer-Verlag Tokyo 2001
Softcover reprint of the hardcover 1st edition 2001
This work is subject to copyright. All rights are reserved, whether the whole or part of the material is concerned, specifically the rights of translation, reprinting, reuse of illustrations, recitation, broadcasting, reproduction on microfilms or in other ways, and storage in data banks.
The use of registered names, trademarks, etc. in this publication does not imply, even in the absence of a specific statement, that such names are exempt from the relevant protective laws and regulations and therefore free for general use.

SPIN: 10773948

Preface

This book is a compilation of papers presented at the International Symposium "Energy and Environment: Technological Challenges for the Future," held in Yokohama and Hakone November 25–27, 1999. The Symposium was the concluding event of the 5-year TEPCO Chairship Program which was established in the Faculty of Science and Technology, Keio University, Yokohama, Japan, in 1994 with a grant from the Tokyo Electric Power Company (TEPCO). Thirteen internationally recognized experts who are actively involved in various aspects of energy and environmental research were invited to the Symposium as guest speakers. Most of them were researchers who had been invited to Keio University either as TEPCO Chair Visiting Professors to conduct courses in the Graduate School of Science and Technology of Keio University or as lecturers for ad hoc TEPCO–Keio Seminars in the Faculty of Science and Technology, Keio University.

Each of the papers contributed by these guest speakers presents an overview, a state-of-the-art review, and/or a perspective on a subject of each speaker's research interest related to a significant aspect of energy and environmental issues. Thus, the book is definitely more than just a guidebook to superficial trends in technology: It is a compilation of the world's top-rank investigations of a wide spectrum of current and future technologies expected to contribute to solving the energy and environmental problems of the next century.

In addition to the technological papers in Parts I–V, the book contains a unique sixth section. Part VI is unique, we believe, because it deals with a subject beyond the confines of *technology*, namely, education—at home, in school, and through the media—about energy and environmental concerns. Part VI is a record of the Panel Discussion held November 27, 1999, in Hakone as part of the Symposium. The Panel Discussion was chaired by Dr. Azusa Tomiura, a member of Science Council of Japan, and was attended by 18 invited guests, the members of the Organizing Committee of the Symposium, and three students from the Graduate School of Science and Technology, Keio University. Compiled in Part VI are the chair's keynote message to the participants in the Panel Discussion, talks by the five panelists, and comments and discussion by the participants.

As the editors of this book and also as the chairs of the Symposium, we greatly thank those who presented their papers in the Symposium and then contributed them to the book. We are also most grateful to Dr. Azusa Tomiura for chairing the Panel Discussion and organizing the resulting documents for publication here. We owe the successful management of the Symposium and that of the subsequent publication of this book to the staff of the Office of Research Administration at Yagami, Keio University, particularly to Mr. Hiroshi Watanabe and Ms. Yukari Mitsuhashi. Finally, we express our heart-felt gratitude to the Tokyo Electric Power Company for their endowment of the chairship to Keio University. We are pleased and proud that the conclusion of the chairship program is marked by the publication of this book.

Yasuhiko H. Mori
Kouhei Ohnishi

Panel Discussion Participants, Hakone Prince Hotel, November 27, 1999

Contents

Organizing Committee of the Symposium

Koichi Hishida, Dept. of System Design Engineering, Keio University

Masanobu Maeda, Dept. of System Design Engineering, Keio University

Yasuhiko H. Mori, Dept. of Mechanical Engineering, Keio University, Symposium Chair

Toshiyuki Murakami, Dept. of System Design Engineering, Keio University

Yuji Nagasaka, Dept. of System Design Engineering, Keio University

Akira Nagashima, Executive Vice-President, Keio University

Kouhei Ohnishi, Dept. of System Design Engineering, Keio University, Symposium Co-Chair

Koichiro Sawa, Dept. of System Design Engineering, Keio University

Naotatsu Shikazono, Dept. of Applied Chemistry, Keio University

Shigeru Tanaka, Dept. of Applied Chemistry, Keio University

Kenji Yasuoka, Dept. of Mechanical Engineering, Keio University

Part I: Production, Transportation and Utilization of Electric Power

Restructuring of the Electric Power Industry in Korea

Jong-Keun Park

Key words. Restructuring—Cost-based pool—CBP—Generation pool—PX

Introduction

In a competitive power market structure environment, markets should be designed to meet local requirements and evolve the plans for transition, training, and change mechanisms. System reliability and market functionality could be sustained by close connection between markets and Energy Management System (EMS).

In this chapter, we introduce the restructuring plan of Korea and explain the detailed organization of the planned market.

Current Electric Industry Structure in Korea

The current electric industry structure in Korea is a vertically integrated system by Korea Electric Power Company (KEPCO). Demand increases by 10% annually, and the total capacity is 43.261 million kW; this figure is the 17th largest in the world. KEPCO has 94.2% of the market share. Plant capacity by plant type and market share are shown in Table 1 and Table 2, respectively.

Table 1. Plant capacity by plan

Type/year	1998	2000	2005	2010	2015
Nuclear	1202	1372	1772	2443	2895
Coal	1140	1410	1902	2130	2172
Liquefied natural gas (LNG)	1222	1344	1735	1845	2025
Petroleum	520	514	560	533	473
Hydraulic	312	387	490	691	691
Total	4396	5027	6459	7642	8256

Department of Electrical Engineering, Seoul National University, San 56-1, Shinrim-Dong, Kwanak-ku, Seoul 151–742, Korea

TABLE 2. Market share of Electric Power Co.

Company	KEPCO	Hanwha	Hydraulic	Total
Capacity (MW)	40760	1500	1001	43261
Share (%)	94.2	3.5	2.3	100

The Restructure Plan

The objectives of restructuring are listed below:

Introduce competition and increase efficiency
Ensure long-term viability of electricity supply
Promote consumer convenience and choice

The restructuring plan contemplates a 10-year implementation period and will progress by four stages. Details are listed below.

Stage 1: December 1999
Stage 2: 2000–2002
Generation competition
Stage 3: 2003–2009
Wholesale competition
Stage 4: 2009–
Retail sale competition

Cost-Based Generation Pool

The objectives of the cost-based generation pool (CBP) are to avoid the market manipulation of the dominating Generation Company (GenCo). and to avoid the obligations in Public Power Association (PPA). CBPs are intermediate instruments to introduce competition in the generation market and are planned to avoid the risks of strong market power and obligation. The principles of CBP are listed below:

System marginal price (SMP) determination in day-ahead market
Each Gensets submits price data by season
Submitted price data be confirmed by each participants
Each Gensets submit availability by hour

Advantages in CBP are that one can avoid risks from strong market power and that it is easy to go to the next competition model without any obligation preventing competition. Disadvantages in CBP are that there is no room for electric supply industry (ESI) business strategy and that there is little incentive to improve management efficiency.

Basic Features of CBP

Pool purchase price (PPP) is determined as system marginal price (SMP) plus capacity payment (CP). SMP is the available cost of hourly marginal Genset. CP is the hourly value of open cycle gas turbine (OCGT) fixed cost.

TABLE 3. Payment to Gensets

Constrained \ Unconstrained	ON	OFF
ON	SMP, CP	Bid, CP
OFF	SMP-Bid, CP	None, CP

SMP, system marginal price; CP, capacity payment

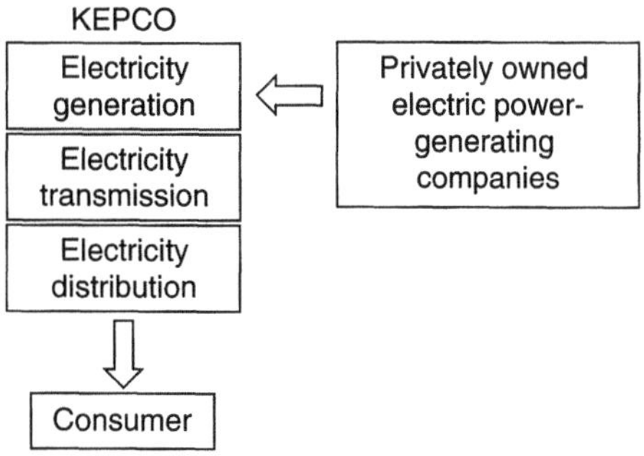

FIG. 1. Stage 1

The system operator makes an unconstrained schedule. When it is calculated, price and availability of each Genset and technical data of each Genset are considered. It determines hourly SMP and generation of each Genset. The system operator also makes a constrained schedule. When it is calculated, price and availability of each Genset, technical data of each Genset, transmission constraint, and fuel constraint including liquefied natural gas (LNG), anthracite coal, and Co-Heat producing (CHP) are considered. It determines the practical hourly generation of each Genset.

The purpose of CP is to compensate the fixed cost of the marginal Genset and to give an appropriate price signal to the market by applying long-run marginal cost. Selection of marginal Genset is based on OCGT.

Actual payment to Gensets is shown in Table 3.

Stage 1: 1999

In Stage 1, KEPCO's generation assets and liabilities will be transferred into new-generation subsidiaries. KEPCO's nonnuclear generating capacity will be separated into five wholly owned generation subsidiaries. Nuclear generation capacity will be separated into one wholly owned generation subsidiary. KEPCO will retain its monopoly position with respect to transmission and distribution/supply.

Stage 2: 2000–2002: Generation Competition

In this stage, generation competition will be implemented. The generation sector will be divided into five nonnuclear companies for sale and one nuclear one. At the beginning,

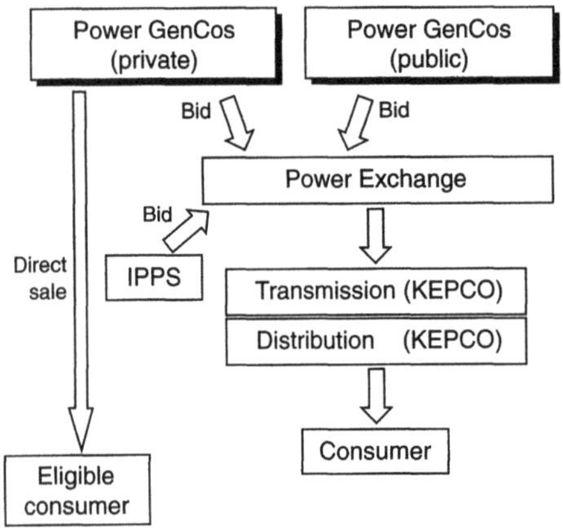

FIG. 2. Stage 2: Generation competition. *IPP*, Independent Power Producer; *GenCos*, Generation Company

a cost-based generation pool will be used but eventually this will be changed to a price-bid generation pool. In this system, vertically integrated Transmission Company (TransCo), Distribution Company (DisCo), Supplier, and Independent System & Market Operator are not biddable and only GenCos can bid to Pool.

In the generation sector, GenCos are divided by fuel type, remaining life, past performance, and present value of the future cash flow division criteria; therefore, a "level playing field" is created. Technical criteria such as geographic location, allocation of base, mid, and peak load units, and allocation of generators for ancillary service could be applied.

Time schedules to introduce the cost-based generation pool are as follows:

October 31: Complete the documents relevant to pool operation
November 15: Complete the equipment installation relevant to pool operation
November 16: Begin test run
January 2000: Actual operation

Stage 3: 2003–2009: Wholesale Competition

In this stage, distribution/supply subsidiaries will be separated. New distribution/supply subsidiaries will initially operate only in specific geographic areas. Several GenCos will compete in the Generation Pool, and several DisCos will purchase electricity from the Pool.

Two-way bidding will be implemented. GenCos will compete in the Pool by bidding with price and availability, and Suppliers will also bid with quantity only or quantity and price. Independent System and Market Operator will be present.

Stage 4: 2009–: Retail Sale Competition

In this stage, full competition will be implemented. Consumers will be able to choose the power source from any distribution/supply company or broker on market terms. KEPCO will continue to act as the sole transmission company.

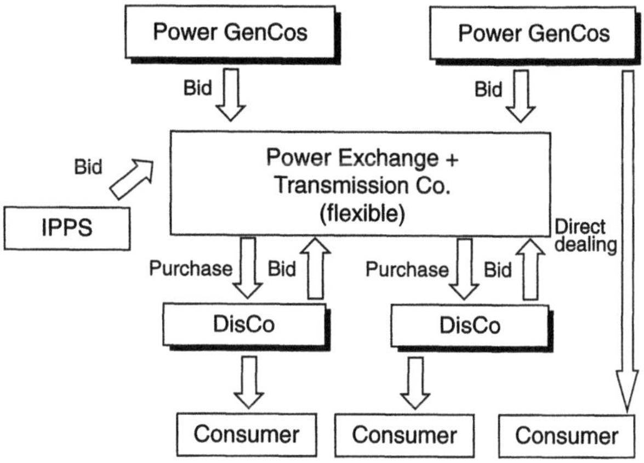

FIG. 3. Stage 3: Wholesale competition. *DisCo*, Distribution Company

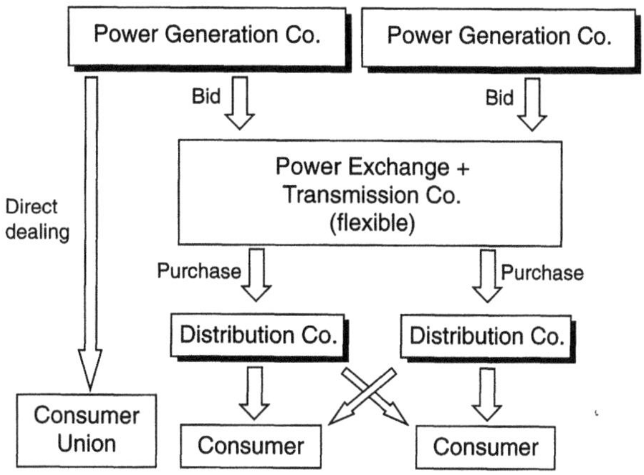

FIG. 4. Stage 4: Retail sale competition

Korean Power Exchange

The tasks of the Korean Power Exchange are listed below:

Pool Governance
Market Operator (MO)
System Operator (SO)

It is planned to open January 2000.
The tasks of Pool Governance are listed below:

Market Opening and Operation
Market Rule Enactment and Revision
Market Monitoring
Dispute Resolution

The tasks of Market Operator (MO) are listed below:

Market Price (SMP) calculation
Metering
Settlement and billing
Fund administration
Support for competition
Members and meeting management

The tasks of System Operator (SO) are listed below:

Economic dispatch
  Load forecast
  Constrained dispatch
Reliable operation of power system
  Analysis and diagnosis
  Protection
  Ancillary service

Member's General Meeting, Director's Meeting, and Committee organize Power Exchange (PX). Generator and supplier (KEPCO) are entitled to power trade party. The transmission & distribution company and the company committed by power exchange could be the market operator. Details of member constitution are explained in Table 4.

The tasks of the Director's Meeting are listed below:

Decision making for operation of power market
Guaranteeing independence of power market
Control of market participants to maintain reliable power system

The Director's Meeting consists of 11 Electricity Business Parties and 2 Non-Electricity Business Parties. Six Generation Subsidiaries Representatives, 1 Retail Co. Representative, 2 Customer Representatives, and 2 Market Operators constitute the Electricity Business Party. Two Public Representatives become the Non-Electricity Business Party.

There are three committees. The Variable-Cost Verification Committee does verification of variable-cost data. The Arbitration Committee arbitrates disputes. The Technical Committee determines price and reviews technical problems. The organization of power exchange (PX) is explained in Fig. 5.

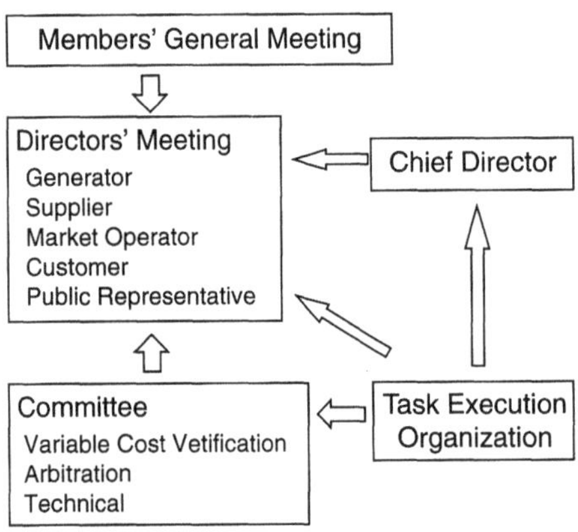

FIG. 5. Organization of PX

TABLE 4. Task commitment

Scenario	Commit	To	PX Task
1	MO SO	Grid Co.	Least organization
2	Nothing	None	MO SO MO
3	SO	Grid Co.	

MO, Market Operator; SO, System Operator

TABLE 5. Merits and demerits

Scenario	Merit	Demerit
1	Stable operation in initial stage SO & MO & grid synergy effect	Unfairness problem arises due to nonseparation of distribution company
2	MO & SO fairness guarantee Match up to restructuring	Trial & error in initial stage
3	SO & grid cooperation	Unfairness problem SO & MO separation (dualization)

Finally, the three scenarios are compared and we summarize the results in Table 4 and Table 5.

Electric Energy and Environment: Some Technological Challenges in Brazil

EDSON H. WATANABE*[1], MAURICIO AREDES[1,2], and CARLOS PORTELA[1]

Summary. Brazil is a country with huge unexplored hydraulic reserves, principally in the Amazon region, that are more than 2000 to 3000 km from the consuming centers. However, the Amazon region has serious environmental concerns, which are imposing many technical challenges. In this chapter, some of the studies being developed at our university with the objective of overcoming these challenges and allowing good use of this renewable energy source with minimum environmental impact is presented. If conventional alternate current (ac) transmission systems were used for these long distances, important severe drawbacks would occur, mainly in relation to the distance, which causes line network behavior quite different from that of lines of the usual length. To overcome this problem, an ac transmission line of little more than half wavelength without fixed reactive compensation but with a small active compensator is proposed to produce a transmission line with excellent performance and low cost per line length. In addition, optimized conductor distribution configuration is being studied to minimize the areas where trees must be cut for the line passage. The dc transmission system is also analyzed, and in this case the unit connected generator/HVDC (high voltage dc system) converter is studied with the objective of minimizing cost as well as maximizing hydraulic turbine efficiency. Also, the possibility of supplying small amounts of power to small communities living near the transmission lines using small HVDC taps based on a high-frequency self-commutated converter is analyzed. Both the unit connection and the small taps have shown to be very interesting solutions for actual applications.

Key words. Long distance transmission of electrical energy—High wavelength transmission—FACT technology—Unit connected HVDC system—Small HVDC tap

Introduction

Electric energy generation in Brazil is based almost exclusively on hydroelectric generation, which is a renewable energy source. In fact, there is some fossil fuel-based generation as well as some based on nuclear power. However, in total they represent less than 4% of the overall electric energy generated in the country.

The Brazilian consumption of electric energy, on average, is still very low as compared with some advanced countries. Although the standard of living and energy consumption

[1]Power Electronics Laboratory, Department of Electrical Engineering, COPPE/Federal University of Rio de Janeiro, Caixa Postal 68504, 21.945-970 Rio de Janeiro, Brazil
[2]Department of Electrotechnics, EE/Federal University of Rio de Janeiro, Rio de Janeiro, Brazil
* Corresponding author

in industry may be quite high in some parts of the country, it is still very low in most of Brazil. To improve the availability of electric energy for industrial development and at the same time improve the general living standard in the country, it is fundamental that electricity becomes available to the consumers. With the objective of increasing electric energy generation, Eletrobras, the Brazilian holding electric power company, plans to double the present generation capacity of 61 GW in the next 10 years. Technical and environmental constraints are some of the challenges to bring this project to reality.

This plan to increase electric power generation in the near future in Brazil will be based on fossil fuels such as natural gas, coal, and oil as well as hydroelectric generation, inclusive in the Amazon region. Today the hydroelectric generation possibilities in the Amazon region are estimated as more than 100 GW. Some of these possibilities are favorable because the power density per inundated area gives an estimate as good as 10 W/m^2. However, in all these areas the distance to the main center of consumption, which may be more than 2000 or 3000 km, represents a technological challenge. Environmental problems involving the inundation area, transmission line construction, and reallocation of population living near the dams are some of the concerns when hydroelectric generation in the Amazon region is discussed. Of course, the studies presented here are not aimed to completely solve these problems. It is intended to help in finding solutions that may minimize their bad effects and improve the country standards, in terms of electric energy generation and use, without depending only on fossil or nuclear fuels.

Alternative generation of electric power based on solar or wind energy is also part of our concern, but this is related more to low levels of power generation and is more appropriate for small communities. In addition, energy conservation is an important issue but is not sufficient to solve the problem of increasing electric power supply. Therefore, these subjects are not discussed here.

It is true that the development of the country will not be accomplished without increasing use of electric energy. The colloquial idea, which is the thinking of most of the common people, as presented by Prof. Werner Leonhard [1], is that "electric energy flows from the wall-socket." This of course is not true and in general people have to be better informed to understand that behind the wall there are many engineering, environmental, and social problems. Thus, several challenges have to be overcome to increase available electric energy.

One of the challenges the country is facing is the development of techniques to transmit electric energy over very long distances, from power plants in the Amazon region to the consuming centers.

Techniques for ac transmission at this long distance are not well developed. If conventional transmission systems were used, severe drawbacks would occur, mainly related to the long distance that causes line network behavior quite different from that of lines of the usual length. These problems would include high transient and sustained overvoltages, resonance effects associated with length and reactive compensation, voltage and transmitted power instabilities, high transmission costs, and significant environmental impact. Several solutions, quite different from conventional transmission systems, may be effective, if specific aspects of long distance transmission systems are taken into account. An optimized transmission system can be defined, without an a priori choice of solution, similar to conventional ones.

It will be shown that a transmission line of a little more than half wavelength [2] can be built, basically without the need of reactive power compensation. Hereafter, for simplicity this special line length will be referred to as "half-wavelength transmission line."

Due to the absence of reactive power compensation, this line presents costs per length less than those for a conventional short line (up to a few hundreds of kilometers). It will also be shown that the use of small FACTS (flexible ac transmission system) controllers can provide efficient control of the power flow through these half-wavelength transmission lines.

The high voltage dc (HVDC) system is also an excellent solution for these very long distance transmissions, especially because it is more efficient and uses more compact transmission lines. However, this choice is not so obvious. The drawback of this kind of electric energy transmission is in the high cost of the ac-to-dc and dc-to-ac converters. Moreover, the difficulty of draining off small amounts of power along the dc line to supply power to small communities is another drawback. Even with these problems it is possible that in some cases dc transmission may be a good choice, especially if the cost of the converter can be reduced and the small HVDC tap becomes a viable solution.

This chapter gives some information about the unit connected generator/rectifier system, which can be much simpler than a conventional generation system because of the reduced amount of equipment employed, which in turn leads to lower costs. In addition, this type of generation system can operate at variable speed [3–5], improving the effectiveness of the generation system, including the dam. This variable speed generation can minimize the inundation area of the dam.

In the event that HVDC transmission is used, a way to drain small amounts of power along the line to feed small communities is of fundamental importance. In this sense, the authors propose the use of high-frequency soft-switching techniques to design a new series tap to drain some few megawatts of power from the dc line.

This chapter introduces some new concepts regarding ac transmission systems with little more than half wavelength and some topics about compact transmission lines. To control these half-wavelength transmission lines, some kinds of FACTS devices are discussed. Finally, two topics related to HVDC systems are discussed: (a) direct connection of synchronous generators to ac-to-dc converters, and (b) small HVDC taps to feed small amounts of power (up to 5 MW) to small communities.

AC Transmission System in the Amazon

If an ac transmission system is used for long distance transmission, possibly the main challenge is to build a transmission line as long as 3000 km. In the Amazon region it is desired that the line should be as compact as possible to allow maximum power transmission occupying a minimum area.

Figure 1 shows the north–south transmission line already in operation in the Amazon region. To construct this line, some trees were cut bellow it, but after this for some time due to environmental restrictions the electric utility that operates the line was not allowed to keep the trees low by cutting them. Consequently, some short circuits occurred between the line and the trees, interrupting power transmission. This kind of restriction did not exist in the past, and it is now a challenge to transmission line designers to make the lines as compact as possible.

Another problem of a long-distance transmission line is that by conventional techniques it should be compensated in such a way as to produce an electrical equivalent short line (less than one-fourth of the wavelength). This would require huge compensation elements (capacitors and reactors). In addition, it would need large active compen-

FIG. 1. A transmission line in the Amazon region

sators, making its cost very high. As a solution to this problem, the transmission system with more than half wavelength lines is proposed and discussed.

The north–south transmission line, without compensation [6], has the following reference conditions: four conductors per phase (ACSR "rail" conductors; diameter, 29.59 mm); nominal voltage, phase to phase, U = 500 kV; characteristic power, $P_c = 1300$ MW; and current density, $J_c = 0.759\,A/mm^2$. This line has 1020 km and connects some power stations and the Northern and Southeast-South Brazilian networks. It has been installed with about 100% shunt compensation (1632 Mvar in twelve 136 Mvar reactors) and about 60% series compensation (966 Mvar in six 161 Mvar capacitor banks), both permanent, and an additional two sets of controlled series compensation based on TCSC (thyristor controlled series capacitors) [7]. Naturally, despite such large reactive compensation additional reactive compensation in both terminal networks is needed to transmit power in this line. To allow 1000 MW transmitted power with this compensated line, an additional fixed reactive compensation, of 2313 Mvar, has been installed at both networks.

This line is presented here only for comparison purposes and to show clearly that if conventional concepts are used the cost of the transmission line becomes very high because of the extremely high level of reactive compensation. The authors present here another concept of transmission system that is more adequate for long distance (around 2500 km) transmission of large power blocks, as in the case of transmission from the Amazon to Southeast Brazil.

Transmission System with a Little More Than Half Wavelength

To clarify the most important physical aspects of the operating conditions of very long lines [2], let us assume a line with no losses, total length L, longitudinal reactance per unit length X, transversal admittance per unit length Y, both for nonhomopolar conditions, and line frequency f. In the case of longitudinal compensation and or transversal compensation, at conventional (not very long) distances along the line, such compensation may be "included" in "equivalent average" X and Y values. The electric length of the line, θ, in radians and treated as nondimensional, at frequency f (being $\omega = 2\pi f$ and v the phase velocity), is

$$\theta = \sqrt{XY}L = \frac{\omega}{v}L \tag{1}$$

$$v = \frac{\omega}{\sqrt{XY}} \tag{2}$$

If X and Y do not include compensation, the phase velocity, v, is almost independent of line constructive parameters, and of the order of 0.96 to 0.99 of the electromagnetic propagation speed in vacuum. At a reference voltage U_0, the characteristic impedance Z_c and the characteristic power P_c are given as

$$Z_c = \sqrt{\frac{X}{Y}} \qquad P_c = \frac{U_0^2}{Z_c} \tag{3}$$

Let us consider eventual longitudinal (series) and transversal (shunt) reactive compensation, along the line, at distances not too long (much smaller than a quarter wavelength at power frequency), by means of "reactive compensation factors," α and β. The "average" per unit length longitudinal reactance and transversal admittance, including compensation, are given by

$$X = \alpha X_0 \quad \text{and} \quad Y = \beta Y_0 \tag{4}$$

where X_0 and Y_0 are the per unit length longitudinal reactance and transversal admittance of the line without compensation.

If no reactive compensation is present, $\alpha = 1$, $\beta = 1$, and, for example, in a line with 30% longitudinal capacitive compensation and 60% transversal inductive compensation, $\alpha = 0.70$, $\beta = 0.40$.

The eventual longitudinal and transversal reactive compensation have the following effects in electric "line" length, θ, characteristic impedance, Z_c, and characteristic power, P_c:

$$\theta = \sqrt{\alpha\beta}\theta_0 \qquad Z_c = \sqrt{\frac{\alpha}{\beta}}Z_{c0} \qquad P_c = \sqrt{\frac{\beta}{\alpha}}P_{c0} \tag{5}$$

where the subscript "0" means values without compensation.

In a line with 600 km, at 60 Hz ($\theta_0 = 0.762$), using 40% capacitive longitudinal compensation ($\alpha = 0.60$) and 65% inductive transversal compensation ($\beta = 0.35$), θ is reduced to 0.349 (equivalent to 275 km at 60 Hz), characteristic impedance is multiplied by 1.31, and characteristic power by 0.76.

In traditional networks, with line lengths of a few hundred kilometers, the reactive compensation is used to reduce θ to "much less" than $\pi/2$ (a quarter of the wavelength) and to adapt P_c, that, together with θ, define voltage profiles, some switching overvoltages, and reactive power absorbed by the line. In case of very long distances (2000–3000 km), to reduce θ to much less than $\pi/2$ would signify extremely high levels of reactive compensation. This fact increases the cost of transmission, and with several technical severe consequences, due to a multitude of resonance type conditions.

The solution proposed by Portela and Gomes [2] is that, for very long distances, it is wise to work with θ a little higher than π (half wavelength), avoiding the need of high levels of reactive compensation, and obtaining a transmission system much more cheaply and with much better behavior.

Neglecting losses at the power frequency and in balanced conditions, the behavior of the line is defined by θ and Z_c. The voltages at the extremities, U_1 and U_2, in complex notation are

$$U_2 = U_0 \quad \text{and} \quad U_1 = U_0 e^{j\delta} \tag{6}$$

Apart from a proportionality factor P_c, the active and reactive power at both extremities and along the line, depends on θ and δ.

Let us consider lines with the following electric lengths:

a. $\theta = 0.05\pi$ (about 124 km at 60 Hz)
b. $\theta = 0.10\pi$ (about 248 km at 60 Hz)
c. $\theta = 0.90\pi$ (about 2228 km at 60 Hz)
d. $\theta = 0.95\pi$ (about 2351 km at 60 Hz)
e. $\theta = 1.05\pi$ (about 2599 km at 60 Hz)
f. $\theta = 1.10\pi$ (about 2722 km at 60 Hz).

Figure 2 shows for these six examples the normalized transmitted active power, P/P_c, the normalized reactive power absorbed by the line (sum of reactive power supplied to the line at both terminals), Q/P_c, and the normalized voltage, at the middle of the line, U_m/U_0 as functions of δ.

Examples a and b correspond to the "usual" lengths of relatively short lines. They must be operated in vicinity of $\delta = 0$, in which an increase in δ increases the transmitted power. The transmitted power may exceed the characteristic power P_c. However, this increases considerably the reactive power absorbed by the line.

Examples c, d, e, and f correspond to very long lines, longer than a quarter wavelength ($\theta = \pi/2$). In examples e and f, the lengths are a little greater than a half wavelength.

For examples c and d, in the vicinity of $\delta = \pi$, the derivative of transmitted power in relation to δ is negative, and so it does not cause the natural stabilizing effect of a positive derivative. The natural effect of a positive derivative, as occurs with short lines, is one of the reasons why alternating current electrical networks are stable (with few exceptions), considering the electromechanical behavior of generator units and loads. Unless extremely complex control systems are considered, affecting all main power stations of the network, it is not adequate to have transmission power corridors with length between a quarter and a half wavelength.

For examples e and f, in the vicinity of $\delta = \pi$, the derivative of transmitted power in relation to δ is positive. Hence, the natural stabilizing effect in a line with length a little more than half wavelength occurs; this is similar to what happens in short lines, near $\delta = 0$. The behavior of this type of line, in the vicinity of $\delta = \pi$, from the aspect of the line terminal, is similar to the behavior of a short line in the vicinity of $\delta = 0$, for a transmitted power in the range of $-P_c < P < P_c$. As shown by Portela and Gomes [2], reactive power consumption of the line is moderate and voltages along the line do not exceed the terminal rated value, U_0. The main difference is related to the voltage in the middle of the line, which is proportional to the transmitted power. If characteristic power is

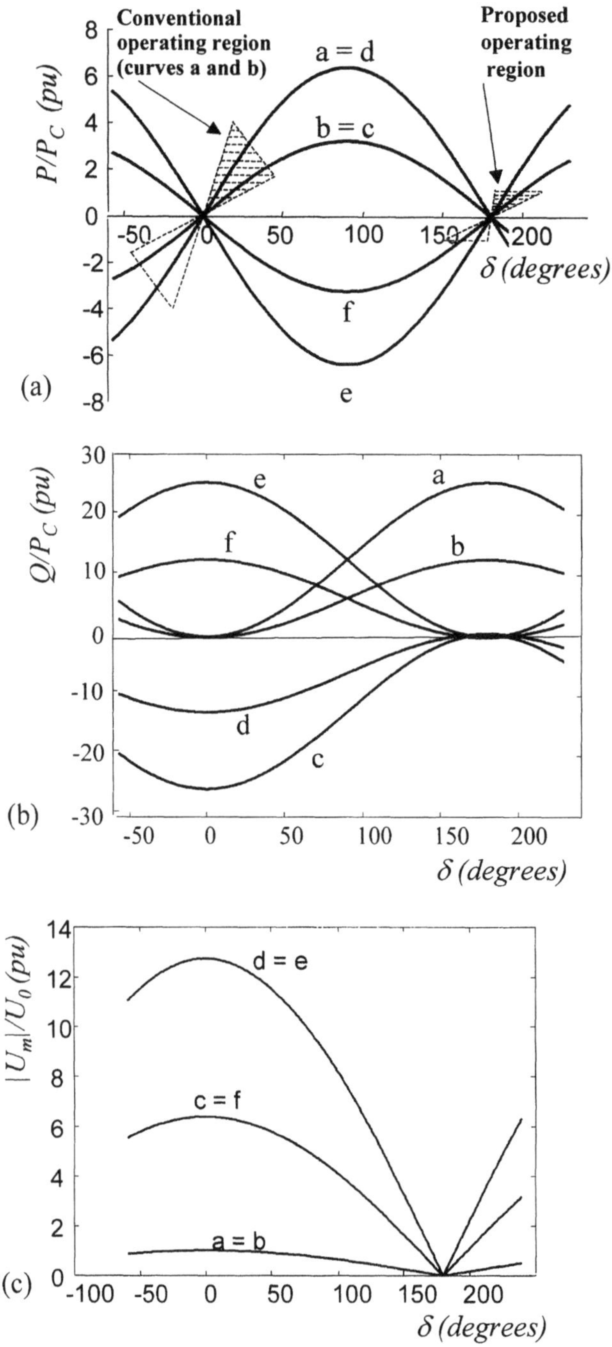

FIG. 2. **a** Normalized transmitted power as a function of δ for different line lengths. **b** Normalized absorbed reactive power as a function of δ for different line lengths. **c** Normalized voltage, at the middle of the line, U_m/U_0, as a function of δ for different line lengths

referred to maximum voltage along the line, the maximum transmitted power is limited to characteristic power (which does not occur in short lines).

At least for point-to-point long distance transmission, the fact that the voltage at the middle point of the line varies between 0 and U_0 constitutes no major inconvenience. If it is wished to connect some relatively small loads in the middle of the line, there are several ways to do so. It is convenient to adopt some nonconventional solution, taking into account that in the central part of the line, the voltage is not "almost constant," but varies according to transmitted power, whereas the current is "almost constant." This problem has not been analyzed yet, but it can be probably solved by using power electronics converters.

The previous short discussion has shown that there are several important reasons to choose an electric length of line, θ, a little greater than half wavelength, for long distance transmission, as concerns normal operating conditions and inherent investment. The "exact" choice is not critical. A range of $1.05\pi < \theta < 1.10\pi$ is a reasonable first approach. In addition, for "slow" and "fast" transient behavior, this choice has very important advantages, as discussed by Portela and Gomes [2]. The solution of long distance transmission with θ a little higher than π ($1.05\pi < \theta < 1.10\pi$, for example) is quite robust for electromechanical behavior and also for relatively slow transients associated with voltage control. For example, a relatively small reactive control, equivalent to a change in θ, allows a fast change in the transmitted power, much faster than the time needed to change the mechanical phase of generators.

Figure 3 shows the elasticity of the transmitted power P with respect to the longitudinal reactive compensation factor α. This figure shows that for conventional short lines, the gain for longitudinal compensation is approximately equal to unity; this means that each percent of series compensation causes an increase of 1% in the transmitted power. However, when the transmission line is long, for instance, $\theta = 190°$, this gain is approximately equal to 10. This means that the compensation effect is, for this line length, about 10 times higher, which allows the use of very small compensators. This effect also appears in the case of shunt compensation.

One example considering a line length $L = 2722\,\text{km}$ ($\theta = 1.10\pi$) without shunt or series fixed or controlled compensation was studied. In this case, a small active series compensator of only ±4%, which produced a longitudinal compensation in the range given by $0.96 < \alpha < 1.04$, was considered. This line, for the case of no compensation ($\alpha = 1$),

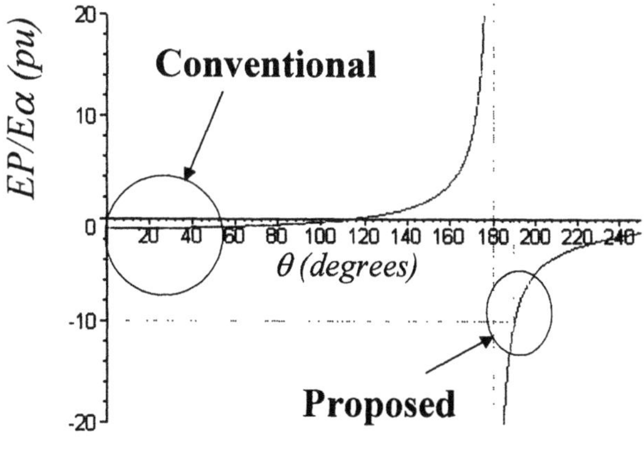

FIG. 3. Elasticity of transmitted power × the line length θ for active series compensation

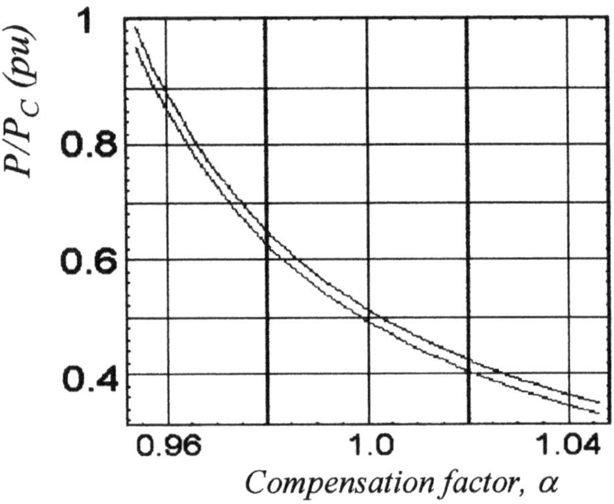

FIG. 4. Normalized transmitted power P/P_C as a function of the series compensation factor α—upper and lower curves apply to the two line terminals

with the phase difference between the terminal voltage equal to $\delta = 1.05\pi$, produced a transmitted power equal $P = 0.506\,P_c$, if no losses are considered. If losses are considered, power at terminals 1 and 2 would be $P_1 = 0.515\,P_c$ and $P_2 = 0.494\,P_c$. Figure 4 shows how the transmitted power varies with respect to the compensation factor α. This figures shows that a controlled series compensator of only about ±4% would allow the control of the transmitted power in the range of approximately 66%.

Transmission System with a Little More Than Half Wavelength: Control Techniques

As shown in the previous section, the transmission system with a little more than half wavelength, as proposed by Portela and Gomes [2], may be an interesting solution for long distance point-to-point power transfer. The operating conditions of this transmission system can be improved with some kind of small controller based on power electronics. This controller can be synthesized based on the FACTS (flexible AC transmission system) concept, which was introduced by Hingorani [8,9] more than 10 years ago and is well condensed in a book written by Hingorani and Gyugyi [10]. These compensators can be connected in series or in shunt with the line and can be based on thyristors or on a new generation of self-commutating power electronics switches, such as an IGBT (insulated gate bipolar transistor), GTO (gate turn-off thyristor), or GCT (gate commutated thyristor) [11].

Active Compensation for Transmission System with a Little More Than Half Wavelength

Various different circuit topologies synthesize the shunt or series FACTS compensators. Their basic description of operation and control can be found in Hingorani and Gyugyi [10]. Some of these compensators, which are more suitable for the long line compensation, are discussed in this section. First, devices based on thyristors are presented and then devices based on self-commutated switches.

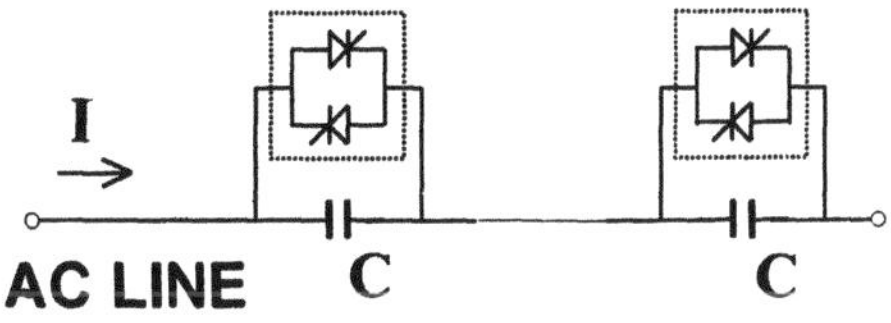

FIG. 5. Thyristor switched series capacitor (TSSC)

FACTS Devices Based on the Thyristor

There are various types of FACTS equipments based on the thyristor. Some of them, like the TSC (thyristor switched capacitor) for shunt applications and the TSSC (thyristor switched series capacitor) for series application do not allow continuous (vernier) control, but in-step changing compensation. These devices are useful to adjust the compensation level to some value close to that which is necessary but they are not appropriate for fine-tuning control. On the other hand, the thyristor-controlled reactor (TCR) and its combination with a fixed capacitor form the SVC (static var compensator) for shunt connection and the TCSC (thyristor controlled series capacitor) for series connection. These last two devices allow continuous control and therefore can be used for fine adjustment of the controlled variable, generally active power flowing through the line.

Our main concern related to the transmission system with a little more than half wavelength is the series compensation, as it is more effective than shunt compensation. Therefore, if thyristor technology is used the TSSC or the TCSC are the natural choices. Figure 5 shows the basic circuit of the TSSC in various modules. These modules should have different values so that they will produce, although discrete, different compensation values.

FACTS devices based on self-commutated switches allows better controllability [10] but the SVC, based on thyristors, presents the great advantage of its relatively long history of actual applications and good results around the world. This fact encouraged the development of the TCSC, which is a SVC connected in series. Today, there are at least three TCSC in commercial operation in the world [7,12,13].

The TCSC circuit is shown in Fig. 6. This compensator can produce continuous impedance control by controlling the firing angle α, as shown in Fig. 7a. As it can be noted, there is a firing angle range around 145° that produces discontinuity in the impedance characteristics. Thus, the TCSC is normally used only in the capacitive characteristic side or in the inductive side close to $\alpha = 0$. Continuous control from capacitive to inductive impedance is not possible.

An analytical model was developed to study the interaction of SVC and the TCSC with other power electronics devices [14,15]. One important conclusion of these works is that conventional models do not represent these compensators accurately enough to study controller interactions in the subsynchronous frequency range.

FACTS Devices Based on Self-Commutating Switches

There are various families of FACTS devices based on self-commutating switches: STATCOM (static compensator) for shunt compensation, SSSC (static synchronous series compensator) for series compensation, and UPFC (unified power flow controller) for shunt and series compensation. All these devices are based on ac-to-dc converters and "minimum" energy storage element (capacitor or reactor) [10]. There are some few

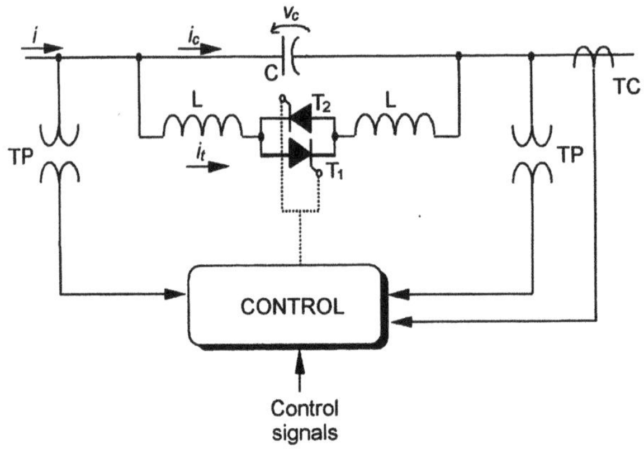

FIG. 6. Thyristor controlled series capacitor (TCSC)

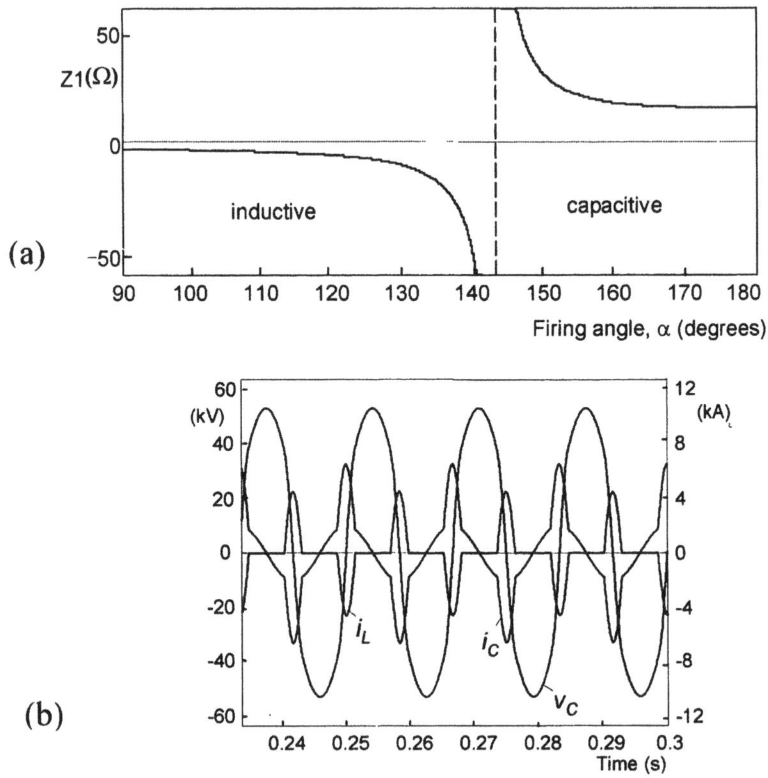

FIG. 7. **a** Equivalent impedance of the TCSC as a function of the firing angle α. **b** Current and voltage waveforms in the TCSC

examples of actual application of these devices [16,17], but they are a serious candidate for wide application in the near future because they present fast and precise response and are very compact as well. Their problems seem to be still the relatively high cost and lack of previous experience in real operating conditions.

Another device based on the self-commutated switch is the GCSC (GTO thyristor controlled series capacitor) that is a dual circuit for the traditional TCR (thyristor controlled reactor). The Power Electronics Group at Federal University of Rio de Janeiro (UFRJ) has been studying the STATCOM [18] the SSSC, the UPFC and the GCSC [10,19,20] as well. These studies are aimed at understanding the behavior of these devices, having in mind the possibility of use in long transmission lines.

The SSSC is a device that would allow easy capacitive or inductive series compensation, being appropriate for a full power flow control like the one shown in Fig. 4. However, this kind of devices would need a relatively complex converter/transformer topology. The STATCOM technology is better known because there are some examples in operation [16–17].

The GCSC is a very simple device with probably the simplest configuration among all the FACTS devices. The basic circuit configuration is shown in Fig. 8. The compensating capacitor is connected in series with the line and two GTO valves (it could be a GCT or

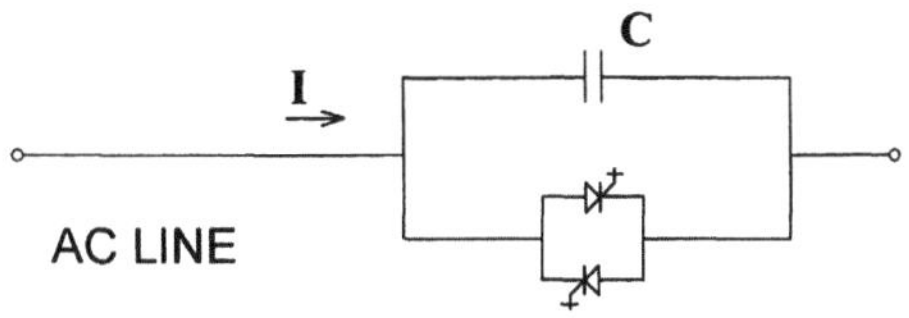

FIG. 8. Basic circuit of the GTO thyristor controlled series capacitors (GCSC)

(a)

(b)

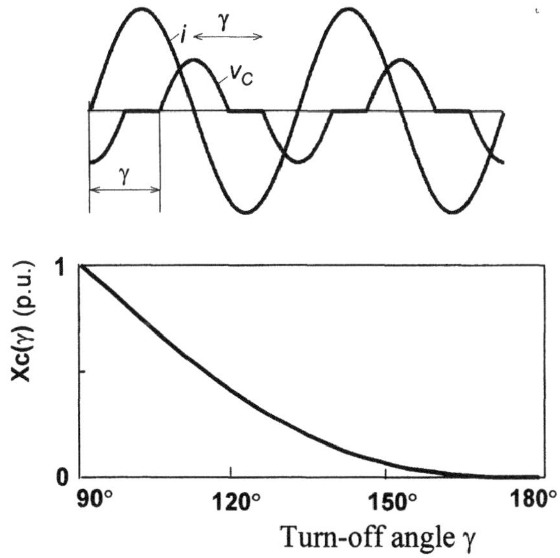

FIG. 9. **a** Voltage and current waveforms in the CGSC. **b** Equivalent impedance as a function of the turn-off angle γ

other self-commutated device) are connected in antiparallel. Figure 9a shows the voltage and current waveforms in the series capacitors. In this case, it is assumed that the current i flowing in a long transmission line is sinusoidal and the period that the capacitor is connected to the line or is short-circuited is controlled by the turn-off angle γ. Note the contrast that in the TCR the control is done by the firing angle and here by the turn-off angle. The voltage waveform on the capacitor in this GCSC is similar to the current waveform in TCR. In the TCSC, the current in the reactor is controlled with the objective to produce variable equivalent capacitive impedance. In the GCSC, the control is done directly on the capacitor without the need of a reactor. Therefore, this is a more compact compensator. Figure 9b shows the equivalent impedance of the GCSC as a function of the turn-off angle γ.

As in the case of TCSC, the harmonic components generated by the switching may be a problem in some applications. For these cases, the six-pulse GCSC (Fig. 10) may be a solution. A higher pulse number GCSC is also possible [19], but for most cases two pulses (as shown in Fig. 8) are enough. Figure 11 shows an example of a transmission line with a GCSC in series, and Fig. 12 shows the transient response when the turn-off angle is varied from $\gamma = 90°$ to $\gamma = 120°$. The controller gains were not optimized, but the power flow response can be classified as very fast.

The authors believe that the GCSC may be a good solution for the active compensator for transmission systems with a little more than half wavelength. If capacitive and inductive compensation is necessary, a reactor can be put in series with the GCSC. In this case, when the capacitor in the GCSC is completely bypassed by the GTO, the equivalent reactance will be inductive. Increasing the amount of capacitor in the circuit is equal to offset the inductive component. For the example shown in Fig. 8, a ±4% variable series compensator can be synthesized using a GCSC. In this case, the GCSC should have a capacitor corresponding to 8% of the total series reactance associated with a reactor corresponding to 4% of the total series reactance. The simplicity of this circuit is one of its strongest points for practical use. Another important thing is that for high-power application normally it is necessary to connect switches in series to produce high-voltage valves. However, because of the high speed of switching, the connection of large numbers of devices in series has been a serious problem. In the case of the GCSC, the switching

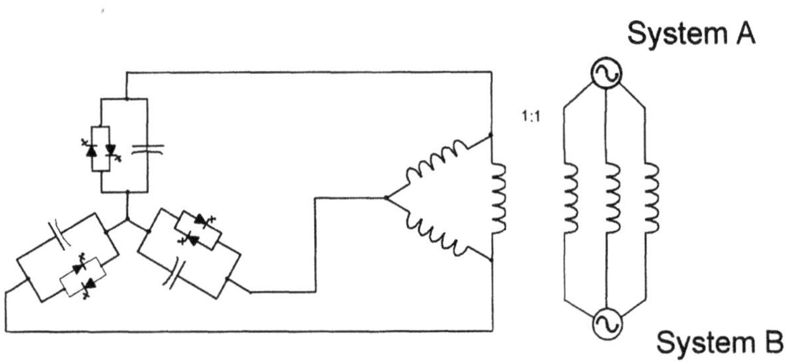

FIG. 10. Six-pulse GCSC

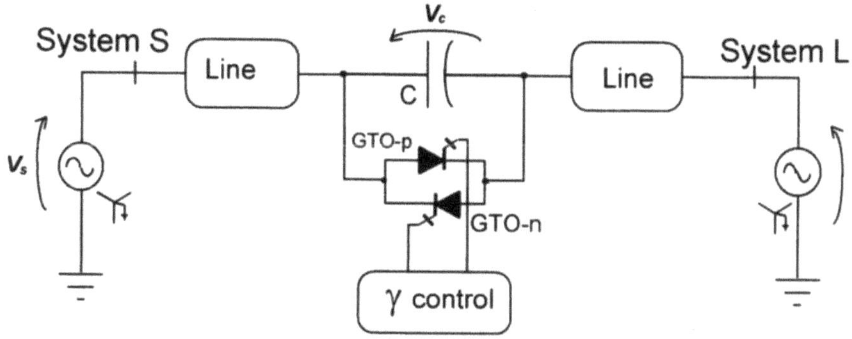

FIG. 11. Example of a GCSC in series with a transmission line

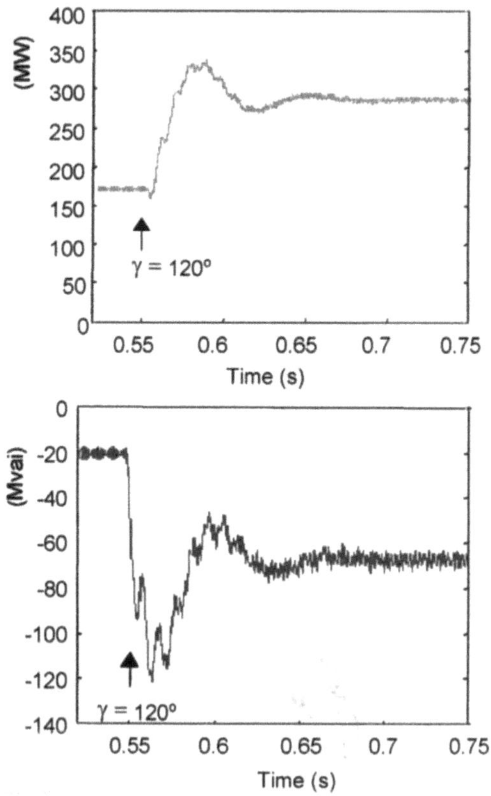

(a)

(b)

FIG. 12. Examples of transient response of a GCSC

is always done at zero voltage (ZVS, zero voltage switching). Therefore, the series association of switches for very high power GCSC applications is much easier than in hard switching converters [21].

Vegetation Clearance Area for ac Transmission Lines

As shown in Fig. 1, the transmission lines in the Amazon region must pass through the forest. The vegetation clearance area is, therefore, a problem, and some kind of optimization to minimize the impact on the trees was analyzed. Figure 13 shows the dimensions of the tower used in the North–South transmission line shown in Fig. 1. These dimensions will be taken as reference for comparison. For the discussion presented next, it is necessary to define some coordinates (Fig. 14). In this figure, it is assumed that the x-axis is along the transmission line and its origin is exactly between two consecutive towers. The distance between these towers is 300 m. The y-axis is perpendicular to the

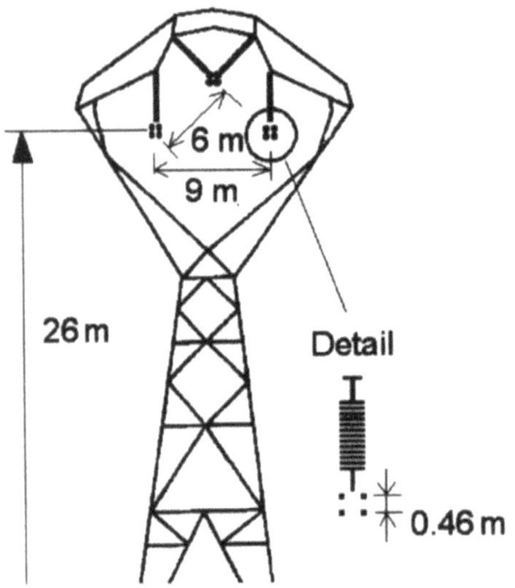

FIG. 13. Schematic representation of conductor position at tower

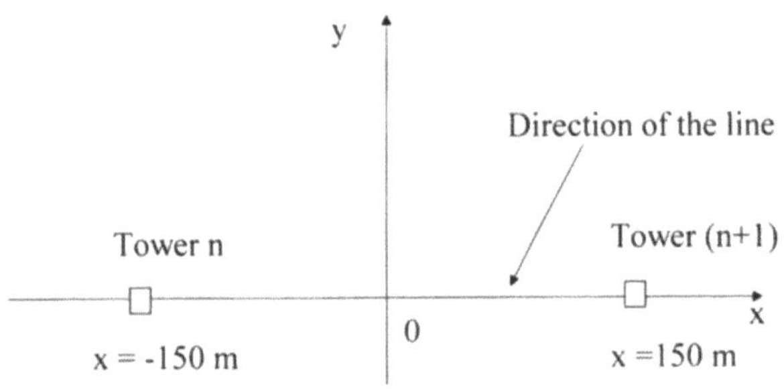

FIG. 14. Definition of coordinates to study vegetation clearance area

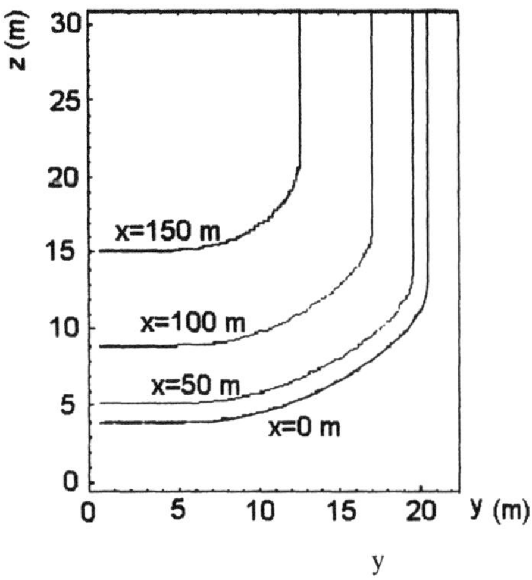

FIG. 15. Limit of vegetation clearance, in vicinity of three-phase line of Fig. 13, in reference conditions, considering conductor displacement due to wind, in flat soil, for various values of x

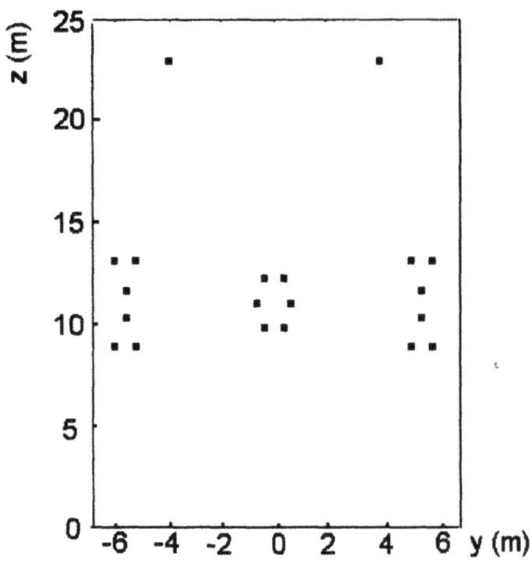

FIG. 16. Schematic representation of conductor position at midspan of a three-phase line with optimized geometry

x-axis and in the horizontal plane. The z-axis is in the vertical direction, and $z = 0$ at ground surface.

Figure 15 shows the limit of vegetation clearance, in the vicinity of the North–South line, in reference conditions, considering conductor displacement due to wind, in flat soil, for some x values. In this case, for instance, for $x = 0$, that is, in the middle of the space between two towers, for $y = 0$, tree height must be limited to 4 m. For $x = 0$, the trees still must be keep low for y up to approximately 21 m.

Figure 16 shows the schematic representation of conductor position at midspan for a three-phase transmission line. This line has six conductors per phase (ACSR "rail"

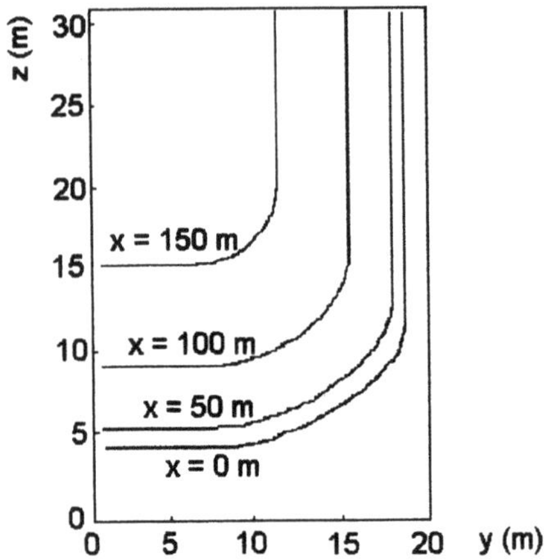

FIG. 17. Limit of vegetation clearance in vicinity of the optimized geometry three-phase line shown in Fig. 16

conductors; diameter, 29.59 mm) with optimized geometry. The optimization was done to obtain maximum transmission capacity, respecting electric field and overvoltage capability. In addition, the following restrictions were imposed: center of the three phases at the same level, conductors of each phase symmetrically disposed in relation to a vertical plane. The reference conditions, for the optimized line, are phase-to-phase nominal voltage, U = 500 kV, characteristic power, P_c = 1910 MW, and current density, J_c = 0.761 A/mm^2. In this situation, Fig. 17 shows the limit of vegetation clearance, in the vicinity of the line, in reference conditions, considering conductor displacement due to wind, in flat soil, for some x values. This figure shows that for $x = 0$, the trees have to be cut for y up to about 19 m.

Another possibility analyzed was the six-phase transmission system. The schematic representation of conductor position at midspan for this case is shown in Fig. 18. This line has five conductors per phase (ACSR "rail" conductors; diameter, 29.59 mm) with optimized geometry, to obtain maximum transmission capacity, respecting electric field and overvoltage capability. In addition, it was imposed that the conductors of each phase were symmetrically disposed in relation to a vertical plane, with conductor centers in an ellipse. The reference conditions, for the optimized line, are nominal voltage, phase-to-ground and between consecutive phases, U = 350 kV, characteristic power, P_c = 3955 MW, and current density, J_c = 0.779 A/mm^2. The characteristic power is about twice as with three-phase lines and, for $x = 0$, the trees have to be cut for y up to 18.5 m (Fig. 19).

Table 1 shows a summary of the comparison of three transmission systems in terms of ecological impact and reactive compensation. As it is evident from this table, and also through a more detailed comparison:

For ecological impact: the transmission system in example 2 has an ecological impact much lower than example 1, and example 3 has a still lower ecological impact

For reactive compensation: example 1 implies a massive reactive compensation, with a high cost, which implies an important increase of transmission system cost; examples 2 and 3 need no fixed reactive compensation

TABLE 1. Summary of comparison of three transmission system examples for ecological impact and reactive compensation

	Mean area, A, measured in a plane perpendicular to line axis, in which trees 20 m high may require height reduction Per unit ecological impact, E, in vegetation, for trees 20 m high, defined as $E = A/P$ Fixed reactive compensation, Q, for a transmission of a power P at a distance L Per unit fixed compensation, q, defined as $q = Q/P/L$	
	With line height without increase ($\Delta = 0$) for reduction of impact on vegetation	With line height with a small increase ($\Delta = 5$ m) for reduction on impact on vegetation
Example 1: Compact 500-kV line and traditional reactive compensation. Transmission capability P = 1000 MW for transmission distance L = 1020 km	$A = 430.84\ m^2$ $E = 0.422\ m^2/MW$ Q = 4911 Mvar q = 4.81 Mvar/MW/(1000 km)	$A = 241.84\ m^2$ $E = 0.237\ m^2/MW$ Q = 4911 Mvar q = 4.81 Mvar/MW/(1000 km)
Example 2: High transmission capability, three-phase, nonconventional 500-kV line. Transmission capability P = 1910 MW for long distance (e.g., L = 2722 km) transmission, without fixed reactive compensation. Eventual small amount of controlled reactive compensation only for control purpose	$A = 379.95\ m^2$ $E = 0.199\ m^2/MW$ Q = 0 Mvar q = 0 Mvar/MW/(1000 km)	$A = 217.83\ m^2$ $E = 0.099\ m^2/MW$ Q = 0 Mvar q = 0 Mvar/MW/(1000 km)
Example 3: High transmission capability, six-phase, nonconventional 350-kV (phase to ground) line. Transmission capability P = 3955 MW for long distance (e.g., 2722 km) transmission, without fixed reactive compensation. Eventual small amount of controlled reactive compensation only for control purpose	$A = 331\ m^2$ $E = 0.084\ m^2/MW$ Q = 0 Mvar q = 0 Mvar/MW/(1000 km)	$A = 188.86\ m^2$ $E = 0.048\ m^2/MW$ Q = 0 Mvar q = 0 Mvar/MW/(1000 km)

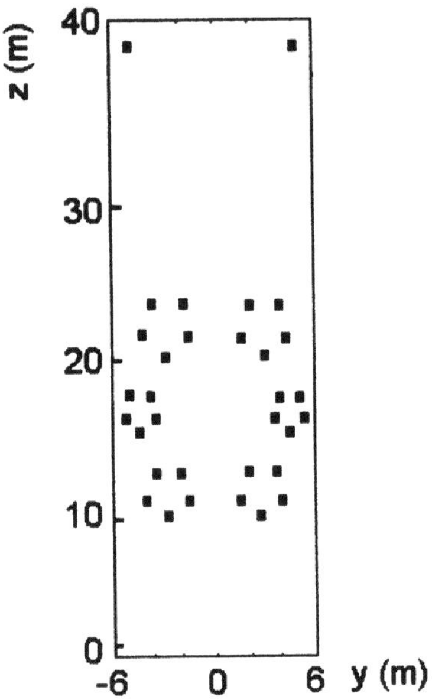

FIG. 18. Schematic representation of conductor position at midspan in a six-phase line

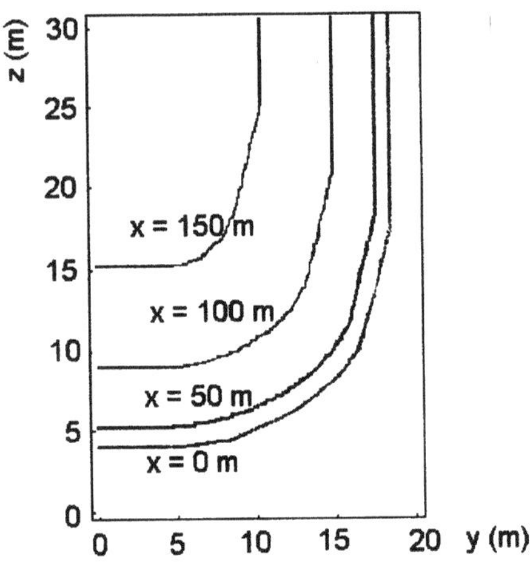

FIG. 19. Limit of vegetation clearance area in vicinity of six-phase line of Fig. 18

dc Transmission System

As mentioned before, when long distances are involved, the HVDC transmission system is one real choice that cannot be ignored. This type of transmission system presents excellent characteristics in terms of efficiency and compactness of the line. In Brazil, one hydroelectric power plant built together with Paraguay generates 12.6 GW of power. Half this power is generated at 60 Hz and is transmitted through ac lines to the consuming center in Brazil. The other half is generated at 50 Hz and is transmitted through HVDC lines to Brazil. This HVDC line is rated at 6.3 GW in two bipoles of ±600 kV. Even with this good example of real application, the costs of these systems, especially the converter stations, are relatively high, and techniques to drain small amounts of power along the line are still not well demonstrated. As an option to minimize cost and improve the effectiveness of the generation system, we discuss the unit connected generator/HVDC converter system, as well as one possible small HVDC tapping converter.

Unit Connected System

The basic configuration of a conventional HVDC generation/transmission system is shown in Fig. 20. The generators are connected to a common ac busbar and therefore their voltage and frequency must be controlled. In addition, at the time these systems were proposed [22], the idea was that the harmonics component in the current should not flow into the machines. Therefore, these systems have always ac filters to eliminate the harmonics component generated by the converter switching. For the case of 12-pulse converters, the ac filters are designed basically in three modules tuned to eliminate the components of the order 11 and 13, 23 and 25, and a high pass. The use of a common ac busbar and the filters presents some advantages, such as in applications where there is an ac load to be supplied by this same busbar. However, when this system is isolated, say in the Amazon region, there would be no need for this controlled busbar.

In a unit-connected generator/converter system, the ac filter and the common ac busbar are eliminated, decreasing the installation costs. Moreover, because the frequency does not need to be controlled, the generator speed can vary and the generation is in variable frequency, to optimize energy use [3–5]. Figure 21 shows one basic configuration of the unit connected system, as studied by Daldegan [23]. In this type of system a converter transformer is used to step up voltage at the generator terminals to a value necessary to produce a correct dc voltage at the converter output.

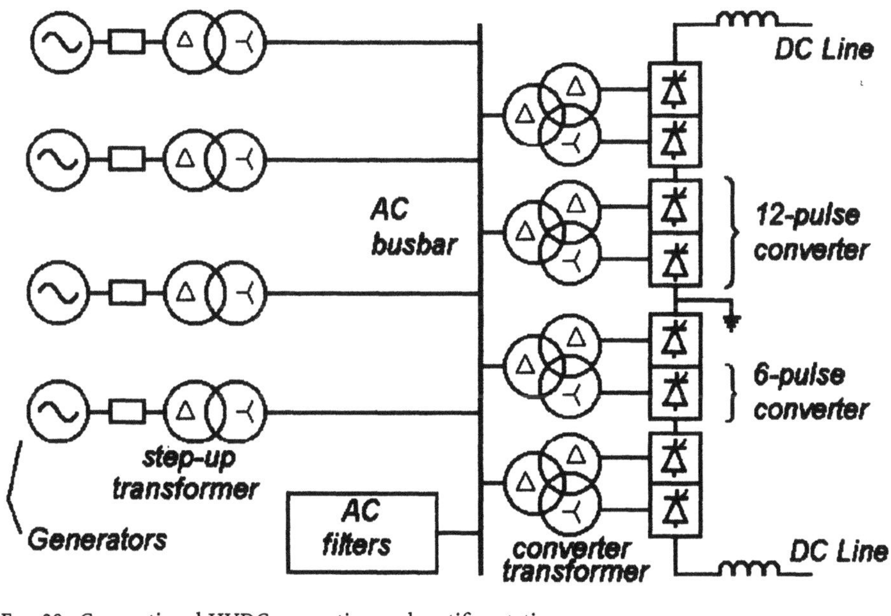

FIG. 20. Conventional HVDC generation and rectifier station

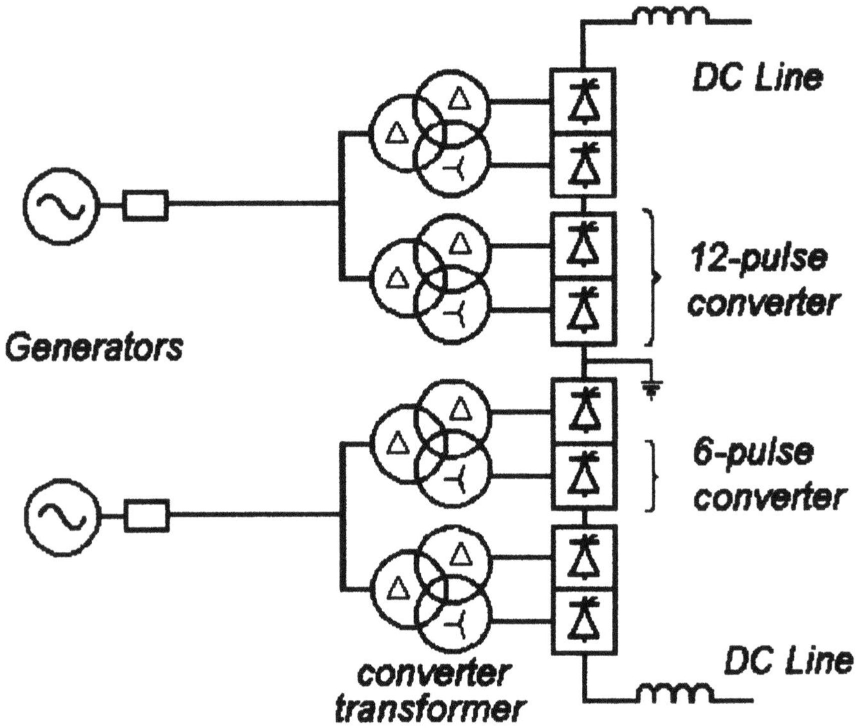

FIG. 21. Generator/HVDC converter unit connected system

In a conventional HVDC converter, the voltage waveforms can be easily derived for each 6-pulse converter knowing the voltage at the common busbar, which is sinusoidal, and the firing angle. The commutations in each 6-pulse converter do not interfere in the other converter commutation. Therefore, the voltage waveforms can be derived independently for each 6-pulse converter and summed up in the dc side to obtain the dc output voltage. However, in the unit connected system, because the voltages at the generator terminals are not filtered, they are not purely sinusoidal. Figure 22 shows a typical simulated generator terminal voltage. The commutation in each 6-pulse converter depends on the generator subtransient reactance (x'', in pu) and the transformer leakage reactance (x_t, in pu). The subtransient reactance is the common reactance for all the commutations in the bridge, and the transformer leakage reactance influences only the commutations related to the converter directly connected to this transformer. This voltage waveform was analyzed in detail by Daldegan [23]. This work shows also that for a simulated typical 415-MW generator the frequencies of the harmonic components produced by the rectifiers are in a range much higher than the machine's operational inductances. Consequently, these harmonic currents flowing in the generator winding do not affect its electromotive force, which is kept sinusoidal for all operating conditions. Therefore, an equivalent circuit composed by a sinusoidal voltage source behind the subtransient voltage can represent the generator.

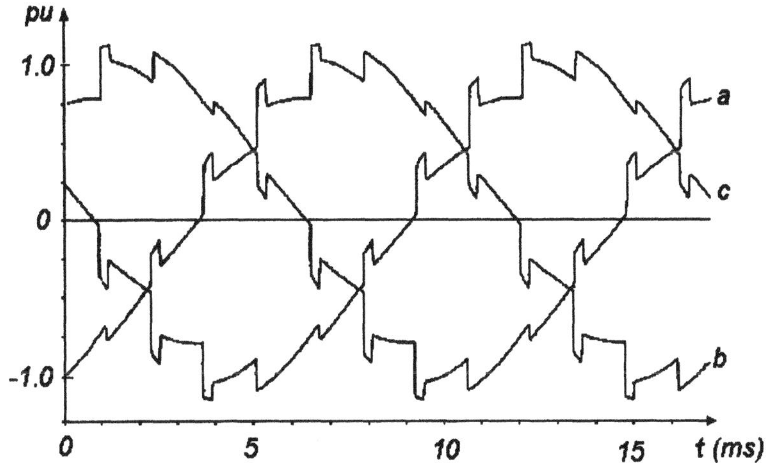

FIG. 22. Unit connected generator terminal voltage

Figure 23 shows an interesting simulation [23] of the voltage waveforms of the generator electromotive force and terminal voltages for the case when the converter current is increased from zero to its short circuit value. The sinusoidal voltage waveforms are the electromotive forces and the terminal voltage waveforms deteriorate as the dc and, consequently, the ac currents are increased. This simulation shows the various modes and regions of operation of the unit connected system. Also, it shows that even with a thyristor firing angle equal to zero (at the beginning), the commutation process naturally imposes a firing angle greater than zero, as the current increases. At the end, when the converter is short-circuited, the voltage waveforms become sinusoidal. All these modes of operation are different from those in conventional HVDC converters and deserve special attention to analyze the system performance. Figure 24 shows the average voltage at the converter dc output. All modes of operation are represented in these curves. The parameter K is defined as

$$K = \frac{x''/2}{x_t + x''/2} \tag{7}$$

The factor 2 dividing the subtransient reactance is because the power of the generator is twice the power of the converter transformers. The horizontal axis is the dc current (I_d) normalized by its maximum short-circuit value (I_{dcc2}), and the vertical axis represents the dc output voltage (V_d) of the two six-pulse converter divided by twice the maximum value of each 6-pulse converter (V_{d0}). The outer curve for $K = 0$ corresponds to the situation of conventional HVDC systems where the ac filter is present and the subtransient reactance does not interfere in the commutation process. The situation for $K = 1$ represents a hypothetical condition where the generator has no subtransient reactance. The curves for actual cases would be for K around 0.5. This figure shows that the output voltage of a unit connected HVDC system is, at a given dc current, smaller than in a conventional HVDC system. However, as said before, the cost of the installation is much lower in the unit connected system. One possibility to decrease the cost of the system even more is to substitute the thyristor valves for diodes in the converters. In this case,

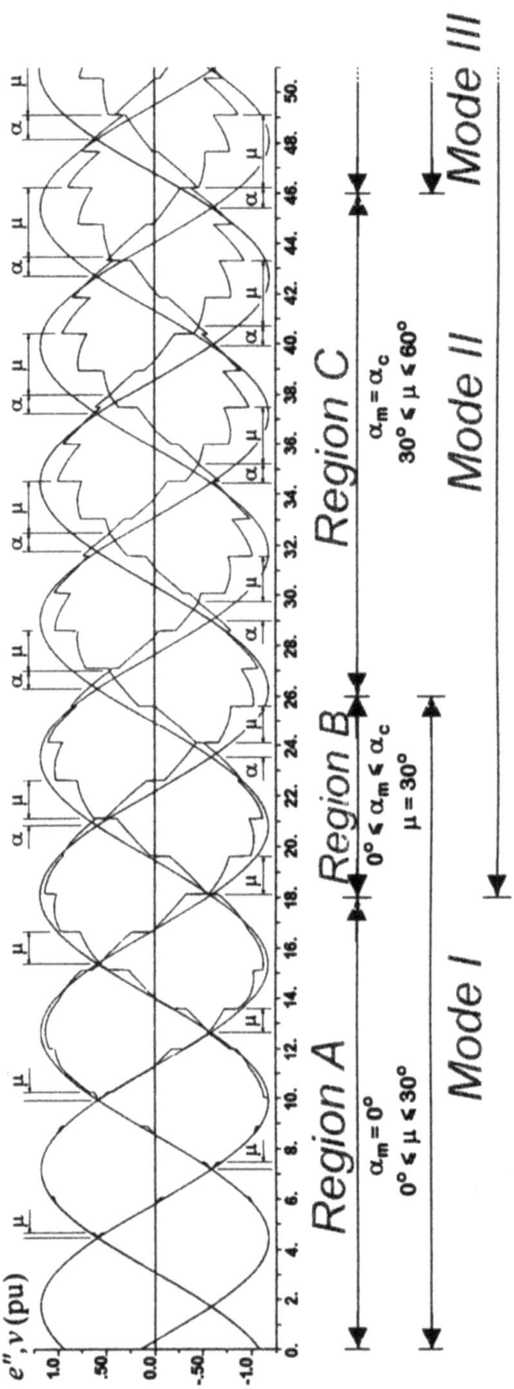

FIG. 23. Different modes of operation of the unit connected generator/converter

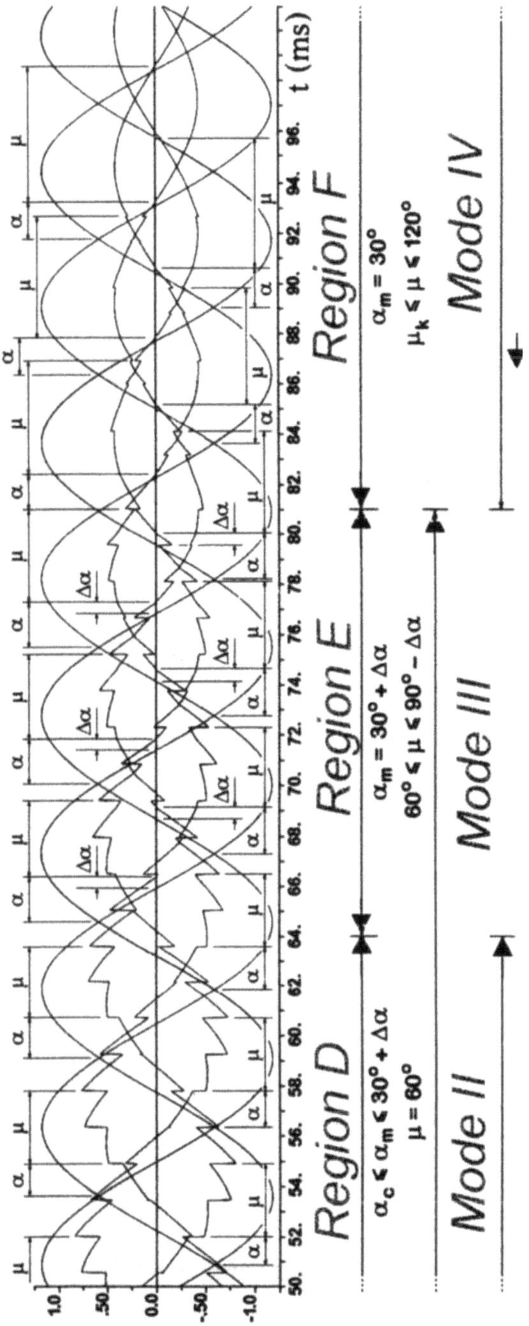

FIG. 23. *Continued*

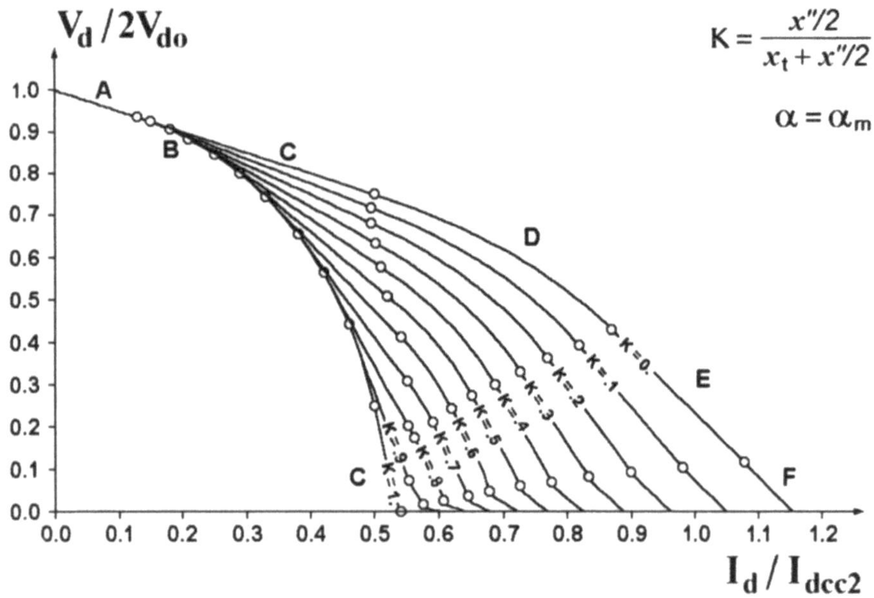

FIG. 24. DC volt/current characteristics in the unit connection

the voltage control has to be done in the machine field, which is somewhat slow but possibly fast enough for the requirements of the transmission system control [23].

The theoretical studies of the unit connected HVDC system show advantages in terms of cost of the converter station with good performance. Another point to be considered is the fact that it can operate at variable speed [3–5], which can improve the efficiency of the generation system. This may be an important way to minimize environmental impact.

Small HVDC Tap

One problem of the dc transmission system is the difficulty of supplying small amounts of power to small communities living along the line [24,25]. The solution to this problem could be the insertion of a tap based on line commutated converters, in series with the dc line. However, this solution would be effective only for relatively high power (few percent of the transmitted power). Another possibility is the use of a voltage-source inverter [26], which is also not suitable for the range of power considered here. The small tap in consideration here can be in the range of up to some few megawatts, representing less than 1% of the total transmitted power in the HVDC line. For this problem, a different nonconventional solution has to be developed. The idea here is to propose a small HVDC tap based on a high-frequency switching converter with the characteristic that if it fails no harm is caused to the main system. In addition, faults in the main system do not affect the integrity of the tap.

Figure 25 shows the basic diagram of a small tapping system for an HVDC transmission line. The HVDC system is equal to any conventional system where the rectifier and the inverter are based on two 6-pulse converters connected in series, forming a 12-pulse

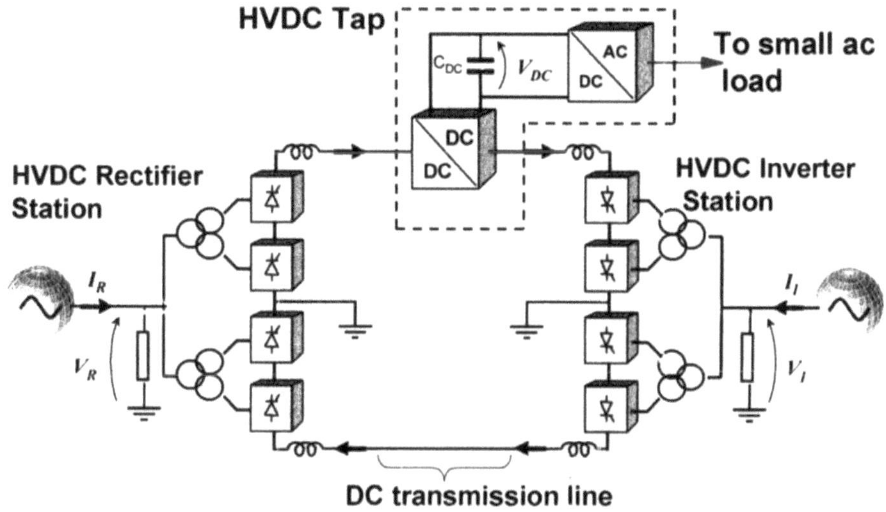

FIG. 25. Small tapping of an HVDC transmission system

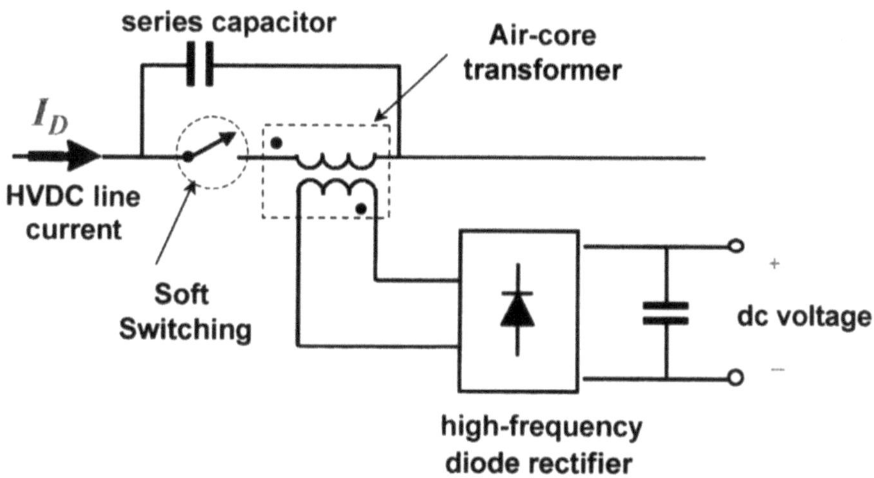

FIG. 26. Basic diagram of the HVDC tap

converter as in Fig. 20. The small dc tap converter is inserted in series with the transmission line and is composed of a dc/dc converter, a dc voltage link, and a dc-to-ac converter to feed the ac load. The dc-to-ac converter can be a well-known voltage source converter used in STATCOM (static compensator) or in UPS (uninterruptable power supply), and therefore is not discussed here. Our main interest is to discuss the dc/dc converter.

Figure 26 shows the schematic diagram of this dc/dc converter. A soft-switching switch in series with an air-core transformer composes the main circuit. In parallel with this switch/transformer, there is a capacitor. In the secondary of the transformer, a high-frequency diode rectifier charges the dc link capacitor where a controlled dc voltage is obtained. The air-core transformer has a very small magnetizing inductance so, if the switch is kept on, the dc/dc converter does not operate and no voltage drop appears for the main transmission. When the switch is turned on and off, there is a resonance between the series capacitor and the transformer inductance. If the switch is turned on and off exactly when the voltage on it is zero, a ZVS (zero voltage switching) is obtained, decreasing the stresses on this switch. Therefore, this technique allows the use of higher switching frequency than when conventional hard switching is used.

The series capacitor and the transformer inductance were chosen such that the switching frequency is around 5 kHz and the switching is always done at zero voltage condition. This high frequency also allows the use of an air-core transformer, which is much simpler to build than the iron-core transformer [27]. Figure 27 shows the details of the circuit. An H-bridge switch synthesizes [28] the soft-switching switch. This type of switch is very well suited for this application because of its low voltage stresses and the possibility to easily design valves based on self-commutating power devices for this kind of application [21].

Figure 28 shows the main voltage and current waveforms in the H-bridge switch. Figure 29 shows the operating characteristics of the dc/dc converter. This figure shows the dc voltage at the converter capacitor, the power delivered to the small ac load, and the duty cycle of the H-bridge switch. From $t = 0$ s to $t = 0.15$ s, no power is supplied to the load and the converter takes about 50 ms to stabilize the output voltage around 22 kV. During this time the duty cycle varies accordingly to control the voltage. At $t = 0.15$ s, a load with 2.5 MW is connected and a slight drop in the dc voltage can be seen followed by the variation of the duty cycle to control this drop. The voltage control system guarantees zero error for the controlled dc voltage and can keep it regulated even when the load is increased to 5 MW at $t = 0.3$ s.

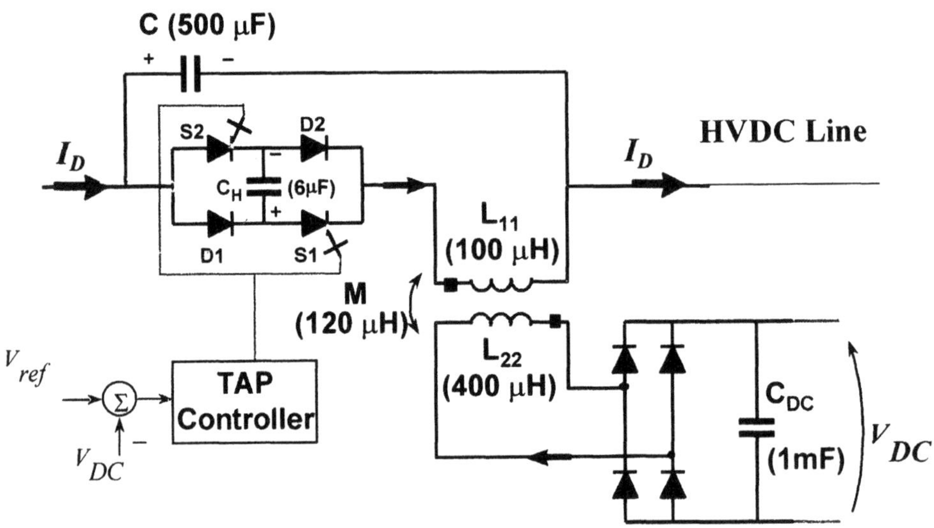

FIG. 27. The dc/dc converter of the HVDC tap

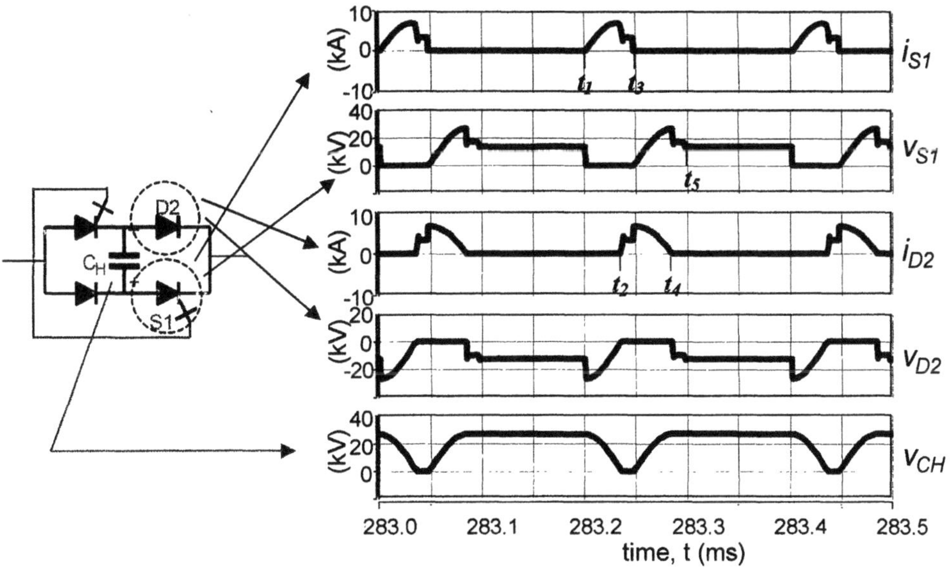

FIG. 28. Soft switching characteristics of the H-bridge

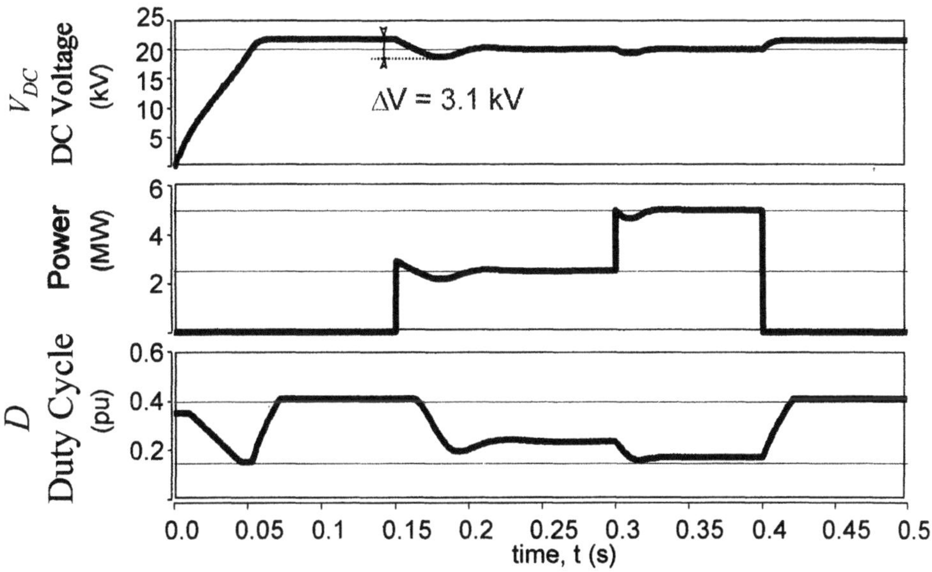

FIG. 29. HVDC tap performance

Further studies were performed, and it was shown that faults in the small tap do not affect the main system and, in the worst case, a bypass switch can be introduced to short-circuit it. In addition, faults in the dc line, rectifier, or inverter of the main HVDC system do not affect the operation of the small tap, if the dc current is maintained. The only obvious problem is that, if the dc current is zero, no power can be drawn to the small load.

This circuit still needs some more studies but seems to be an interesting solution for actual application, where small communities live spread along a vast area below a dc transmission line.

Discussion

The work developed by the authors and presented here is not aimed at finding a final solution to the problem of electric energy transmission from the Amazon or any other long distance transmission system, but to offer a wide range of possible solutions for different cases. Therefore, if ac transmission is chosen for a specific case, the use of the concept of transmission system with little more than half wavelength associated with optimized transmission lines and the six-phase system should be considered. On the other hand, if dc transmission is chosen, the unit connected system has to be considered together with a small tap to supply power to small communities along the transmission line.

It is clear in the mind of the authors that the use of energy from the Amazon region may cause some environmental impact and it is also clear that this impact must be as minimal as possible. Also, what is more important, the environmental impact caused by a hydroelectric power plant in this region should be carefully analyzed and compared with impacts caused by other generating systems such as those based on fossil or nuclear fuel. Rosa et al. [29] show that hydroelectric reservoir may be a significant source of greenhouse gases due to the bacterial decomposition of the submerged organic material. However, the effects of these gases are much smaller than those caused by the CO_2 emission in a thermal plant as the result of fossil fuel combustion.

Conclusions

This chapter introduced some work being developed by the authors with the objective of offering various solutions for very long distance transmission of electric energy. The basic concern in this development is the optimization of the electric system and minimization of environmental impact. In this sense, the transmission system with little more than half wavelength was presented and was shown to offer a real solution for very long distance point-to-point ac transmission. It was also shown that the behavior of this type of transmission system could be improved, in several aspects, with a small power electronics reactive controller. The optimized transmission line, in either three- or six-phase systems, can minimize the necessity to cut trees below the transmission lines, thus minimizing the environmental impact. The unit connected transmission system was analyzed, showing that the direct connection of generator to converter can operate without the conventional ac busbar and without ac filters. It was shown that, in principle, the harmonic components in the current do not affect the generator operation, and the use of diodes instead of thyristors may produce a unit connected system with low cost. Also, generation at variable speed is possible with this system and can optimize the efficiency of the energy conversion in hydraulic turbines. A new series dc tap was proposed as a solution to drain small amounts of power along a HVDC transmission line. This tap has been shown be very reliable and has the advantage of not interfering in the main transmission system if faults occurs on it.

References

1. Leonhard W (1998) Electrical energy. Continuity and progress in fast-moving world. Invited Paper, Proceedings of the Brazilian Conference on Automatic Control, September, Uberlandia, Brazil
2. Portela C, Gomes S Jr (1998) Analysis and optimization of non-conventional transmission trunks considering new technological possibilities. CD-ROM of the VI Symposium of Specialists in Electric Operational and Expansion Planning, VI SEPOPE, paper # BSP-032.pdf, Salvador, May 24–28, Brazil
3. CIGRÉ Joint Working Group 11/14-09 (1997) Guide for preliminary design and specification of hydro stations with HVDC unit connected generators. Report #116. CIGRÉ, Paris
4. Barros JGC (1989) Direct connection of generators to HVDC converters: main characteristics, comparative advantages and prospectives in Brazil. II Symposium of Specialists in Electric Operation and Expansion Planning, II SEPOPE, paper #IP-17, August, Sao Paulo, Brazil
5. Dinh QN, Arrilaga J (1999) Factors affecting the commutation reactance of direct connected generator-converter units. IEEE Transactions on Power Delivery 14(3):1090–1095
6. Gama C (1999) North-south interconnection—general characterization of the project—transmission aspects (in portuguese). Presentation at Federal University of Rio de Janeiro, July
7. Gama C, Leoni R, Gribel JB, Fraga R, Eiras MJ, Ping W, Ricardo A, Cavalcanti J, Tenório R (1998) Brazilian north-south interconnection—application of thyristor controlled series compensation (TCSC) to damp inter-area oscillation mode. 37th Session, CIGRÉ, Paris, August
8. Hingorani NG (1993) Power electronics in electric utilities: role of power electronics in future power systems. Proceedings of IEEE (Special Issue on Power Electronics), April
9. Hingorani NG (1988) Flexible AC transmission. IEEE Spectrum, April, pp 40–45
10. Hingorani NG, Gyugyi L (1999) Understanding FACTS: concepts and technology of flexible AC transmission systems. IEEE Press, New York
11. Baliga BJ (1995) Power ICs in the saddle. IEEE Spectrum, July, pp 34–40
12. Christl NR, Heidin R, Sadek K, Lützelberger P, Krauze PE, Mckenna SM, Montoya AM, Torgeson D (1992) Advanced series compensation (ASC) with thyristor controlled impedance, 34th Session, CIGRÉ, paper 14/37/38-05, Paris, France
13. Zhou X, Guo J, Liang J, Zhang C, Yin Y, Hu X, Chen Q, Wang W, Li B, Chen Y, Li Z, Tao J, Zhang Z, Jiang J (1998) Analysis and control of Yimin-Fengtun 500kV TCSC system. Electric Power System Research 46:157–168
14. Pilotto LAS, Alves JER, Watanabe EH, Hammad AE (1995) A nonlinear switching function model for static var compensators. IPEC—Yokohama 1995, April
15. Alves JER, Pilotto LAS, Watanabe EH (2000) A nonlinear generalized switching function model for SVC and TCSC devices. Proceedings of the International Power Electronics Conference—Tokyo 2000, April
16. Mori S, Matsuno K, Takeda M, Seto M (1993) Development of a large static var generator using self-commutated inverters for improving power system stability. IEEE Transactions on Power Delivery 8(1):371–377

17. Schauder C, Gernhardt M, Stacey E, Lemak T, Gyugyi L, Cease TW, Edris A (1995) Development of a ±100 Mvar static condenser for voltage control of transmission system. IEEE Transactions on Power Delivery 10(3):1486–1493
18. Barbosa PG, Misaka I, Watanabe EH (1994) Shunt-PWM advanced var compensators based on voltage source inverters for FACTS applications. Proceedings of IV Symposium of Specialists in Electric Operational and Expansion Planning—IV SEPOPE, paper #SP-11, Foz do Iguassu, Brazil, May
19. Souza LFW de, Watanabe EH, Aredes M (2000) Multi-module and multi-pulse compensators using the GTO controlled series capacitor. IEEE Transactions on Power Delivery 15(2):725–731
20. Karady GG, Ortmeyer TH, Pilvelait BR, Maratukulam D (1993) Continuously regulated series capacitor. IEEE Transactions on Power Delivery 8(3):1348–1354
21. Watanabe EH, Aredes M, Souza LFW de, Bellar MD (2000) Series connection of power switches for very high power applications and zero voltage switching. IEEE Transactions on Power Electronics 15(1):44–50
22. Kimbark EW (1971) Direct current transmission. Wiley-Interscience, New York
23. Daldegan J Jr (1999) Static converters and synchronous generators in unity connection (in Portuguese). Doctorate thesis, Federal University of Rio de Janeiro
24. Bahrman MP, et al (1995) Integration of small taps into (existing) HVDC links. IEEE Transactions on Power Delivery 10(3):1699–1706
25. Bowles JP, Nakra HL, Turner AB (1981) A small series tap on an HVDC line. IEEE Transactions on Power Apparatus and System 100(2):857–862
26. Zhao Z, Iravani MR (1994) Application of GTO voltage-source inverter for tapping HVDC power. IEE Proceedings on Generation Transmission and Distribution 141(1):19–26
27. Aghaebrahimi MR, Menzies RW (1997) Small power tapping from HVDC transmission systems: a novel approach. IEEE Transactions on Power Delivery 12(4):1698–1703
28. Salazar LD, Joos G, Ziogas PD (1992) A low loss soft switching PWM CSI. Proceedings of IEEE—Power Electronics Specialist Conference, New York, pp 1098–1104
29. Rosa LP, Tolmasquin MT, Rovere E, Legey LF, Miguez J, Schaeffer R (1996) Carbon dioxide and methane emissions: a developing country perspective. COPPE/Federal University, Rio de Janeiro

Superconducting Magnet Technology: A Magnet Engineer's View

Yukikazu Iwasa

Summary. This chapter covers a magnet engineer's view of superconducting magnet technology. The coverage begins with a review of low- and high-temperature superconductivity and proceeds with (1) a discussion of technical issues for applications that rely on superconducting magnets; (2) a description of the Superconductivity Partnership Initiatives (SPI), a U.S. Department of Energy program on electric power applications of high-temperature superconductivity; and (3) a description of the current activities on superconducting magnet technology at the MIT Francis Bitter Magnet Laboratory. The chapter concludes with the author's assessment of the present status and future prospects for superconducting magnets.

Key words. Superconductivity—High-temperature superconductors—Super conducting magnet technology—Solid nitrogen—Flux pump

Superconductivity

Zero Resistance

Discovered by Kamerlingh Onnes in 1911, superconductors have one key property that makes them absolutely appealing for magnet technology applications: "zero" electrical resistance ($R = 0$). However, this unique feature exists only under dc conditions; under time-varying (ac) conditions, superconductors are dissipative. The top drawing of Fig. 1 shows a normal metal conductor carrying dc or ac current; in either case the conductor dissipates heat (red) that emanates from the conductor as indicated by the red arrows. The middle drawing depicts a superconductor with dc current; no heat is generated in the conductor. With ac current (bottom drawing), however, the superconductor generates heat. The red dots within the superconductor are "normal islands" in a sea of superconductivity (blue) [1]; the normal islands are responsible for the dissipation.

Francis Bitter Magnet Laboratory, Massachusetts Institute of Technology, Cambridge, MA 02139-4307, USA

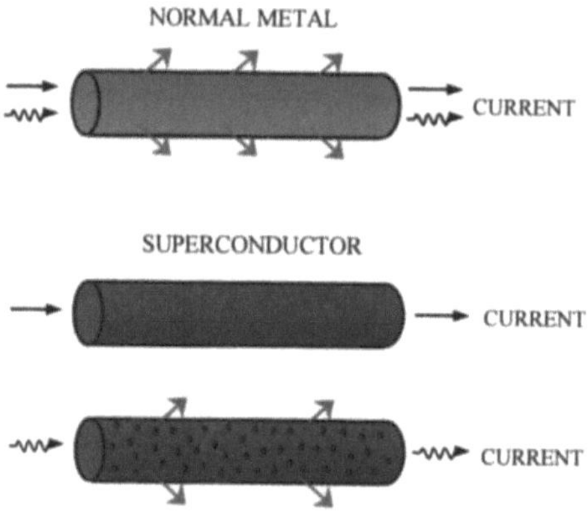

FIG. 1. *Top*: Normal metal conductor carrying current, dc or ac. *Red arrows* denote heat that emanates from the conductor. *Middle*: Superconductor carrying DC current; here, zero dissipation. *Bottom*: Superconductor carrying AC current; *red arrows* signify heat emanating from the conductor, which contains "normal islands" (*red dots*) in a sea of superconductivity

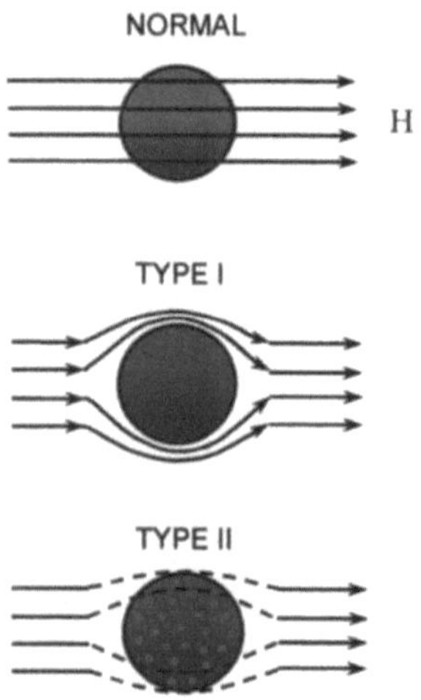

FIG. 2. Spherical conductors each in a uniform magnetic field (*H*). *Top*: Normal metal conductor; the *H* field passes through the sphere. *Middle*: Type I superconductor exhibiting the Meissner effect; the *H* field is completely expelled from the sphere. *Bottom*: Type II superconductor showing a partial penetration of the *H* field. The red dots in type II superconductor are the same "normal islands" of Fig. 1 responsible for dissipation under ac conditions

Meissner Effect

Another important property of the superconductor is the Meissner effect, discovered by Meissner and Ochsenfeld in 1933. In a superconductor exhibiting the Meissner effect the magnetic field is completely expelled, as depicted in the middle drawing of Fig. 2, which shows a superconducting sphere placed in a uniform magnetic field. As indicated in the drawing, this superconductor is of type I. The top drawing of Fig. 2 shows a normal metal sphere under the same condition; the field passes through the sphere. The bottom drawing of Fig. 2 shows a superconducting sphere of type II in which a partial penetration of field is indicated. The red dots are the same normal islands responsible for dissipation under ac conditions: the dissipative superconductor of Fig. 1 is thus of type II.

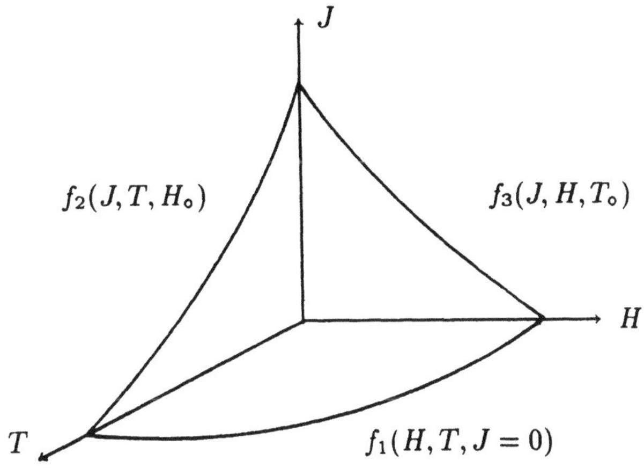

FIG. 3. Critical surface of a typical superconductor [1]

Three Critical Parameters

Superconductivity is a phase bounded by three critical parameters: (1) temperature, T_c; (2) magnetic field, H_c; and (3) current density, J_c. Figure 3 shows a schematic diagram for the critical surface; superconductivity can exist within the phase volume defined by the critical surface [1].

Two Kinds of Superconductors

There are "metallic" superconductors, known also as low-temperature superconductors (LTS), with critical temperatures below 25 K, and "oxide" superconductors, known also as high-temperature superconductors (HTS), discovered in 1986 by Bednorz and Müller. Presently, two HTS appear useful for superconducting magnets: (1) copper oxides of yttrium and barium (YBaCuO) with $T_c = 93$ K; and (2) of bismuth, strontium, and calcium (BiSrCaCuO) with $T_c \sim 110$ K.

Figure 4 shows the chronological progression of the rise in T_c. It began in 1911 with $T_c = 4.2$ K (Hg) and climbed in the early 1970s to $T_c = 23.2$ K (metallic compound, Nb_3Ge), and, after the breakthrough of 1986, reaching the today's highest T_c of nearly 140 K (HgBaCaCuO under a high pressure).

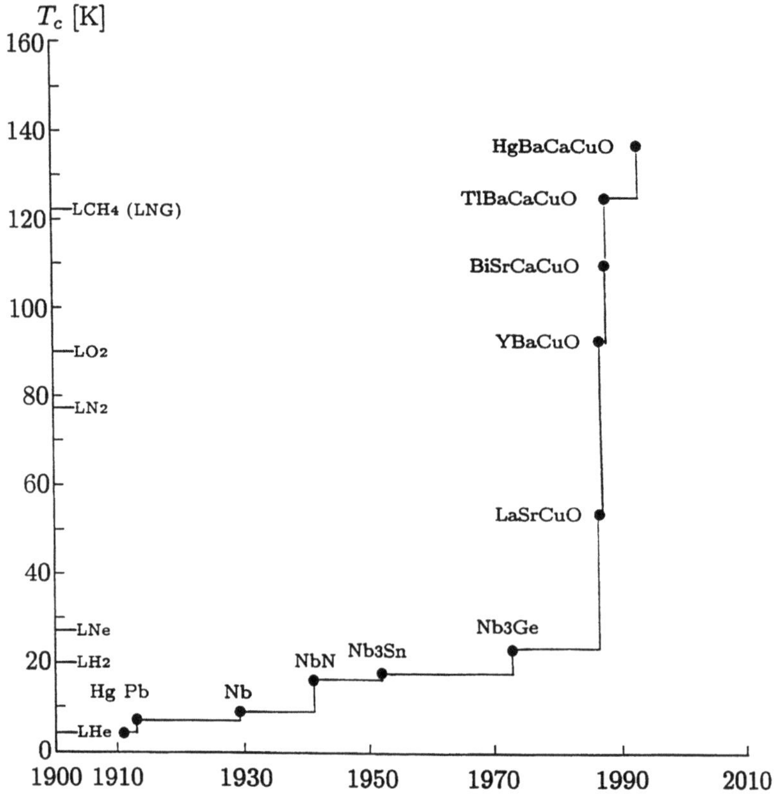

FIG. 4. Chronological progression of the rise in T_c

Superconducting Magnet Technology

Applications and Interdisciplinary Engineering

Superconducting magnets (SCM) are major components in some of the following systems: energy (nuclear fusion); medicine (magnetic resonance imaging, MRI); research (nuclear magnetic resonance, NMR; high-field magnets); transportation (maglev); and space exploration. Superconducting magnet technology comprises engineering aspects associated with the design, manufacture, and operation of superconducting magnets. Superconducting magnet technology is interdisciplinary in that it requires knowledge and expertise in the following fields of engineering: mechanical, electrical, materials, cryogenics, and nuclear [1].

Positive Aspect of Superconductivity

For SCM, superconductivity offers only one positive aspect: $R = 0$ under dc conditions, as stated above. This $R = 0$ feature permits two possibilities for SCM operation: (1) because the joule dissipation is zero ($I^2R = 0$), it is possible to generate a large (>2 T) magnetic field over useful volumes; and (2) because the system's total joule dissipation energy can be made arbitrarily small in comparison to its total stored magnetic energy, it is possible to operate an SCM in "persistent" mode, i.e., $dB/dt = 0$.

Negative Aspects of Superconductivity

There are three negative aspects in superconductivity: (1) T_c is much below room temperature; (2) type II superconductors, used for SCM, are dissipative under ac conditions; and (3) superconductors themselves are expensive compared with ordinary materials used in devices—copper, steel, insulating materials.

The condition $T_c \ll$ room temperature means operation in the cold environment or the need for refrigeration and thermal insulation, which makes it difficult to produce light and compact superconducting magnets. The dissipation in type II superconductors not only requires refrigeration for magnet windings but also imposes heat transfer considerations in the winding designs. Expensive superconductors make these magnets less competitive against normal metal magnets even before the expenses associated with low-temperature operation are added.

Generation of Magnetic Field

We consider power requirements for the generation of a magnetic flux density, $\vec{B}$, with a copper magnet and its superconducting counterpart.

Copper Magnet

As illustrated in Fig. 5a, the copper magnet requires an input power, P_{joule}, which must be balanced by a cooling power, P_{cool}.

SCM

The superconducting magnet (Super, Fig. 5b) requires a refrigeration power, P_{refg}, to balance out a thermal heat input, P_{thrm}, to the magnet.

The SCM is viable when P_{joule} is much greater than $[P_{comp}]_{300\,K}$, the compressor power required to produce P_{refg}. $[P_{comp}]_{300\,K}$ is proportional to P_{refg}, where the proportionality constant varies from a low of 10 for an SCM system operating at ~80 K (T_{op}) to 500 for large systems and 5000 for small systems, both operating at ~4 K. That is, SCM is viable when

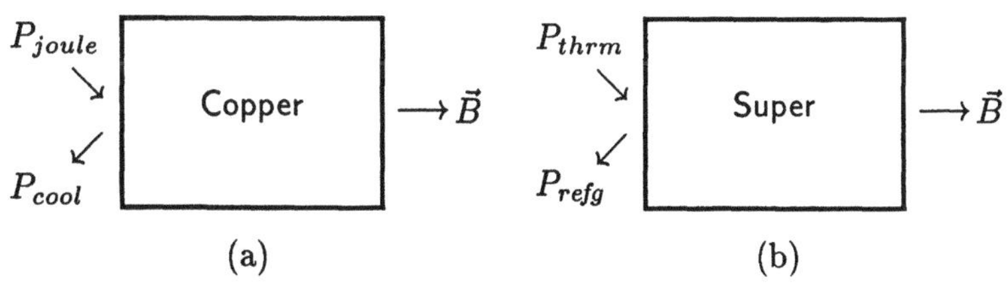

FIG. 5a,b. Power balances for the generation of a magnetic field, $\vec{H}$, (a) with a copper magnet and (b) with a superconducting magnet

$$P_{joule} >> \left[P_{comp}\right]_{300K} = (10-500-5000) \times P_{refg} \tag{1}$$

$T_{op} \sim 80\,K$

"large" units "small" units

$T_{op} \sim 4\,K$

Joule Dissipation

It may be shown that P_{joule} may be given by [1]

$$\begin{aligned} P_{joule} &\propto \rho_{cond} \times a \times H^2 \quad (H \propto a \times J) \\ &\propto \rho_{cond} \times a^3 \times J^2 \end{aligned} \tag{2}$$

where ρ_{cond} is the conductor resistivity; a is the magnet characteristic dimension, e.g., diameter; and J is the conductor current density.

Equation 2 shows that SCM is generally viable for a magnet system that would otherwise require large P_{joule} or that requiring large field volume, large B (>2 T), or a combination of both. As noted, SCM is also viable when the condition of $dB/dt = 0$ must be met when $B > \sim 0.5\,T$.

Areas Firmly Established for SCM

The following areas may be considered to be firmly established for SCM.

- MRI: medicine
- NMR: chemistry, biotechnology, pharmaceutical
- High-field research (HFR) facilities
- High-energy physics (HEP) atom smashers

Works in Progress: Acceptance Still in the Future

In the following areas, it is generally agreed that acceptance of SCM is still in the future.

- Energy
  - ◇ Fusion; superconducting magnetic energy storages; generators; transmission cables; fault-current limiters; transformers; motors
- Maglev (in Japan)
- Quantum Technology ⇒ "Theory of Everything"
  - ◇ NMR for quantum computation; NMR, although an excellent R&D technique for QC, is considered unsuitable in the "real" machine

Criterion for Acceptance

A criterion for acceptance in the marketplace for a given device may be defined in an expense versus uniqueness phase diagram (Fig. 6): the cost limit line defines the device's acceptance (green region) or rejection (red region). SCM devices are expensive, occupying the upper region, independent of uniqueness. To date, as stated earlier, among SCM devices only MRI, NMR, HFR, and HEP offer sufficient uniqueness to offset their high expense to gain acceptance. These devices are thus in the green region.

FIG. 6. Expense versus uniqueness diagram for marketplace acceptance

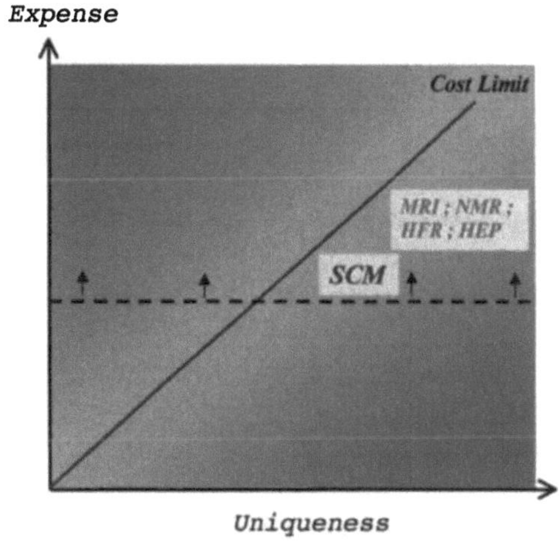

TABLE 1. SPI Projects

Application	Team leader	Start (m/yr)	End (m/yr)	Budget ($M)
Motors	Rockwell/Reliance	1/96	8/01	20.5
Fault-current limiters	General Atomics	3/96	9/99	9.2
Transmission cables (A)	Southwire	4/98	3/01	7.9
Transformers (A)	Waukesha Electric	5/98	5/01	5.9
Flywheel storage	Boeing	5/98	1/01	7.5
Transmission cables (B)	Pirelli Cable	9/98	3/01	5.4
Transformers (B)	ABB	9/98	12/01	10.4
Magnetic separators	DuPont	12/98	12/01	10.4

Applications to Electric Power in the United States

Superconductivity Partnership Initiatives

The Superconductivity Partnership Initiatives (SPI) has been established by the U.S. Department of Energy (DOE) to foster integrated development efforts to accelerate the development of HTS application prototypes that can then be commercialized by U.S. industry. The SPI is a collaborative superconductivity program among (1) DOE, (2) industries—utilities and HTS manufacturers, (3) national laboratories, and (4) universities.

The SPI has two projects each on transmission cables and transformers, and one each on motors, flywheel storage, and magnetic separators. The fault-current limiter project was completed in 1999. Table 1 presents each project's relevant information, including the total budget (R. Eaton, DOE [Golden Office], personal communication, 1999).

TABLE 2. Project specifications

Application	Specs	HTS (supplier)	T_{op} (K)
Motors	1000/5000 hp	Bi-2223 (ASC)	20–30
Fault-current limiters	15-kV "class"	Bi-2223 (IGC)	20–35
Transmission cables (A)	30-m; 3ϕ 12.5 kV/1.25 kA	Bi-2223 (IGC)	70 → 77
Transformers (A)	5/10 MVA	Bi-2212 (IGC)	20–30
Flywheel storage	10 kWh	Bulk YBCO (Prax Air)	77
Transmission cables (B)	120 m; 3ϕ 24 kV/2.4 kA	Bi-2223 (ASC)	70 → 77
Transformers (B)	10 MVA	Bi-2223 (ASC)	77
Magnetic separators	~1.5 T 20–25 cm in diameter	2212 or 2223[a]	20–30

[a] IGC, ASC, or Oxford

FIG. 7. One of the three HTS magnets built by IGC and tested successfully in 1999 for the fault-current limiter project led by General Atomics. Each magnet, 0.75 m high, is wound with Bi-2223/Ag composite tape (total, 17 km long)

Table 2 presents specifications, HTS and supplier, and operating temperature (T_{op}) for each project (R. Eaton, personal communication). Note that both transmission cable projects have T_{op} in the range 70 to 77 K; this is because in each project the superconductor is cooled by liquid nitrogen, subcooled (70 K) at the inlet, and boiling (77 K) at the exit. Also noteworthy is the fact that if cooling is not by liquid nitrogen then the operating temperature is in the range 20 to 30 K, which is provided by having the superconductor coupled to cryocoolers. With the exception of the flywheel storage project, the HTS of choice is bismuth based, either Bi-2223 or Bi-2212.

Figure 7 shows a picture of one of the three superconducting magnets built by IGC and successfully tested in 1999 for the fault-current limiter project led by General Atomics. Each magnet, 1 m in diameter and 0.75 m in height, is wound with Bi-2223/Ag composite tape (total, 17 km long) (R. Eaton, personal communication). These magnets, believed to be the largest HTS coils built to date, were each tested up to 3 kA DC current at 77 K and 7 kA pulse current at 23 K.

Activities at Francis Bitter Magnet Laboratory (FBML)

In this section we describe briefly the superconducting magnet technology activities at FBML, ongoing (1–4) and scheduled to start in 2000 (5–7):

1. Experimental and numerical simulation study of the recovery process in an HTS coil driven normal by transient heating
2. Development of an HTS magnet/solid nitrogen ($SolN_2$) system
3. Levitation in YBCO/Nd-Fe-B permanent magnet systems
4. Development of an HTS current lead ("MIT Lead")
5. $SolN_2$ in the HTS winding under transient heating
6. Development of a flux pump for an HTS insert in an NMR magnet
7. Development of HTS inserts for 1-GHz NMR magnets

Recovery Process

HTS magnets used in fault-current limiters, transformers, motors, and other electric power devices would, under fault-mode conditions, be subjected to transient heating. In a fault-current limiter, for example, a transient current much greater than the superconductor's rated critical current will pass through the HTS magnets, although briefly, quenching the magnets. The magnets are designed to recover from this kind of transient heating. Understanding of this recovery process is important for protection of the magnets.

The FBML study focuses on this recovery process primarily through an experiment with a small HTS coil subjected to an overcurrent pulse that momentarily drives the coil into the normal state. The coil is composed of two sections, inner and outer, each wound with Bi-2223/Ag composite tape; the outer section is wound directly over the inner section. Table 3 lists key parameters of the coil and tapes. Note that the sections are wound with different tapes, the tape for the inner section having slightly better critical current performance than the tape for the outer section.

TABLE 3. BSCCO-2223 coil parameters

Parameters	Inner	Outer
Sections		
Winding i.d. (mm)	100.0	105.8
Winding o.d. (mm)	104.6	110.4
Winding height (mm)	76.0	76.0
# layers	8	8
# turns per layer	20	20
# turns	161	157
Tapes		
Overall width (mm)	3.5	3.5
Thickness (mm)	0.23	0.23
Conductor length (m)	52	69
Ag/SC ratio	2.33	2.33
R_{ag} @RT (Ω)	1.57	1.60
R_{ag} @20 K (mΩ)	6.6	6.7
Measured I_c @77 K (0 T) (A)	23.5[a]	19.9[a]
(A)	27.5[b]	23.7[b]
Projected I_c @20 K (0 T) (A)	127[a]	107[a]
	148[b]	128[b]

[a] At 0.1 μV/cm
[b] At 1.0 μV/cm

Experimental Setup and Procedure

The coil is placed inside a cryostat and coupled to a cryocooler that maintains the coil at an operating temperature in the range 20–60 K. The procedure consists of the following sequence: (1) set the coil at an operating temperature; (2) apply a pulse current; and (3) record coil voltage and temperature.

Figure 8a shows a circuit model for the test coil. The solid and dotted lines in Fig. 8b represent, respectively, the pulse current applied to the coil and the steady current flowing through a device modeled by the test coil under normal operating conditions.

Overcurrent Dissipation in BSCCO-2223/Ag Tape

Figure 9 shows a circuit model for BSCCO/Ag composite tape. The BSCCO carries current up to its critical current, $I_c(T)$, which depends on the conductor temperature. The Ag sheath carries the difference between the total current, in this case a peak current I_{pk}, and $I_c(T)$. The resistive voltage, $V_r(T)$, across and joule heating over the conductor, $P(T)$, are given by

$$V_r(T) = R_{ag}\left[I_{pk} - I_c(T)\right] \quad (3)$$

$$P(T) = V_r(T) I_{pk} \quad (4)$$

Because $I_c(T)$ is a decreasing function of T, for a constant I_{pk}, $V_r(T)$ and $P(T)$ are increasing functions of T, which can lead to a thermal runaway situation if $P(T)$ is not properly balanced by cooling.

Figure 10 shows $V_r(t)$ traces for the BSCCO/Ag composite carrying I_r of Fig. 8b with I_{pk} slightly greater than I_c. At $t = 0$, the superconductor is only slightly resistive; with $P(T)$

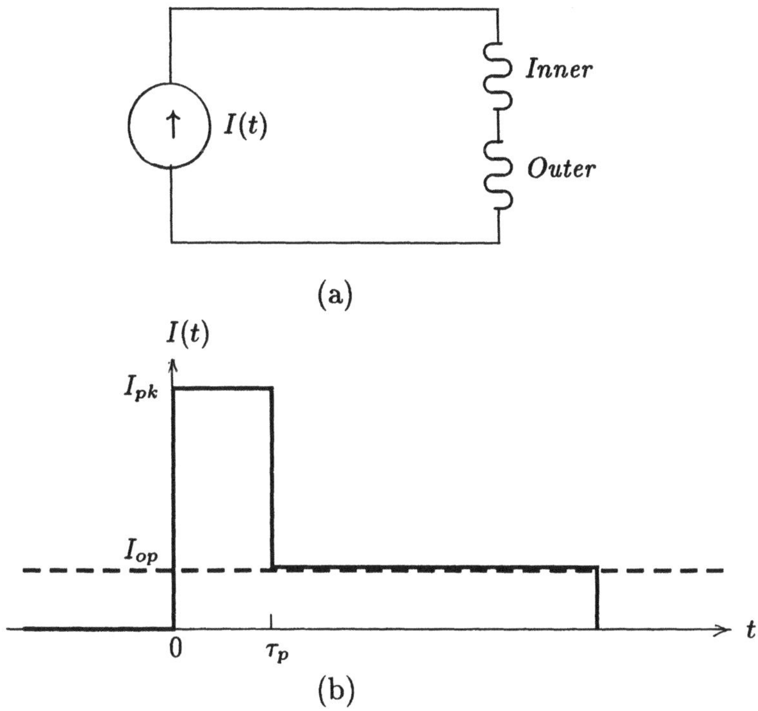

FIG. 8. **a** Circuit model for a two-section test coil used in the experiment. **b** *Solid lines*, pulse current applied to the test coil; *dotted lines*, steady current in an HTS magnet modeled by the test coil under normal operating conditions

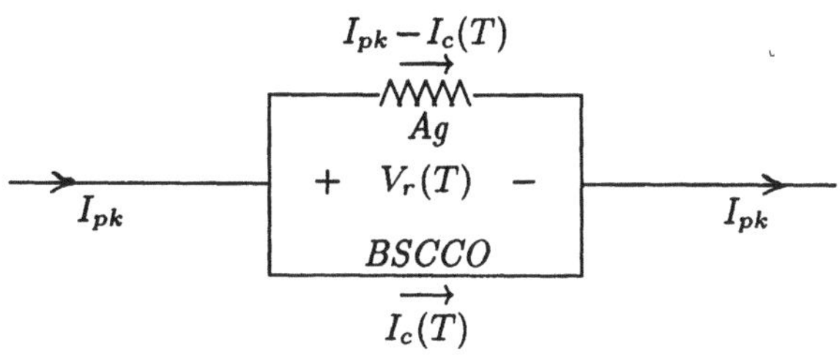

FIG. 9. Circuit model for BSCCO/Ag composite tape

heating up the superconductor, $I_c(T)$ decreases, raising $V_r(T)$, which in turn increases T further. The thermal runaway process is slowed down at $t = \tau_p$ when I_r is decreased to I_{op}.

Whether the conductor continues to heat up (quench curve in Fig. 10) or recover (recovery curve) depends on the balance between dissipation and cooling. Parameters such as I_{op}, T_{op}, I_{pk}, τ_p, and the cooling condition are key to dissipation and cooling.

Figure 11 presents a typical set of current (top) and voltage (middle and bottom) traces recorded in the experiment [2]. In this example, the coil was at $T_{op} = 39\,\mathrm{K}$ with $I_{op} = 28\,\mathrm{A}$, and curves are shown for three values of I_{pk}: 120 A (blue), 128 A (green), and 131 A (red).

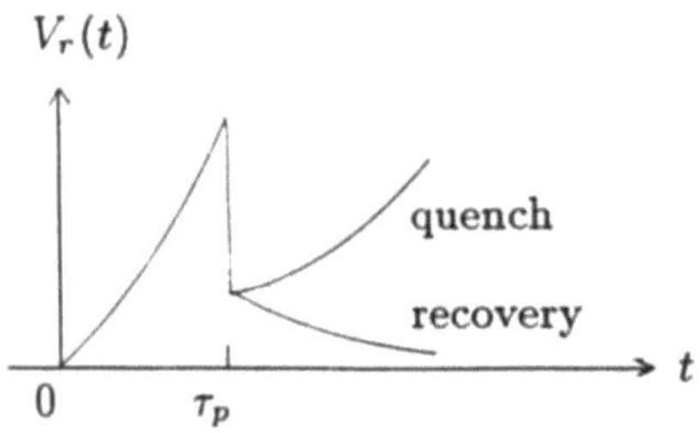

FIG. 10. $V_r(t)$ traces in response to $I_r(t)$ shown in Fig. 8b

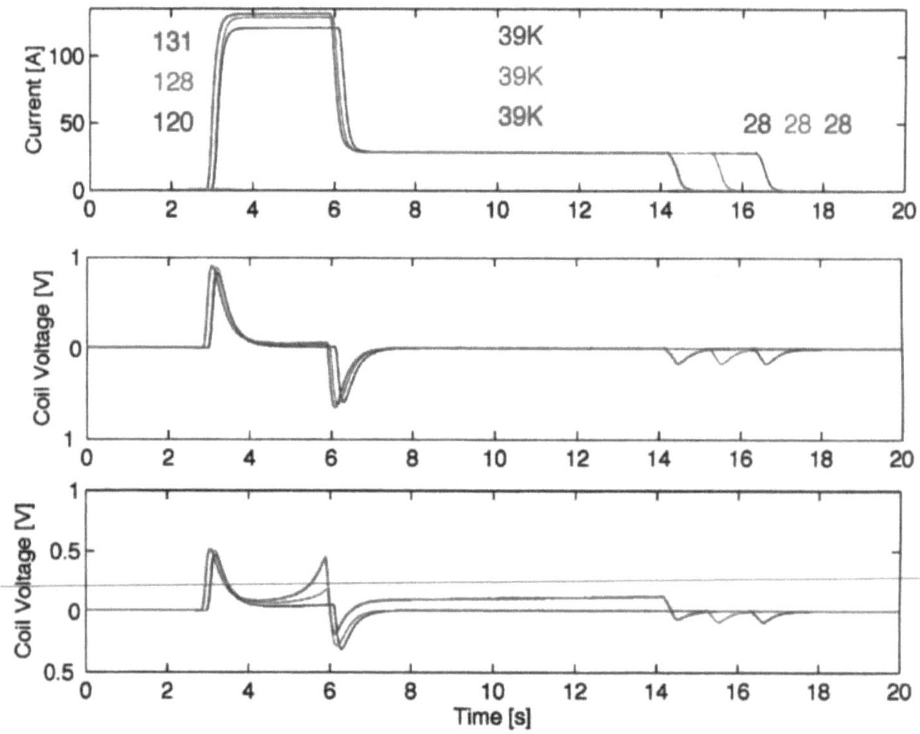

FIG. 11. Current (*top*) and voltage (*middle*, inner section; *bottom*, outer section) traces at $T_{op} = 39\,K$ [2]

The middle voltage traces (inner section), with the exception of inductive pulses, indicate that the inner section remains completely superconducting.

The bottom traces (outer section) show both inductive and resistive voltages. At $I_{pk} = 123\,A$, the outer section remains superconducting. At $I_{pk} = 128\,A$, the outer section becomes resistive when it is carrying 128 A but recovers when the current is reduced to 28 A. At $I_{pk} = 131\,A$, the outer section remains resistive throughout the pulsing.

HTS Magnet/Solid Nitrogen ($SolN_2$) System

A new HTS magnet system that incorporates solid nitrogen ($SolN_2$) combines (1) simplicity and ease of operation of a ferromagnetic permanent magnet and (2) strength and versatility in field generation of an electromagnet. Once energized and producing a desired field, the system, without being coupled to a cryocooler for refrigeration, can maintain the field for long periods [3]. This *cryocoolerless* "permanent" magnet system is particularly appropriate for magnetic devices on-board vehicles (ground, water, space) where "permanence" in practice means durations ranging from a few hours to a year. For on-board applications, $SolN_2$ also eliminates "sloshing" associated with liquid coolants. Two other innovative ideas are "recooling" and "recharging" capabilities; that is, the system is designed to be recooled while it maintains its constant field so as to make the field literally permanent. Like rechargeable batteries, it may also be recharged if its upper operating temperature is exceeded and the field is destroyed. The new system is feasible because of a capability unique to high-temperature superconductors: their ability to remain superconducting over a *wide* temperature span.

Innovative Design Features

The innovative design features incorporated in this system that permit it to be practical are (1) incorporation into the cold system of a volume of solid nitrogen of about the same size as the magnet winding, roughly tripling the system's heat capacity over that of the magnet's mass, with an added increase in system weight of only about 10%; (2) variation in the magnet's operating temperature over a wide span, from about 20 K at the initial stage to 40–60 K at the final stage—heretofore, superconducting magnets have always been designed to operate at a virtually constant temperature; and (3) stringent minimization of total heat input to the cold system.

Figure 12 shows heat capacity, C_p, versus T plots for $SolN_2$ and other solids in the low temperature range. Note that $SolN_2$ undergoes a phase change at 35.6 K, absorbing an energy density of 8.1 J/cm^3 as it warms up. Because the specific density of $SolN_2$ in this temperature range is about 1/10 of those of silver and lead, in terms of specific heat [J/g K], $SolN_2$ is far superior to silver and lead. Figure 13 presents predicted warming trends for each solid with a heat input of 0.11 W and an initial temperature of 20 K; Fig. 13a is for each solid with a volume of 2000 cm^3, and Fig. 13b is for each solid with a mass of 2 kg (B. J. Haid, personal communication, 1999).

Recooling-to-Recooling Time Period (RRTP)

RRTP is the duration over which the system may be operated decoupled from a cooling source: system RRTP should be as long as possible. Figure 14 shows a typical set of cycles, temperature (green) and cooling (blue). In the figure, RRTP corresponds to a time duration of 6 h, 6 days, 6 weeks, or 6 months. The power balance for the cold body of $C_{cd}(T)$ subjected to heat input Q_{in}, is given by

$$C_{cb}(T)\frac{dT_{op}}{dt} = Q_{in}(T) \tag{5}$$

Equation 5 may be integrated to obtain an expression for RRTP:

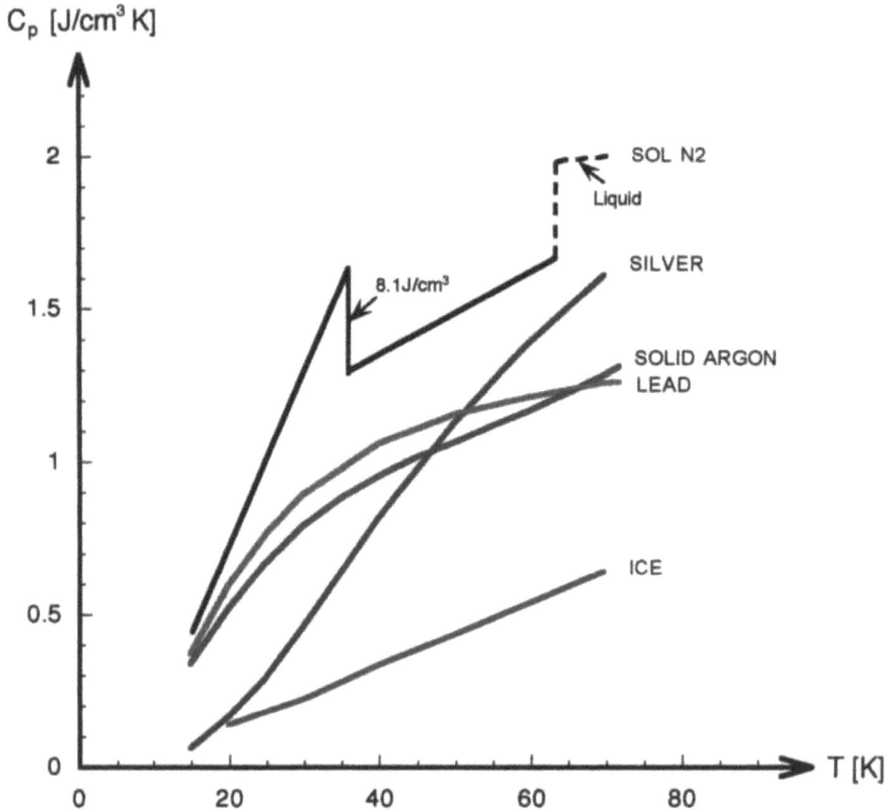

FIG. 12. C_p versus T plots for $SolN_2$, silver, solid argon, lead, and ice

$$RRTP \equiv \Delta t_{op} = \int_{T_{opmn}}^{T_{opmx}} \frac{C_{cb}(T)}{Q_{in}(T)} dT_{op} = \frac{\Delta H_{cb}(T_{opmn}, T_{opmx})}{\langle Q_{in} \rangle} \tag{6}$$

T_{opmn} and T_{opmx} are, respectively, the minimum and maximum operating temperatures, e.g., 20 K and 40 K; $\langle Q_{in} \rangle$ is the average heat input.

RRTP may be increased by (1) widening $\Delta T_{op} = T_{opmx} T_{opmn}$; (2) enhancing $\Delta H_{cb}(T_{opmn}, T_{opmx})$; and (3) minimizing $\langle Q_{in} \rangle$ with innovative cryostat designs (mechanical, cryogenic, electromagnetic).

Figure 15 shows the operating range of an HTS magnet in terms of its I_c versus T plot. The magnet is operated at 100 A in the temperature range $T_{opmn} = 20$ K and $T_{opmx} = 40$ K. Note in this temperature range that the magnet remains well within the superconducting state (blue); only when its temperature exceeds about 50 K at 100 A does the magnet enter the normal state (red).

Figure 16 shows the new system, drawn schematically, in which an HTS magnet is surrounded by $SolN_2$. Liquid helium flowing through a copper coil condenses nitrogen gas to liquid, changes the liquid to $SolN_2$, and cools it and the HTS magnet to T_{opmn}. The helium leaves the system as vapor. The HTS magnet is energized through a pair of current

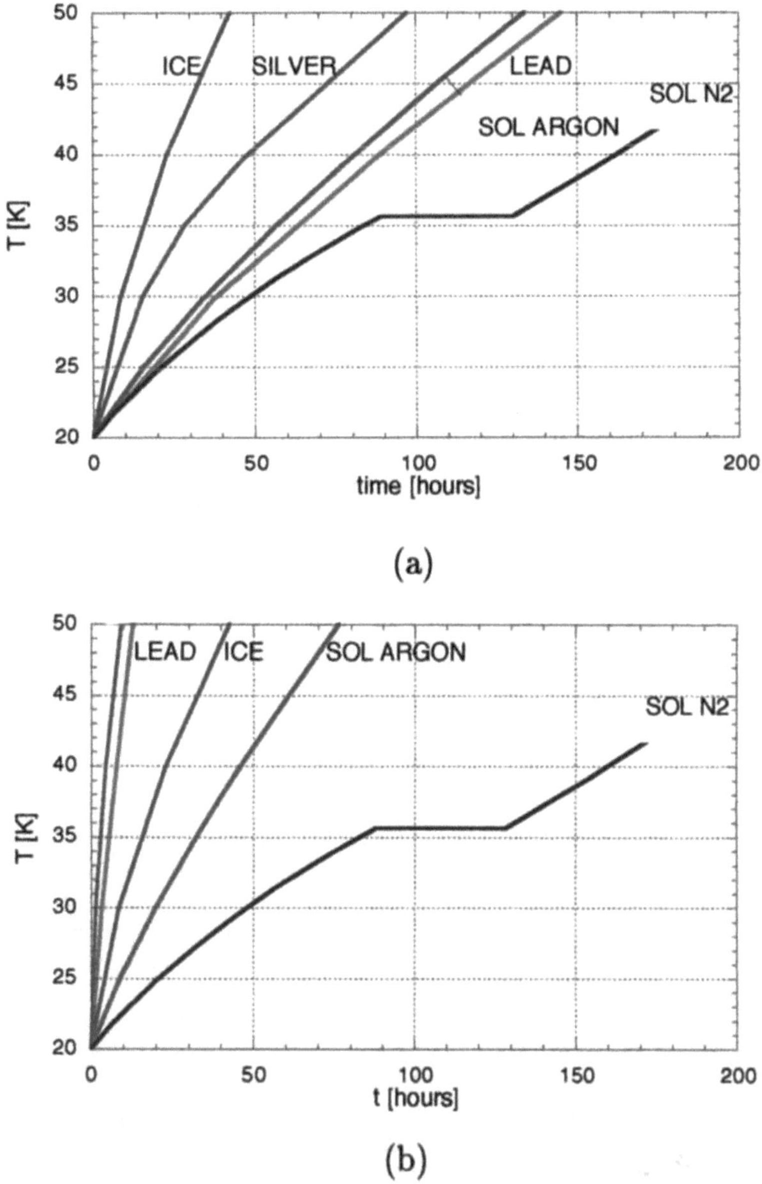

FIG. 13a,b. Predicted warming trends for each solid from 20 K with a heat input of 0.11 W (**a**) with a volume of 2000 cm^3 and (**b**) with a mass of 2 kg

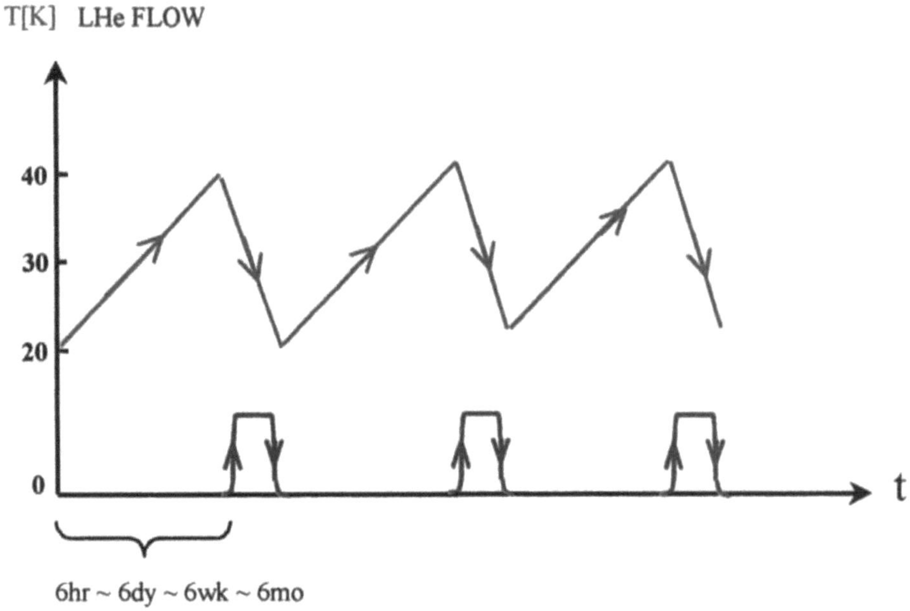

FIG. 14. Operating cycles, temperature (green) and cooling (blue); RRTP can be 6 h, 6 days, 6 weeks, or 6 months

leads; afterward the leads are disconnected from the system to minimize conduction heat input.

Figure 17 shows a schematic drawing of a system presently being developed at the FBML; the system is scheduled to be completed in the fall of 2000 (B. J. Haid, personal communication). The system's HTS magnet is composed of six double pancake coils, each wound with Bi-2223/Ag composite tape. The magnet is expected to generate a magnetic flux density in the range of 0.5–1 T in a room temperature bore of 12 mm. As in the schematic system of Fig. 16, the system is cooled by liquid helium forced through a copper coil; its current leads are also disconnectable. The system, once pumped down to create a vacuum environment to minimize convective heat input, will be decoupled from the pumping station. A getter attached to the bottom (not shown in the figure) maintains a good vacuum.

Levitation in YBCO/Nd-Fe-B Permanent Magnet System

This project studies levitation of a bulk YBCO disk or a set of disks above an Nd-Fe-B permanent magnet (PM) or a set of permanent magnets, focusing on repulsive forces and supercurrent—magnitude and distribution—induced in the bulk YBCO disk(s). Applications of this levitation include "human levitation" [4] and low-boiloff cryostats.

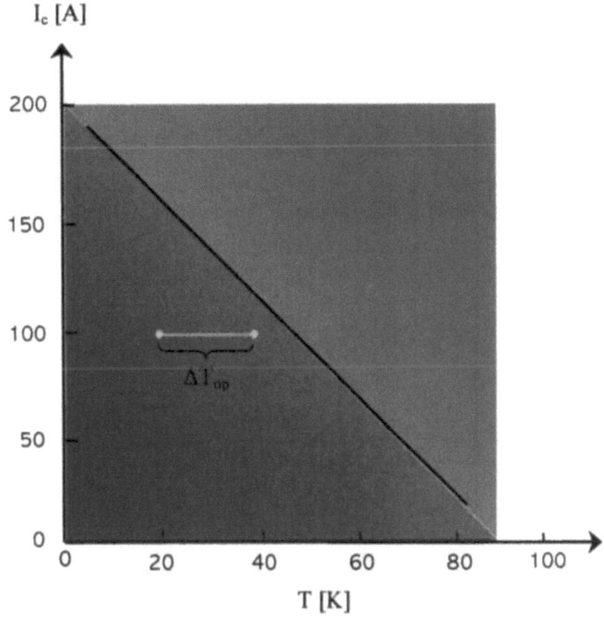

FIG. 15. I_c versus T plot of an HTS magnet operated at 100 A in the temperature range $T_{op_{mn}}$ = 20 K and $T_{op_{mx}}$ = 40 K

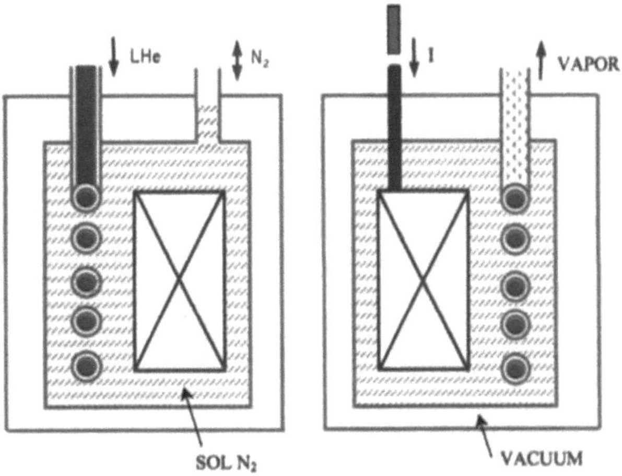

FIG. 16. New system, drawn schematically

System Model

The system, studied theoretically by Kordyuk [5] using a "frozen-image" method, models the bulk YBCO disk as a combination of type I (perfect diamagnet) and type II (hysteretic). Figure 18 illustrates a levitation system in which a permanent magnet disk is levitating an YBCO disk, with each disk schematically represented by a dot, magnetized in the direction indicated by the arrow. In the initial state (Fig. 18a), the induced magnetization of the YBCO and the magnetization of the permanent magnet disk are identical but point in the opposite directions so that the field within the YBCO disk is zero.

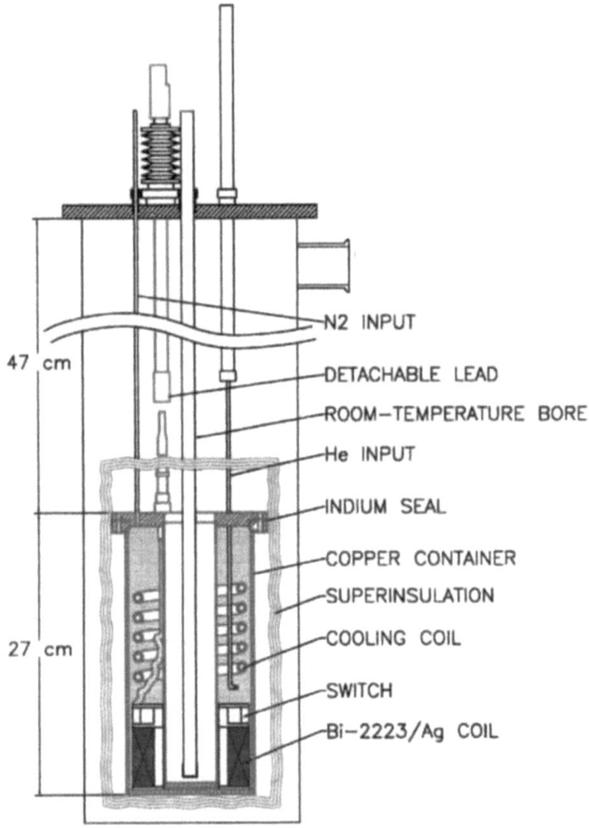

FIG. 17. Schematic drawing of a new system being developed at the FBML

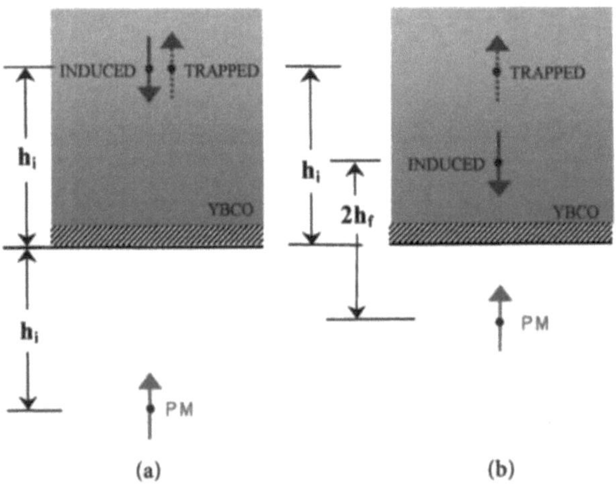

FIG. 18a,b. Schematic representation of a levitation system of a permanent magnet disk and a bulk YBCO disk, at two states: (**a**) initial and (**b**) final

This part of the model assumes that the YBCO is a type I superconductor. The magnetization pointing upward and labeled "trapped" models the YBCO as a type II superconductor with a trapped magnetization that is identical in magnitude but pointing opposite to the induced magnetization, a result of which is a zero net force between the two disks. We thus have:

$$F = f(m_{pm}, m_{id}, h_i) - f(m_{pm}, m_{tf}, h_i) = 0 \tag{7}$$

where F is the levitation force between the two disks; $f(m_{pm}, m_{id}, h_i)$ is the force due to the magnetization of the permanent magnet disk, m_{pm}, and the induced magnetization of the YBCO disk, m_{id}, at a distance of $2h_i$ (a factor of 2 because the YBCO disk is a mirror image of the permanent magnet disk); and $f(m_{pm}, m_{tr}, h_i)$ is the force due to m_{pm} and the trapped magnetization of the YBCO disk, m_{tr}, at a distance of $2h_i$. Because $|m_{id}| = |m_{tf}|$, we have $F = 0$.

Figure 18b represents the final state in which the permanent magnet disk is moved upward so that the distance between it and the induced magnetization of the YBCO disk is $2h_f$. According to this model, the trapped magnetization remains stationary, resulting in a net levitation force between the two disks because $|f(m_{pm}, m_{id}, h_f)| > |f(m_{pm}, m_{tf}, h_i)|$.

Experimental Setup

Figure 19 is a schematic drawing of the experimental setup. An YBCO disk is attached to the tip of an aluminum beam, the other end of which is held clamped. The force on the disk produces a bending moment in the beam, the magnitude of which is measured by two strain gauges attached to the surfaces of the beam near the clamped end. When the top surface gauge is in tension, the bottom surface gauge is in compression and vice versa. Coaxially below the YBCO disk at distance h (between the bottom YBCO disk surface and the top permanent magnet disk surface) is a permanent magnet disk placed on the surface of a steel plate. The steel plate enhances the field of the permanent magnet disk in the space above the disk and almost eliminates that in the space below the disk. Both disks are placed in a styrofoam "cryostat" that is filled with liquid nitrogen to transform the YBCO disk from normal in the initial state to superconducting in the final state. The styrofoam cryostat is placed on a jack whose vertical location is adjustable to within ±0.5 mm.

Procedure

In the initial state, the styrofoam is empty and the YBCO is in the normal state and the jack is set so that $h = h_i$. The styrofoam is filled with liquid nitrogen and magnetic moments m_{id} and m_{tr} are induced in the YBCO (Fig. 18a). In the final state, the permanent magnet disk is moved vertically so that $h = h_f < h_i$ (Fig. 18b). The levitation force, F, generates a bending moment in the beam that is measured by the two strain gauges.

Experimental Results

Kordyuk's model implies that F is independent of the thickness of the YBCO disk. Figure 20 presents h_f versus F plots for two 25.4-mmϕ YBCO disks, 2 mm thick (blue) and 5 mm thick (green), both at $h_i = 10$ mm and with a 12.7 = mmϕ permanent magnet disk. At the

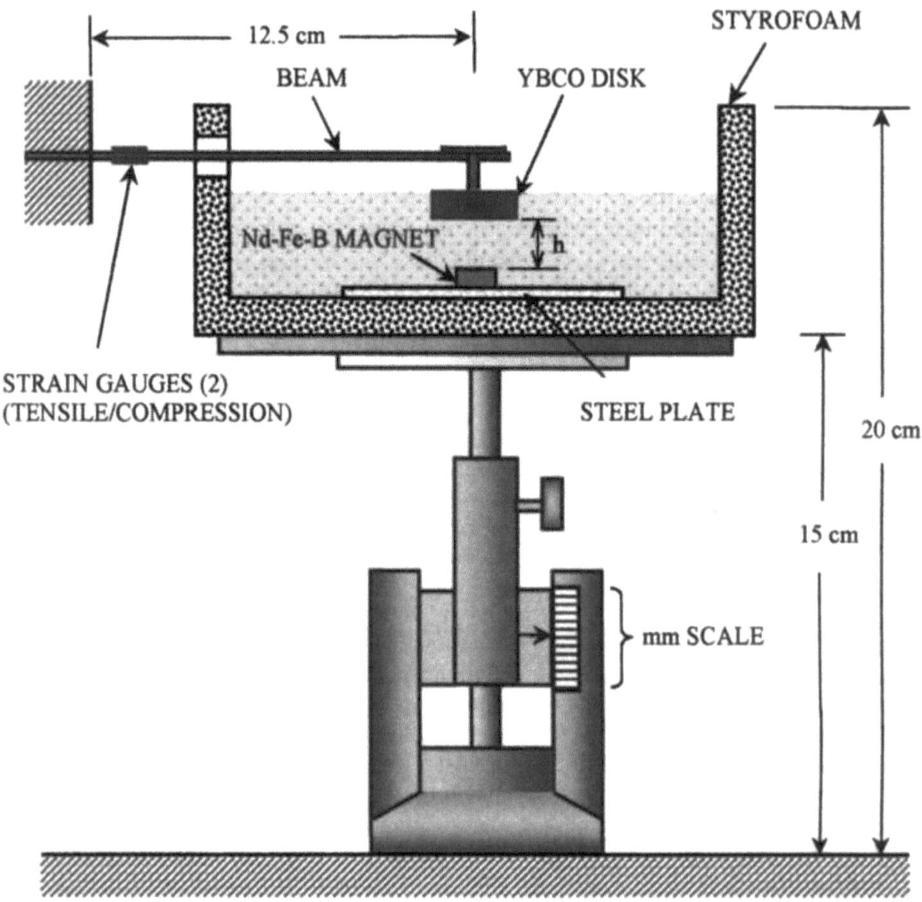

FIG. 19. Schematic drawing of the experimental setup for measuring levitation force between an YBCO disk and a permanent magnet disk

initial state ($h_f = h_i = 10$ mm), there is, as expected from the model, no force ($F = 0$). Indeed one condition that might result in a slight levitation force even initially ($h_f = h_i$) is $|m_{id}| > |m_{tf}|$ for an YBCO disk. The condition of $|m_{id}| > |m_{tf}|$ may exist in "defective" YBCO disks.

Figure 21 shows "Bitter" field patterns of two magnetized 31-mmϕ YBCO disks. The top pattern is symmetrical and corresponds to a 4-mm-thick "clean" disk, while the bottom pattern is asymmetrical and corresponds to a 2-mm-thick "defective" disk. It is possible that the condition of $|m_{id}| > |m_{tf}|$ is satisfied with the defective disk and hence $F > 0$ even at the initial state. Figure 22 presents h_f versus F plots for the two disks, clean (blue) and defective (green). The green plot clearly shows $F > 0$ at $h_f = h_i = 10$ mm, implying that the condition of $|m_{id}| > |m_{tf}|$ is satisfied in the defective disk. Because F is greater with the defective disk than with the clean disk, for creating a greater levitation force for a given set of YBCO and permanent magnet disks, it appears that YBCO disks should have defects that create the condition of $|m_{id}| > |m_{tf}|$.

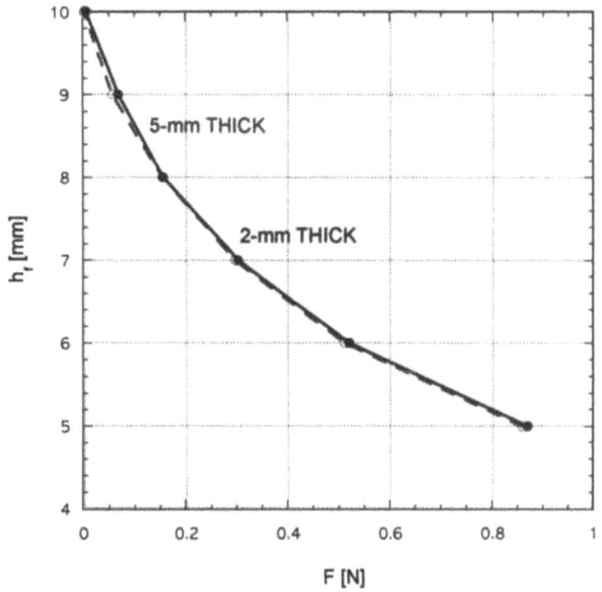

FIG. 20. h_f versus F plots for two 25.4-mmϕ YBCO disks, 2-mm thick (*blue*) and 5-mm thick (*green*), both at h_i = 10 mm and with a 12.7 = mmϕ permanent magnet disk

"MIT" HTS Current Lead

In this project a new concept for the design of a high-temperature superconducting (HTS) current lead, called here the "MIT Lead," is under development. Over a short length at the warm end (e.g., 80 K), the MIT Lead, composed of paralleled BSCCO-2223/Ag-Au tapes, is operated in the *current-sharing* mode [6]. Because MIT Leads rely on cooling by the effluent helium vapor, they are applicable only to devices such as superconducting magnets that operate in a bath of liquid helium. The current-sharing mode operation results in a significant saving of the superconducting materials (BSCCO, Ag, and Au) needed to construct an MIT Lead as compared with a "conventional" lead of the same configuration that is rated for the same transport current and operates fully superconducting over the same temperature range.

Figure 23 shows a schematic drawing of the basic configuration of a current lead operating between room temperature and 4.2 K (liquid helium). The lead is composed of an HTS lead and a gas-cooled lead, the two leads coupled at 80 K. Figure 24 illustrates I_c versus T plots for two HTS leads, both rated at 6 kA. The "IGC" line corresponds to a lead manufactured by the Intermagnetics General Corp., designed to have the HTS section operate fully superconducting over the entire temperature range, 4.2–80 K. Note that at 80 K, the IGC lead has I_c of 6 kA. The "MIT" line corresponds to an MIT Lead that operates in the current-sharing region (red) in the range 73.3 to 80 K, with I_c of 3 KA at 80 K. Because its I_c is one-half that of the IGC lead, the MIT Lead consequently requires one-half of HTS tapes required by the IGC lead. Detailed analysis of the IGC and MIT leads demonstrates that despite joule heating generated in the current-sharing region, the cold-end heat input of the MIT Lead can be equal to or be optimized to be smaller than that of the IGC lead [7]. The analysis also shows that a protection criterion under fault mode or heat exchange consideration ultimately determines the extent to which the material saving can be achieved in MIT Leads.

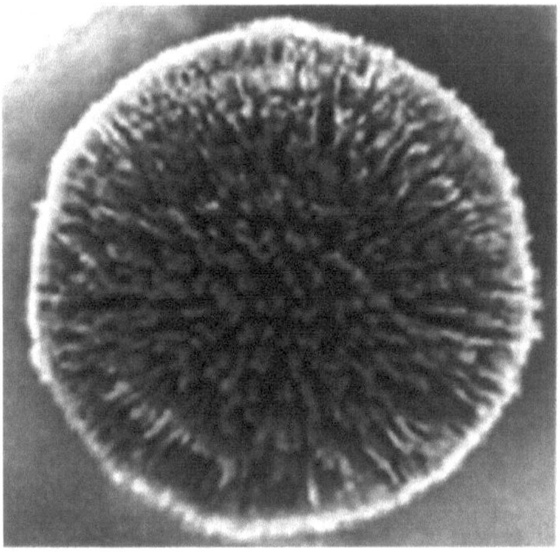

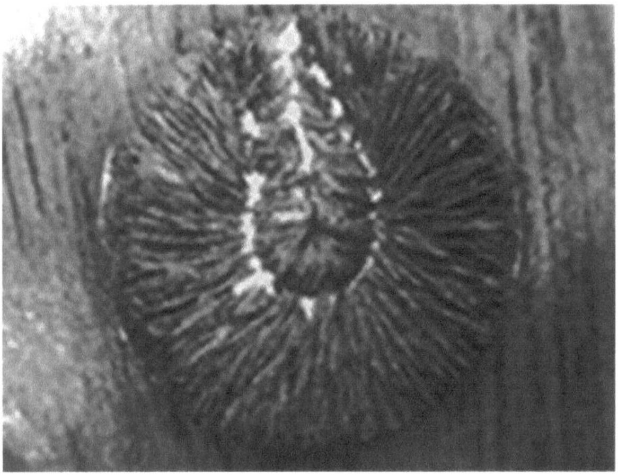

FIG. 21. "Bitter" field patterns for two magnetized 31-mmϕ YBCO disks: "clean" (4 mm thick) (*top*) and "defective" (2 mm thick) (*bottom*)

$SolN_2$ for HTS Winding Under Transient Heating

The large heat capacity of solid nitrogen ($SolN_2$) in the temperature range 20–60 K may also be used to limit the temperature rise of HTS magnets subjected to overcurrent pulses in electric power devices such as fault-current limiters, transformers, motors, and transmission cables. One difficulty in using $SolN_2$ as a heat absorber under transient heating is the extraordinary small thermal diffusivity of $SolN_2$, $[D_{th}]_{sn2}$, compared with that of silver, $[D_{th}]_{ag}$: it is about 1/6300, as seen below:

$$\frac{[D_{th}]_{sn2}}{[D_{th}]_{ag}} = \frac{\left[\dfrac{k}{C}\right]_{sn2}}{\left[\dfrac{k}{C}\right]_{ag}} = \frac{\left[\dfrac{2\,\mathrm{mW/cmK}}{1.39\,\mathrm{J/cm^3K}}\right]_{sn2}}{\left[\dfrac{8\,\mathrm{W/cmK}}{0.90\,\mathrm{J/cm^3K}}\right]_{ag}} \sim \frac{1}{6300} \tag{8}$$

FIG. 22. h_f versus F plots for two 31-mmϕ YBCO disks, clean (*blue*) and defective (*green*), both at h_i = 10 mm and with a 12.7 = mmϕ permanent magnet disk

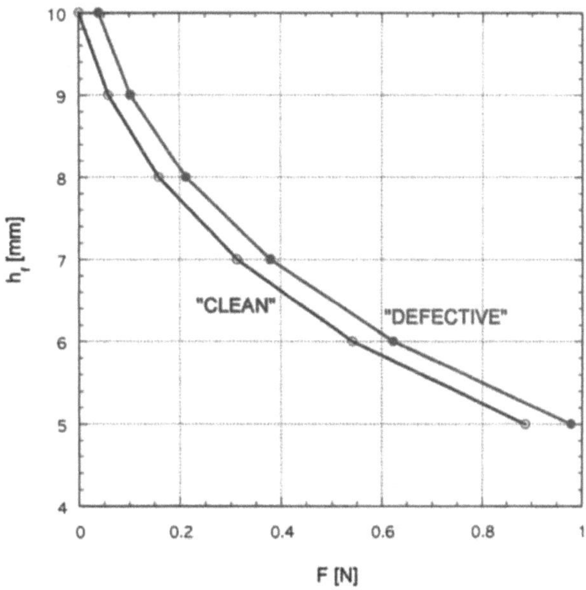

Figure 25 shows a schematic drawing of the experiment setup for this project. Two principal components—a cryocooler and a test coil—are housed inside a cryostat. A test HTS magnet, a single pancake coil as illustrated in the bottom part of the figure or a double-pancake coil, is anchored to the second stage of a cryocooler that maintains the HTS magnet at a given operating temperature. Nitrogen gas, supplied (and relieved) by the "N2 Gas" line in the figure, fills the empty space of the test coil and is condensed and solidified by the cryocooler.

Figure 26 shows typical current (blue) and temperature (green) waveforms for the test coil. The test coil, initially at 40 K, is subjected to an overcurrent pulse, which raises the current through the test coil momentarily (~0.1 s) from 60 A to 200 A. Under transient heating, the thickness of $SolN_2$ in the test coil must be (1) thick enough to contribute heat capacity; (2) thin enough not to degrade λJ, overall current density; and (3) thin enough to make $[\tau_{th}]_{sn2}$ less than heating durations (>100 ms). Thus for a $SolN_2$ thickness, δ_{sn2} = 0.12 mm, one-half of a typical turn-to-turn gap distance in the test coil, we have

$$[\tau_{th}]_{sn2} = \frac{1}{[D_{th}]_{sn2}}\left(\frac{\delta_{sn2}}{\pi}\right)^2 = \frac{(0.012\,\text{cm})^2}{(1.4\times10^{-3}\text{cm}^2/\text{s})\pi^2} = 10\,\text{ms} \qquad (9)$$

That is, for the test coil with a turn-to-turn gap distance of about 0.25 mm, $SolN_2$ would be effective against heating pulses lasting longer than about 10 ms.

Flux Pump for HTS Inserts

This project focuses on the development of a flux pump to be used in high-resolution, high-field NMR magnets, which will most likely be composed of a 900-MHz (21.1-T) LTS

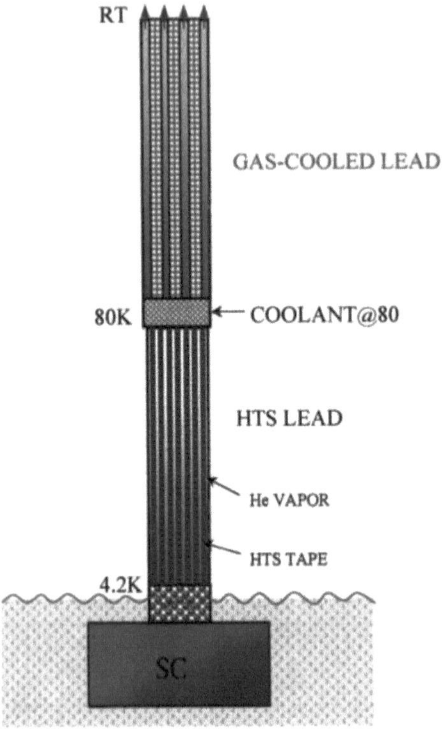

FIG. 23. Schematic drawing of a current lead composed of an HTS section and a gas-cooled lead, coupled at 80 K

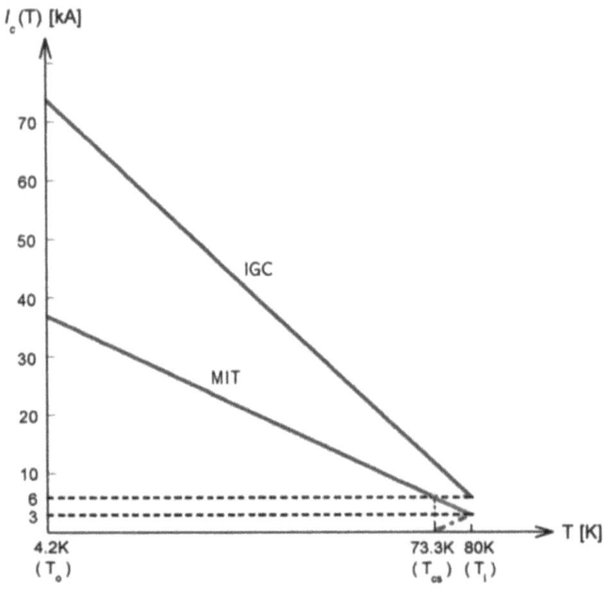

FIG. 24. I_c versus T plots for 6-kA leads. The *IGC* line corresponds to a lead manufactured by the Intermagnetic General Corp

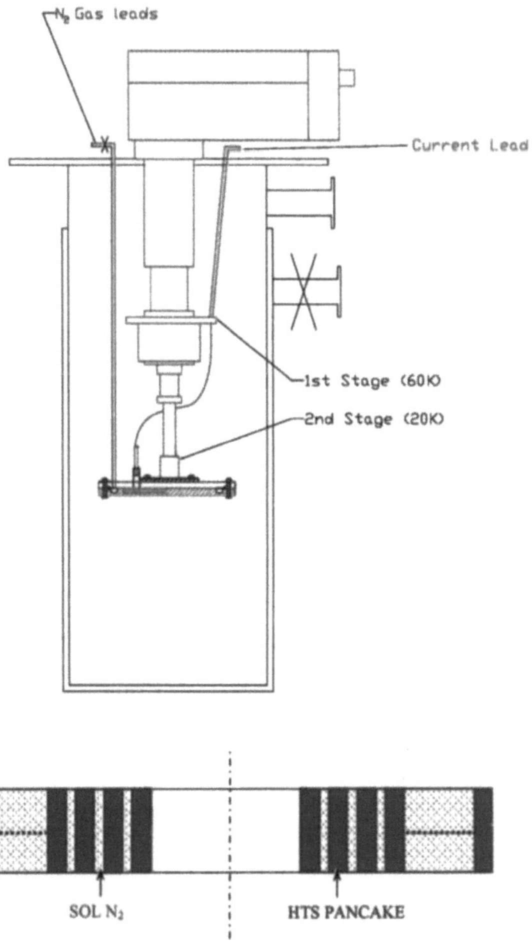

FIG. 25. Schematic drawing of the experimental setup to study the effect of $SolN_2$ on HTS magnet under transient heating. The *bottom* drawing shows a single-pancake test coil with its winding impregnated with $SolN_2$

background field magnet and a 100-MHz (2.4-T) HTS insert. However, because of the HTS poor value of *index* (n), it will be difficult to operate the HTS insert in persistent mode.

Figure 27 shows E versus J curves for three superconductors each with a different index, n. The one with $n = \infty$ corresponds to an ideal superconductor: $E = 0$ for up to J_c, the critical current density at a criterion of $E = E_c$. In the other superconductor there will be an electric field, E_{op}, at an operating current density of J_{op}. The integral of E field along the entire length of the magnet gives rise to the magnet's total dissipative voltage, which causes a field decay if the magnet is operated in persistent mode.

Examples

Typical values of n for Nb-Ti range from 30 to 80, for Nb_3Sn, from 20 to 50, and for Bi-2223, from 10 to 12, although there are a few isolated cases of n as high as 20. Table 4

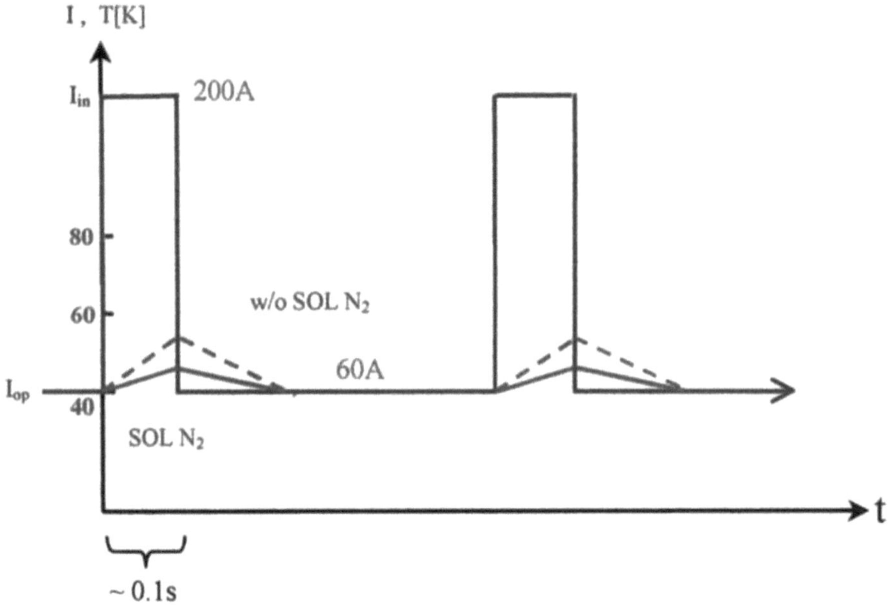

FIG. 26. Current (*blue*) and temperature (*green*) waveforms for the test coil in the presence and absence of $SolN_2$

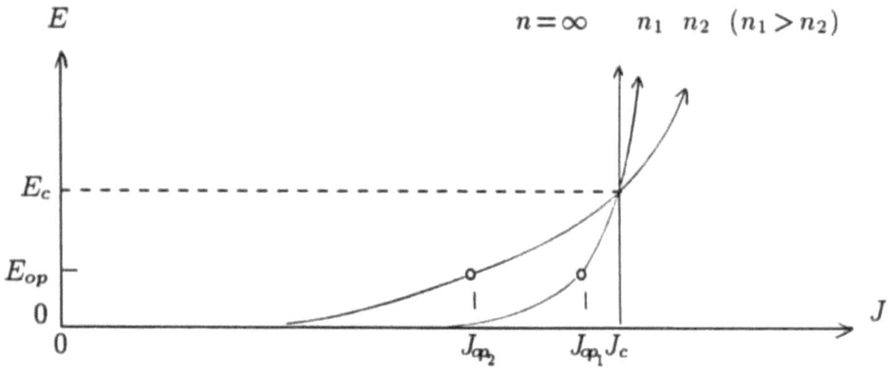

FIG. 27. E versus J plots for superconductors each with a different index, n

presents parameters of LTS and HTS coils, each similar in size and field performance to a typical HTS insert coil for a 1-GHz NMR magnet. Each coil contains a superconductor 2000 m long, of which only 5% (100 m) is in the highest field region of the coil and subject to E_{op} of Fig. 27, because in this region the superconductor J_c is smallest. Each coil is assumed to have a self-inductance of 2 H, and operate at 75 A, which is 75% of I_c. Beyond these parameters, the two coils have different parameters, leading to a significant difference in the decay time: 10^{11} s (32 000 years) for the LTS coil and 3.6×10^6 s (6 weeks) for the HTS coil. A decay time constant of 6 weeks is much shorter than at least approximately 10 000 years required for an NMR-magnet grade HTS insert.

Hydraulic Analogy

Figure 28 shows two hydraulic systems analogous to operating modes of an insert coil for an NMR magnet: (a) persistent and (b) "nearly" persistent. Fluid flow is persistent in Fig. 28a, whereas it is not in Fig. 28b because of a small leak in the system.

Figure 29 shows two hydraulic analogies that would convert a leaky system into effectively persistent: (a) driven, in which fluid circulation is maintained with an external pump and (b) "pumped," in which fluid circulation is made persistent with a pump that matches the leak rate. Q and $q < Q$ each indicates heat input to the "cold" system: the pumped system would receive less heat input than the driven system. Figure 30 shows circuit models. Here Fig. 30a represents a driven system, while Fig. 30b is a nearly persistent system with a flux pump.

TABLE 4. Comparison of LTS and HTS coils: an illustration

Parameters	LTS	HTS
Total conductor length, l (m)	2000	
5% l (peak field), l_{pk} (m)	100 (10^4 cm)	
Self inductance, L (H)	2	
I_c; I_{op} (75% of I_c) (A)	100; 75	
E_c (V/cm)	10^{-9}	10^{-7}
$V_c = l_{pk} E_c$ (V)	10^{-5}	10^{-3}
n	30	10
$V_{op} = V_c(I_{op}/I_c)^n$ (V)	1.8×10^{-9}	5.6×10^{-5}
$R_{op} = V_{op}/I_{op}$ (Ω)	1.8×10^{-11}	5.6×10^{-7}
Time constant (L/R_{op}) (s)	10^{11}	3.6×10^6

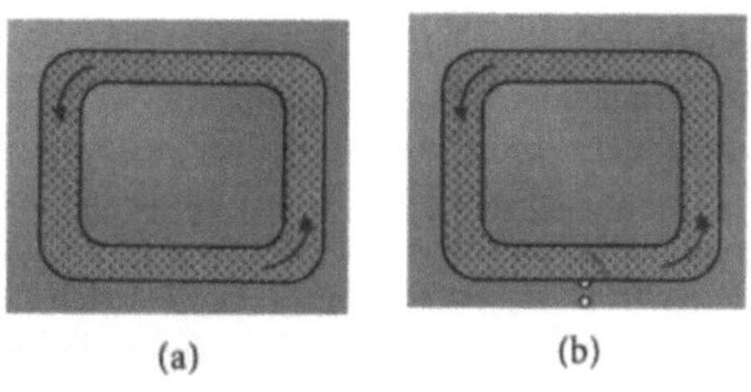

FIG. 28a,b. Hydraulic systems analogous to operating modes of an insert coil for an NMR magnet: (a) persistent mode and (b) "nearly" persistent mode

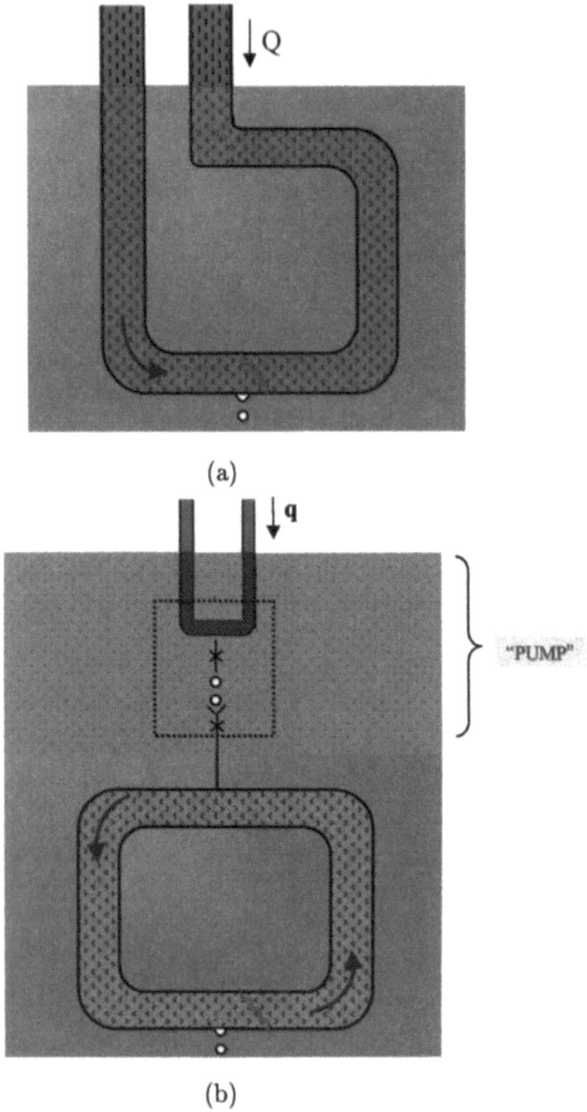

FIG. 29a,b. Hydraulic analogies to convert a nearly persistent system to effectively persistent: (**a**) driven mode and (**b**) "pumped" mode

Beside keeping the heat input to the cold environment smaller than the driven system that must rely on a very stable external current supply, a flux pump can inject precisely metered amounts of flux into the system. Figure 31 illustrates the sequence of operation for a magnet circuit with a flux pump. At first with switches S1 and S2 closed and S3 open, the magnet current, I_1, decays slowly owing to the circuit resistance.

In step 1, switch S3 is closed and the primary of the flux pump is energized to I_p, inducing I_2 in the secondary. At this point I_2 is slightly greater than the intial value of I_1. In step 2, switch S2 is open, forcing $I_1 = I_2$, which itself is an inductance-weighted value of the magnet circuit. (Note that the magnet inductance is much greater than the secondary

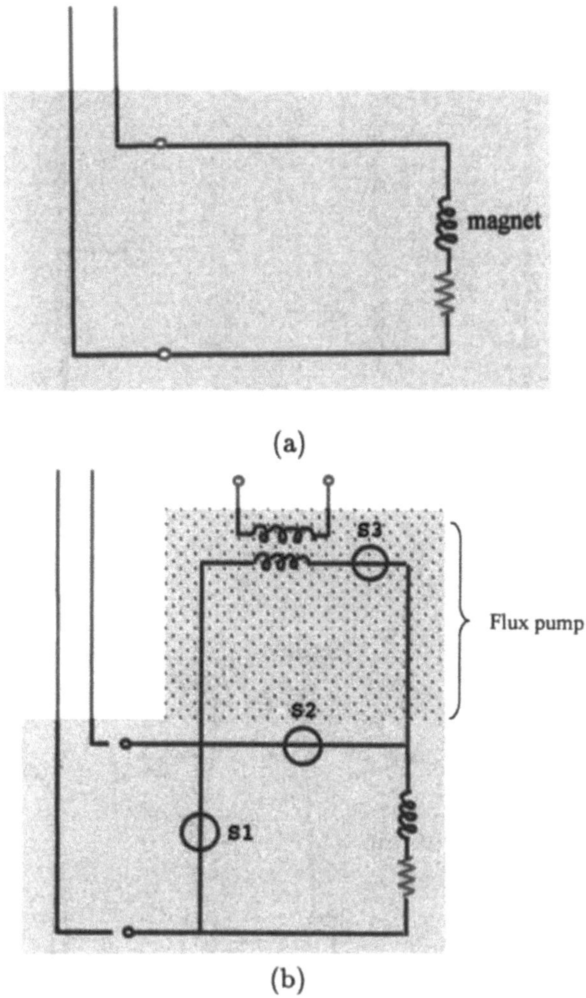

FIG. 30a,b. Two circuits that would convert a nearly persistent circuit to effectively persistent: (**a**) driven mode and (**b**) "pumped" mode

inductance.) In step 3, switch S2 is closed and then switch S3 is opened, returning to the initial condition.

Pulsing frequency depends on the circuit decay time constant and the level of $\Delta I_1/I_1$ that must be maintained. For a typical HTS insert that contributes about 10% of the total field, $\Delta I_1/I_1 \sim 10^{-7}$ is believed adequate.

HTS Inserts for 1-GHz NMR Magnets

This project is the Phase 1 program of a three-phase project at FBML to complete in 2005 a 52-mm RT-bore high-resolution 1 GHz NMR magnet. Table 5 summarizes the three-phase project. In phase 1, a 450-MHz high-resolution LTS/HTS NMR magnet will be operated in 2002. It is a combination of an existing LTS background magnet and an HTS insert

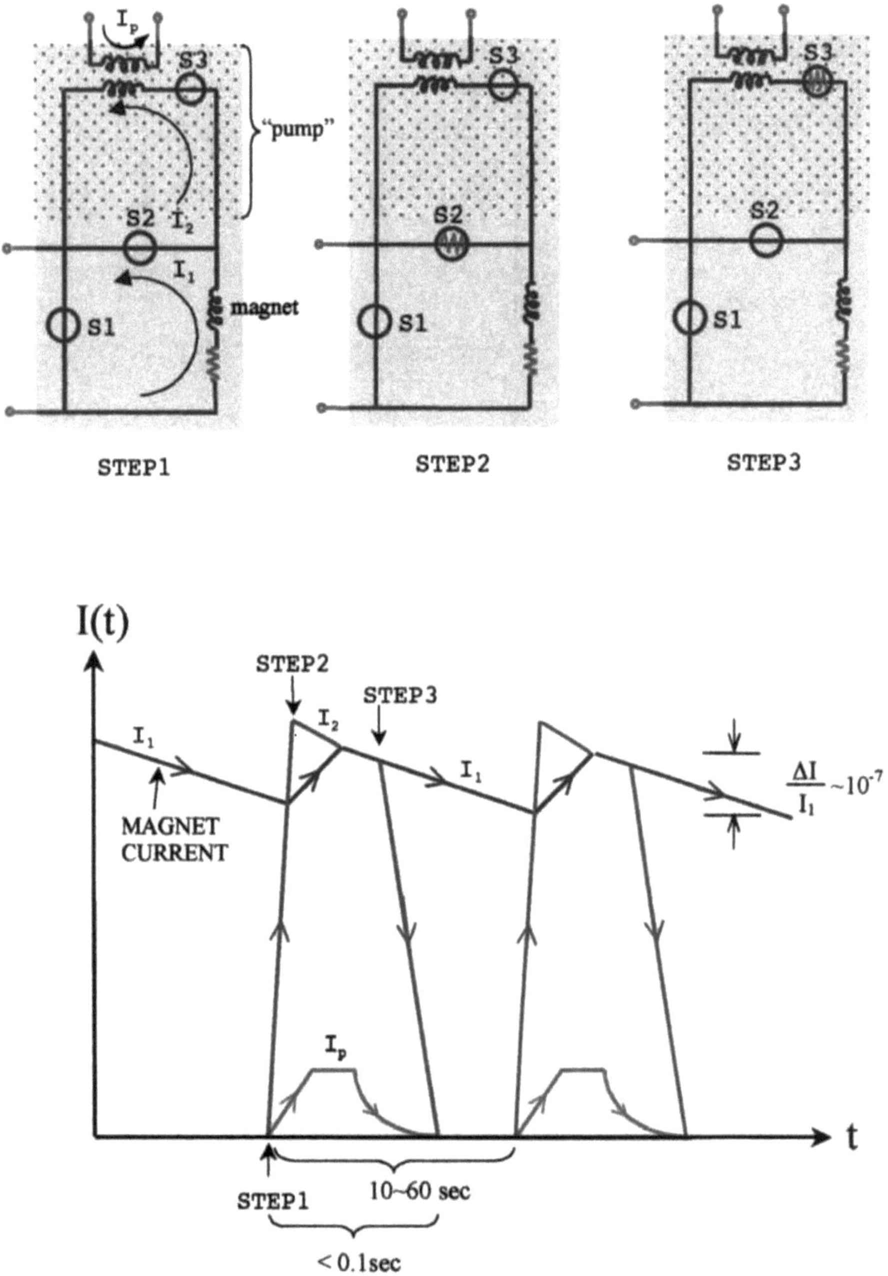

FIG. 31. Sequence of operation: magnet current, $I_1(t)$ (*blue*); the flux pump primary current, $I_p(t)$ (*red*); and secondary current, $I_2(t)$ (*green*)

TABLE 5. Summary of the FBML three-phase 1-GHz NMR magnet project

	Phase 1	Phase 2	Phase 3
Upgraded system (52-mm RT bore)			
Frequency (MHz)	450	700–800	1000
Operating temperature (T_{op}) (K)	2.3	4.2	4.2 (1.8)[a]
Background system			
Frequency (MHz)	372	600–700	900
Cold bore (mm)	140	155–160	
HTS insert (T_{op} range: 1.8–4.2 K)	Insert #1	Insert #2	Insert #3
Winding i.d. (mm)	79.2	80	
Winding o.d. (mm)	127.2	144	
Overall winding length (mm)	343.8	511.80 (tentative)	
Field contribution (T)	1.83	2.35	
Tentative timetable			
Starting date	10/01/00	04/01/02	10/01/03
Completion date	03/31/02	09/30/03	03/31/05

[a] If in 2005, the 900-MHz background magnet still must operate at 1.8 K

(insert #1) that is basically identical to those proposed for phases 2 (insert #2) and 3 (insert #3). We believe that this ultimate goal is attainable only through the successful completion of phases 1 and 2. In each phase an LTS/HTS unit will be built based on essentially the same design principles and philosophy. The present state of the art of superconducting magnet technology as well as that expected in the next few years for both LTS and HTS makes this assiduous step-by-step approach almost mandatory.

The LTS background magnet in the Phase 1 program will be modified from the existing *first* high-resolution LTS (Nb-Ti) 500-MHz NMR magnet completed at FBML in the late 1970s and successfully operated, at 2.3 K, for the next 14 years as a research tool at FBML [8]. In the phase 1 modification, insert #1 will replace the original Nb-Ti high-field insert. Although rated only at 450 MHz and to be operated for a brief period, just long enough to generate NMR signals, this high-resolution LTS/HTS system is expected to validate our basic design principles and philosophy to achieve the 1-GHz goal at the conclusion of the phase 3 program.

Figure 32 shows a cross-sectional view of a high-resolution 450-MHz NMR magnet to be completed in phase 1. As stated earlier, an HTS insert (*green*) is replacing the high-field Nb-Ti insert in the original high-resolution 500-MHz NMR magnet, which, without the insert, is shown blue in the figure. The HTS insert consists of 43 double pancakes, each wound with BSCCO-2223/Ag composite tape. Because of index-induced dissipation, the HTS insert will be operated in driven mode.

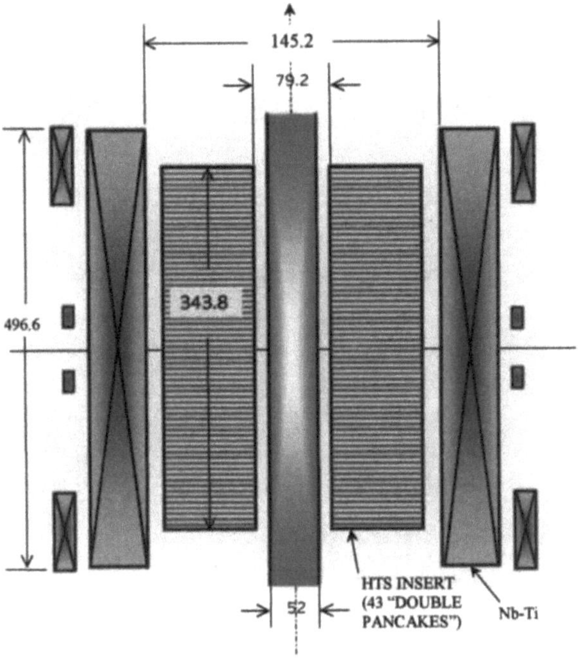

FIG. 32. Cross section of the phase 1 high-resolution 450-MHz NMR magnet. HTS insert (*green*) is replacing the high-field Nb-Ti insert in the original 500-MHz NMR magnet, which, without the insert, is shown in *blue*

Conclusions: Present Status and Future Prospects

LTS Magnets

LTS magnet technology is essentially established, although adiabatic magnets are still vulnerable to *premature quenches*. LTS magnets for NMR, MRI, and HEP applications have been accepted as enabling, while those for energy devices have yet to achieve general acceptance.

HTS Magnets

The greatest deterrent for wide acceptance of HTS magnets is the high cost of HTS conductors [9]. It is also desirable to improve T_c [9]. HTS magnets *do* offer wide operating temperature ranges and are almost immune from disturbances that afflict adiabatic LTS magnets. Remaining technical issues include protection against fault-mode conditions and AC dissipation that must be minimized. Another key challenge is identification, and technical and economic proof, of large-quantity applications where HTS magnets can excel and dominate.

Figure 33 shows the chronological progress of superconductivity from 1911 to 2000 and beyond. Items in red are important events in physics; those in green are in engineering. Superconducting magnet technology began in the 1960s. Applications masked in green are well established for superconductivity, while those in yellow are yet to be established.

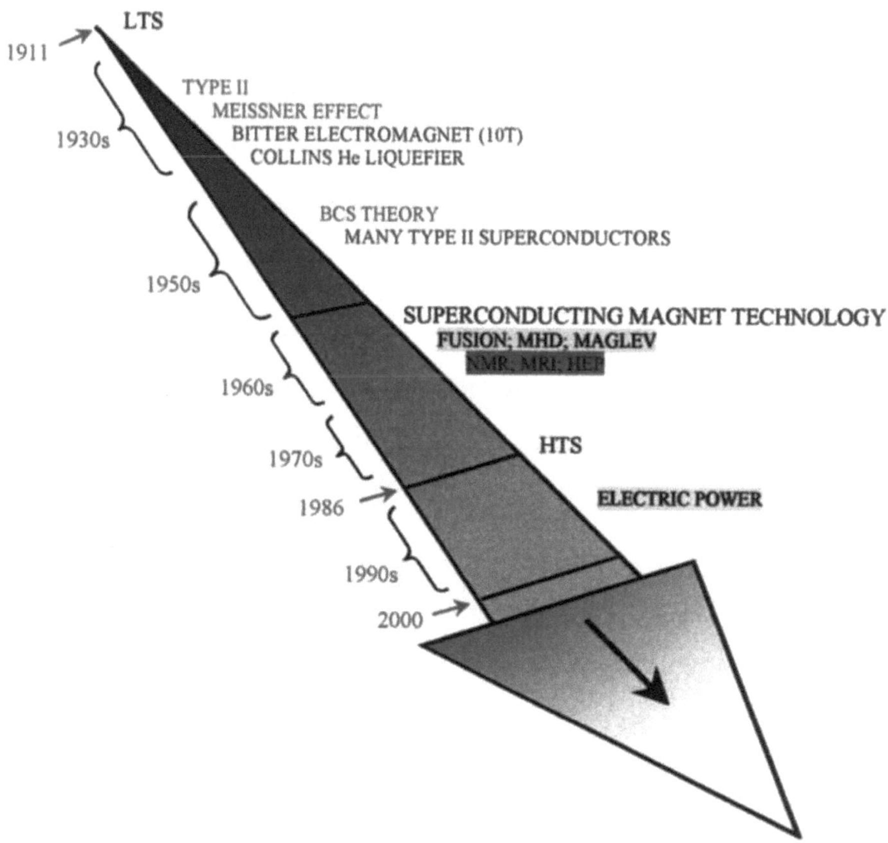

FIG. 33. Superconductivity—past, present, future

Acknowledgments. I express my deep gratitude to the Tokyo Electric Power Company and Keio University for the opportunity to spend a part of each year in 1996–1998 at the university. I thank Profs. Yasuhiko Mori and Kouhei Ohnishi for coordinating my visitation and organizing this Symposium on Energy and Environment based on which this article is prepared; and Prof. Koichiro Sawa for having me in his group during my stay at Keio. I am also grateful to Dr. Haigun Lee for valuable contributions to the ongoing projects at FBML and also for creating the color figures used in my symposium lecture and this article; and to graduate students Benjamin Haid and Karl Kowallis for their contributions to the ongoing projects.

References

1. Iwasa Y (1994) Case studies in superconducting magnets. Plenum Press, New York
2. Kowallis KW (2000) Transient thermal recovery of a pulsed BSCCO-2223 superconducting magnet. M. S. thesis, Dept. of Mechanical Engineering, Massachusetts Institute of Technology, Cambridge

3. Iwasa Y (1998) Permanent magnet system based on high-temperature superconductors with recooling and recharging capabilities. U.S. patent 5,724,820 (March 10, 1998)
4. Murakami M, Yoo SI, Sakai N, Sen SJ, Otsui T, Miyamoto T, Chikumoto N, Mochida T, Higuchi T, Nagashima K (1999) Applications of bulk high temperature superconductors. Proceedings, ICEC 17 (1998). Institute of Physics Publishing, Bristol, p 23
5. Kordyuk AA (1998) Magnetic levitation for hard superconductors. Journal of Applied Physics 83:610
6. Iwasa Y (1999) Method for current sharing in a superconducting lead. Filed for a U.S. patent to the U.S. Patent & Trademark Office, Washington, D.C. (June 9, 1999)
7. Iwasa Y, Lee H (2000) High-temperature superconducting current lead incorporating operation in the current-sharing mode. Cryogenics 40:200–219
8. Williams JEC, Neuringer LJ, Bobrov ES, Weggel RJ, Ruben D, Harrison W (1981) Magnet system of a 500 MHz NMR spectrometer. Part I: Magnet development. Review of Scientific Instrumentation 52:649
9. Iwasa Y (1998) Major obstacle and R&D initiatives for high-temperature superconducting devices. ISTEC Journal 11(3):1

Part II: Thermal Science and Engineering for Energy Conversion/Utilization Processes

Novel Optical Instrumentation Meeting the Challenges for Energy and the Environment

Pavlos G. Aleiferis, Ilias Prassas, and Alex M. K. P. Taylor*

Summary. This chapter is composed of two parts. The first describes novel optical instrumentation, based on the latest image-intensified CCD technology. It has direct storage of the video images in computer memory, applied to the initial flame kernel growth in a Honda VTEC-E optically accessed single cylinder lean-burn engine operating at 1500 rpm and an air/fuel ratio of 22:1. The images are complemented by three-dimensional renditions of the in-cylinder velocity measured by laser Doppler anemometry, which allows visualisation of the mean convection pattern of the rich fuel–air cloud from its inception by evaporation at the cylinder liner to ignition in the pentroof. The combined swirl and tumble vortex is shown by these images to be a single vortex pattern with a strongly inclined axis of rotation. The second part describes work in progress in the development of the shadow Doppler anemometer. This sizing anemometer is based on laser Doppler anemometry, which provides temporally and spatially resolved simultaneous measurements for particles of arbitrary shapes and optical properties. The review encompasses the latest developments in sizing accuracy and the lower limits of measurable size, use of the shadow Doppler anemometer to establish the accuracy of flux measurements made by the phase Doppler anemometer, trajectory bias effects of the shadow Doppler anemometer, and future detector developments.

Key words. Optical techniques—Particle sizing—Shadow Doppler anemometer—Spark-ignition engines—Cyclic variability

Preamble

The themes of this volume are energy and the environment, reflected by the two parts to this contribution. The first and larger part concerns our research into lean-burn piston engines. The advantage in terms of both fuel efficiency and lower emissions compared with stoichiometrically fired engines is well described in standard textbooks and needs

Department of Mechanical Engineering, Imperial College of Science, Technology and Medicine, Exhibition Road, London SW7 2BX, UK
* Corresponding author

no repetition here. The work we describe in this chapter makes extensive use of optical techniques and the novelty they involve. The purpose of using them is explained in the first part of the chapter. The second, smaller part is a description of the "work in progress" in developing a spatially and temporally precise "sizing velocimeter," the particular novelty of this laser-based instrument being that it sizes particles of arbitrary shapes and optical properties. The instrument was originally developed for laboratory-scale experiments involving "pulverised fuel," which is usually in the form of milled hard, or brown/lignite, coals but has application to a wide range of industrial processes that have consequences particularly for emissions. The common theme that links these two otherwise separate contributions is that the work has been performed in conjunction with Japanese companies and colleagues, as becomes apparent below.

Part I. Lean-Burn Piston Engines

Introduction

Scope and Background

The scope of this paper is our recent efforts to elucidate of the mechanisms of charge preparation and the *i*nitial *f*lame *k*ernel *g*rowth (IFKG) of a "lean-burn" spark ignition piston engine. These engines are an attractive means for powering vehicles from the environmental point of view because of their high efficiency—and hence lower specific CO_2 emissions and lower NO_X emissions—compared to a stoichiometrically charged piston engine [1]. Broadly, it is advantageous to make such engines operate as leanly as possible; but as the *a*ir/*f*uel (A/F) ratio of the mixture is increased above stoichiometric levels the *c*ycle-*b*y-*c*ycle *v*ariations (CBCV) in flame development and consequently in *i*ndicated *m*ean *e*ffective *p*ressure (IMEP) become more prominent and eventually limit the range of the lean-burn operation because they cause poor car driveability [2]. This means that maintaining "smooth" engine operation under lean-burn conditions is not an easy task, and the ultimate aim of our work is to provide the understanding necessary to extend the current limits of the A/F ratio at which engine designs operate.

The optical instruments which we use are primarily (1) the *p*hase (PDA) or *l*aser (LDA) *d*oppler *a*nemometer (or both) for simultaneously measuring fuel droplet size and velocity or gaseous velocity and (2) *c*harge *c*ouple *d*evice (CCD) cameras for imaging the IFKG. Both of these approaches rely on the use of an optically accessible piston engine.

The PDA and LDA rely on detection and measurement of the interference pattern that arises when a droplet crosses two intersecting laser beams. The CCD camera is "image-intensified" to detect the low levels of natural emission from the flame. Because of the excellent optical access provided by the engine, we have been able to record two simultaneous orthogonal views of the flame, which allows us to study the flame in greater detail than would have been the case otherwise. It is the CCD camera optical instrumentation that constitutes one novel aspect, mainly in terms of the speed and the ease with which images can be stored and manipulated in a personal computer's (PC's) digital memory and on CD-ROM, by-passing the need to use video tape, which was the case up to a few years ago. The other novel aspect is the comparative ease with which complicated flow patterns can be depicted as a movie made by appropriate commercial software (e.g., QuickTime Player, Apple Computer, Inc.).

Lean-Burn Engines Considered

It has been known for some time that lean-burn engines must have charge stratification, in the sense that the mixture near the spark plug must be richer than average [3]. However, it was only recently that research has demonstrated the flow structures that lead to successful lean-burn engine designs. In this context, and to keep the account brief, we refer to the work of Berckmüller et al. [4], Hardalupas et al. [5], and Carabateas et al. [6]. The former used *l*aser-*i*nduced *f*luorescence (LIF) to measure the A/F ratio in planes near the spark plug throughout the intake and compression strokes of an optically accessed single-cylinder research engine. The latter two investigative groups used PDA and LDA, respectively, to measure the in-cylinder droplet size and velocity and in-cylinder gas flows in nominally identical engine blocks. Hardalupas et al. [5] limited their measurements to planes similar to those of Berckmüller et al. [4] and used a cylinder head nominally identical to that used by Berckmüller et al. [4]. Carabateas et al. [6], in contrast, had access to a completely transparent optical liner and was able to measure three essentially orthogonal components of velocity along a large fraction of the length of the cylinder. Carabateas et al. [6] used a modified cylinder head that employed a passage between the two inlet ports upstream of the plane containing the inlet valves (in conjunction with a twin-jet air-assisted injector). Their experiments showed that the in-cylinder gas motion was similar for both cylinder heads, although the droplet distribution and the lean stability limits were modified. A qualitatively correct impression of the flow of the droplets and the motion of the (gaseous phase) fuel–air mixture has emerged from these studies. However, the flow pattern is strongly three-dimensional; and so far the pattern has been reported in terms of two-dimensional plots that show the expected "tumble" and "swirl" circulatory patterns about axes, respectively, perpendicular to and parallel to the axis of the cylinder. It is nevertheless different to visualise the resulting three-dimensional flow pattern and its implications for improving the extension of the lean-burn limit of the engine. The *first* purpose of this chapter is therefore to present some three-dimensional representations of the flow pattern in an example of a commercially successful lean-burn engine and thereby to facilitate discussion of the implications for improved engine design.

The combustion process in spark-ignition engines consists of four main stages: spark and flame initiation, initial flame kernel development, turbulent flame propagation, and flame termination. It is the first two stages of these four that are the most important in terms of the CBCV during combustion [2]. More specifically, the correlation between the IMEP and the maximum *m*ass *f*raction *b*urned (MFB) or between the IMEP and the *c*rank *a*ngle (CA) corresponding to 5% MFB ($\theta_{Xb5\%}$) under low-load engine conditions has been found to be quite strong [7–9]. The implications of these findings is that the overall "strength" of a cycle is generally determined by the speed of combustion to, at most, 5% MFB.

The 5% MFB figure is arbitrary in the sense that it has been determined by experimental convenience: It is difficult to measure low MFB values from the indicator diagram because of uncertainties in the measured in-cylinder pressure. It is certainly of interest to know how early a correlation is established between $\theta_{Xb5\%}$ and IMEP; consequently, various techniques have to be employed to investigate the initial flame kernel growth at a CA that is degrees closer to ignition timing than $\theta_{Xb5\%}$. Note that $\theta_{Xb5\%}$ corresponds to about a 40° CA *a*fter *i*gnition *t*iming (AIT) for an A/F ratio of 22, a low load (30% volumetric efficiency), and 1500 revolutions per minute (rpm) in the engine investigated.

Imaging the IFKG has proven particularly useful in this context because the minimum flame volume that can be adequately resolved by imaging corresponds to only tenths of one percent, but most of it has been mainly conducted under skip-firing conditions in either square-piston engines or two-valve, headed ones with disk-shaped combustion chambers, based mainly on the Schlieren technique [10–12]. Combustion in pentroof-head geometries has been investigated by directly imaging the light radiated by the flame using image-intensified CCD cameras [13,14]. These studies presented some limited relations between calculated flame areas and maximum pressure (P_{max}), combustion duration, and $\theta_{Xb5\%}$ or $\theta_{Xb10\%}$.

The aim of the recently published work by Aleiferis et al. [15] was to determine how early in the IFKG stage fast cycles arise (i.e., to identify when the relation between IMEP and $\theta_{Xb5\%}$ begins to be established) and, if possible, what causes it. This investigation was concerned with extensive *stereoscopic* CCD imaging of the IFKG, which permitted (among other things estimation of flame volume which is closely related to MFB) at CAs between 10° and 40° AIT to provide insight into the relation between the IFKG and the subsequent level of IMEP or $\theta_{Xb5\%}$ for the critical early stage of 0%–5% MFB.

The *second* purpose of this chapter is to present flame images for spark-plug orientations different from those reported by Aleiferis et al. [15], compare them explicitly with images for stoichiometric flames, and present correlations between $\theta_{Xb5\%}$ and IFKG based on projected flame areas (rather than volume, as reported by Aleiferis et al. [15]). The motivation behind this analysis is the fact that stereoscopic imaging, although clearly desirable, is not always possible. It is of interest to establish the extent to which the projected flame area is an acceptable surrogate for the flame volume.

Engine and Experimental Techniques

A single-cylinder research engine with optical access to the combustion chamber was used for the present work. The engine employed the Honda single *o*ver*h*ead *c*amshaft (SOHC) *v*ariable valve *t*iming and lift *e*lectronic *c*ontrol (VTEC) mechanism for the inlet valves that was introduced by Horie et al. [16] and is known as the VTEC-E mechanism. The research engine retained most of the characteristics of the Honda VTEC-E 1.5-litre, four-valve-per-cylinder, pentroof-type geometry mass production engine that operates typically with an A/F ratio of 22 under 3000 rpm at low power. For those conditions, one of the two intake valves is party deactivated so swirl and tumble motions are introduced into the flow to promote axial stratification of the in-cylinder mixture. The stratification is achieved with sequential injection into the inlet stroke, so a rich mixture is present near the spark plug at ignition timing [5]. At a full load above 2500 rpm and a partial load above 3200 rpm, both intake valves open fully to increase volumetric efficiency; and either a stoichiometric or slightly rich mixture is used to provide high-power output. The research engine operated only in the VTEC-E secondary-valve deactivation mode and produced a swirl ratio of 2.5 and a tumble ratio of 1.9 [5,16]. Optical access to the combustion chamber was available through two orthogonal directions: one other through triangular windows on the cylinder head and the other through the piston crown.

The fuel used, isooctane, was injected by a commercial twin-jet air-assisted injector. The ignition system used was a fully transistorised coil ignition system with electronic ignition available in the Honda mass-production engines. A commercial J-type ground electrode, V-grooved central electrode spark plug was employed. The A/F ratio was adjusted by altering the injection duration, and it was continuously monitored by a high-speed linear A/F ratio sensor mounted in the exhaust manifold. The in-cylinder pressure measurements were conducted with a water-cooled piezoelectric pressure transducer connected to a charge amplifier and digitised with 12-bit resolution. The pressure traces were used to calculate the gross IMEP and its *c*oefficient *o*f *v*ariation

$$COV_{IMEP} = 100\sigma_{IMEP}/IMEP_{mean}$$

where σ_{IMEP} is the standard deviation of IMEP. (It has been found that vehicle driveability problems usually occur when the COV_{IMEP} exceeds 10% [2].) A two-zone thermodynamic model based on the formulation of Lavoie et al. [17] was employed to calculate the MFB from the pressure traces. The operating conditions used throughout all the experiments are as follows: engine speed 1500 rpm, volumetric efficiency 30%, A/F ratio 15 or 22 ($\phi = 1.00$ or 0.68), injection at 30° CA after *i*ntake *t*op *d*ead *c*enter (ITDC), and ignition at 40° CA before *f*iring TDC (FTDC) with no skip-firing. Three-component in-cylinder velocity measurements were obtained by the LDA at various axial distances from the TDC, from 60°–300° CA after ITDC during the intake and compression strokes. Various locations were used on each plane, spanning the entire cylinder cross section. A CCD camera coupled with an ultraviolet (UV)-sensitive variable-gain gated image intensifier and a UV-sensitive lens was used for flame imaging. The camera was connected to a monochrome frame grabber triggered to synchronise image acquisition with the CA position. The images were digitised with 8-bit resolution into 256 gray scales with a typical size of 768 × 576 pixels. Flame images were acquired at various CA positions from 10° to 40° AIT through the piston crown (plane of in-cylinder swirl motion) and a pentroof window (plane of in-cylinder tumble motion) individually or through both of these views simultaneously by employing an array of mirrors to focus the two views on the CCD array of a single video camera. The 10°–20° CA AIT images corresponded to CA positions at which the spark still existed; the latest timing at which flame images were acquired (40° CA AIT) corresponded to the FTDC. Details about the engine test bed arrangement, the experimental techniques, and the processing methods of the acquired data have been described in detail elsewhere [5,6,15].

Fluid Particle Tracking in a Honda VTEC-E Engine

Method

The in-cylinder velocity measurements of Carabateas [18] were stored on a numerical grid and were used to calculate the trajectory of fluid particles inside the cylinder. The particles were released at given locations and were assumed to have the velocity of the measured local flow; they travelled distance *S* during a time (or crank angle) step (Δt). Analytically, the location of a fluid particle was calculated from:

$$\vec{x}_i = \vec{x}_{i-1} + \vec{u}\Delta t \tag{1}$$

where i corresponds to the current time (or crank angle degrees in engine terms), i − 1 is the previous time step, and $\vec{x}$ is the position of a fluid particle, i.e.:

$$\vec{x} = \begin{pmatrix} x \\ y \\ z \end{pmatrix} \tag{2}$$

where the (x, y) pair are the particle coordinates on a cylinder cross section, and z is the axial distance. In a similar fashion:

$$\vec{u} = \begin{pmatrix} u \\ v \\ w \end{pmatrix} \quad (3)$$

where, equivalently, (u, v) are the *mean* velocity components on a cross section, and w is the *mean* axial component. The single most difficult aspect of the calculation was the grid irregularity of the velocity measurements (the mesh was structured along the axis of the cylinder but not in the cross-bore planes); hence advanced interpolation schemes suitable for unstructured meshes were used [19,20]. This irregularity also made it difficult to assess when a particle was about to approach a cylinder wall, adjacent to which only few measurements existed. Readers should also note that this data set did not extend into the pentroof, although such data do exist [21].

Trajectories of typically 100 fluid particles were calculated with a time/crank angle step equivalent to 1° CA. Smaller and larger time steps were also tried, but this one was found to be a good compromise between numerical and experimental accuracy. It should be noted that the measurements were temporally windowed (i.e., a measurement reported at, for example, 500° CA was in fact the average of all measurements obtained for 500° ± 5° CA).

Visualisation

After all of the trajectories were calculated and stored on a hard disk, a postprocessing software, written in Mathematica (Wolfram Research), was used to visualise the results. For this purpose, a true three-dimensional model cylinder with a conical top was designed, resembling the Honda VTEC-E cylinder and cylinder head. A section of the cylinder wall was cut away to allow observation of the visualised flow inside it. The piston motion was simulated by plotting a circular disk with its upper surface at locations that corresponded to those of the piston crown as a function of time. All of the drawings were produced to scale. The current location of a fluid particle was denoted by a coloured sphere, and the "tail" of the sphere was a representation of the locations of the same sphere during the previous time steps and hence represented the time history of the trajectory. The trajectories are *pathlines* (*not* streaklines) of the flow. Snapshots of the flow were created every 5° CA. The snapshot sequence between 0° and 360° CA after ITDC (corresponding to the intake and compression strokes) was integrated in a QuickTime video with Sorenson compression (Sorenson Vision), which realistically depicted the airflow motion inside the engine based on experimental data.

Figure 1 shows eight snapshots from QuickTime animation at selected crank angles between 70° and 270° after ITDC, as indicated in Fig. 1. A small drawing at the bottom shows the relative position of the intake and exhaust valves in *plan* view (i.e., looking from the cylinder head towards the piston crown). In this particular simulation two batches, or "rafts," of fluid particles were released in the flow: (1) 100 particles 10 mm below the TDC at 70° CA after ITDC (colour-coded as blue spheres) and (b) 100 particles 50 mm below TDC at 140° CA after ITDC (colour-coded as red spheres). Note that a proportion of the blue and red particles were lost during the course of the integration because some entered the region adjacent to the wall, for which the experimental grid was too sparse to prevent particle loss from the calculation domain.

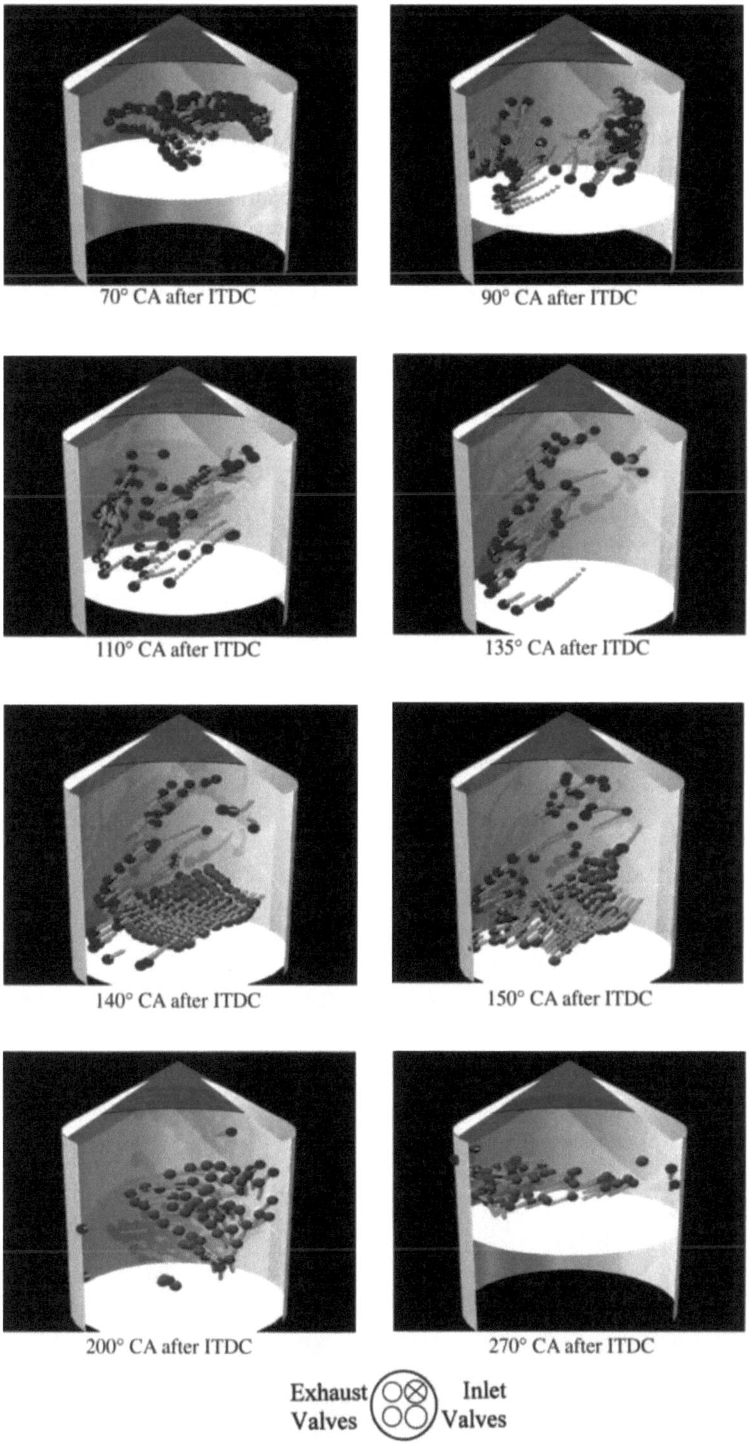

FIG. 1. Locations of fluid particles for various crank angle (*CA*) degrees after intake top dead centre (*ITDC*)

For comparison, Fig. 2 shows the two-dimensional representation of the same velocity field, as measured by *p*ulse *c*ode *p*article *v*elocimetry (PCPV) [6]. The vectors in Fig. 2 give a clear impression of the existence of a swirl motion (at least beyond about 90° CA after ITDC), and the colour-coded contours show the existence of a tumble motion.

For reference, Fig. 3 shows the cycle-averaged droplet volume flux at 16 mm below the spark plug, as measured by Hardalupas et al. [5]. These workers argued, on the basis of

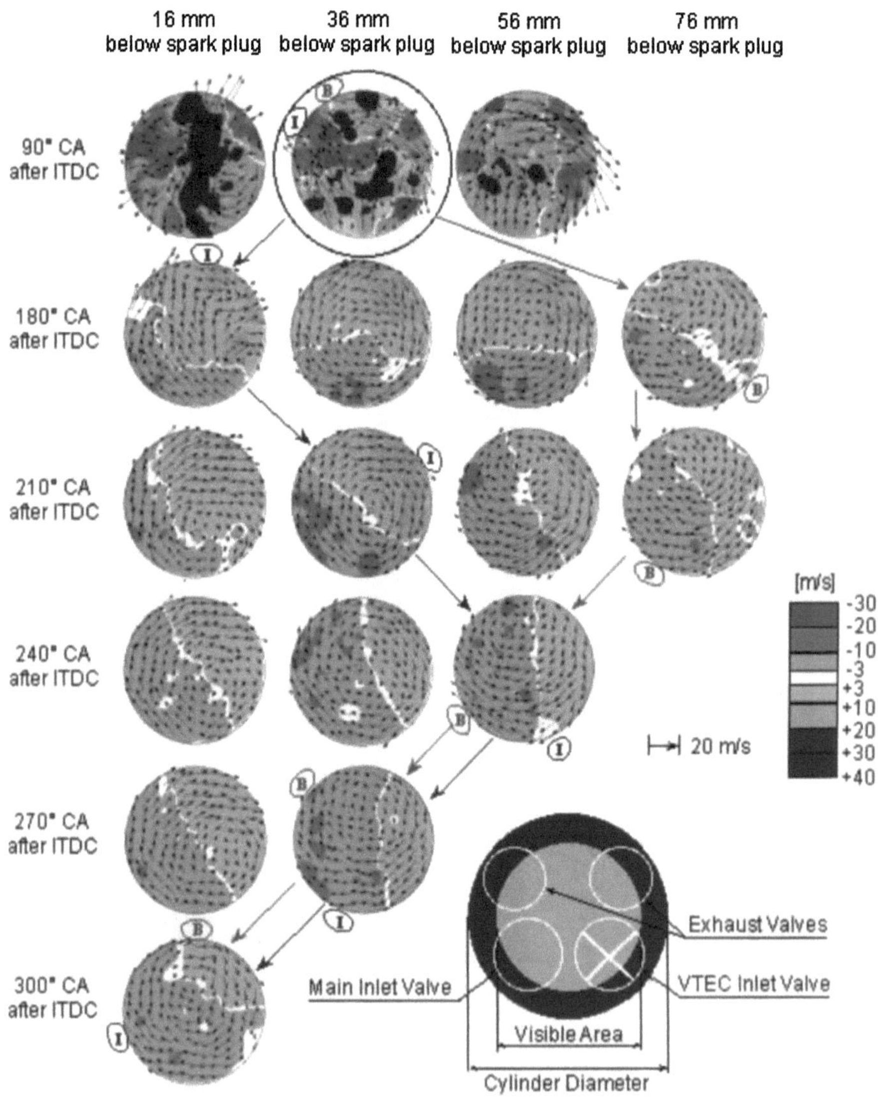

FIG. 2. Contours of instantaneous air velocity on different planes below the spark-plug gap for different crank angle (*CA*) degrees after intake top dead centre (*ITDC*). *VTEC*, variable value timing and lift electronic control

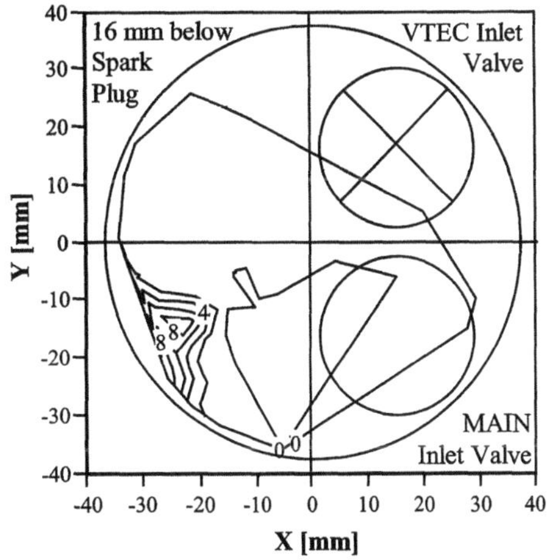

FIG. 3. Cycle-averaged droplet volume flux on a plane 16 mm below the spark plug for start of injection at 30° crank angle after intake top dead centre

Fig. 3 and other information, that a large fraction of the injected fuel entered the cylinder in liquid form (not vapour form), having first struck the inlet valve as a spatially concentrated droplet stream that struck the cylinder liner about 30 mm below the spark plug in the primary exhaust valve quadrant. There the liquid evaporated into a rich vapour cloud, which they argued (based almost exclusively on the evidence of the tumble motion only) was convected by a vortex (with its axis implied to be horizontal) to appear on the cylinder half underneath the inlet valves and then into the pentroof and past the spark plug. They went on to argue that appropriate timing of the injection ensured that this cloud was in the vicinity of the spark plug at ignition. Their arguments were qualitative but were consistent with the LIF measurements of the fuel vapour reported by Berckmüller et al. [4].

In the work of Carabateas et al. [6], a semiquantitative "graphical integration" procedure was adopted, which is shown in Fig. 2 as trajectories I and B. Trajectory I in Fig. 2 corresponds to the suggested trajectory of the vapour cloud produced by the impingement of the droplet stream produced by the modified (or *i*mproved) cylinder head of Carabateas et al. [6], and this stream was circumferentially slightly displaced relative to that measured by Hardalupas et al. [5] and indicated by B ("*b*asic"). The thrust of the work of Carabateas et al. [6] was that even apparently minor displacement in the trajectory of the initially concentrated droplet stream can result in profound differences in the convection of the subsequent vapour cloud, which in this case resulted in favourable improvement of the lean stability limit of the engine.

Figure 4 shows the flow pattern found in the engine and identifies four zones (A, B, C, D) in the flow. Vapour released in zone A does not move forward towards the piston but is drawn to the pentroof, at least initially, by the local reversed air motion. At least part of this reversal is associated with the wake of the main inlet valve. As no measurements are presented in this work for the pentroof, it is not known what trajectory this vapour would follow and where it would be at ignition. However, it is this structure that interacts, at least initially, with the droplet stream that evaporates on the cylinder liner in the case described by Carabateas et al. [6]. Zone B is the main forward flow that carries the vapour around the cylinder, and it is this flow that seems to carry the vapour, from

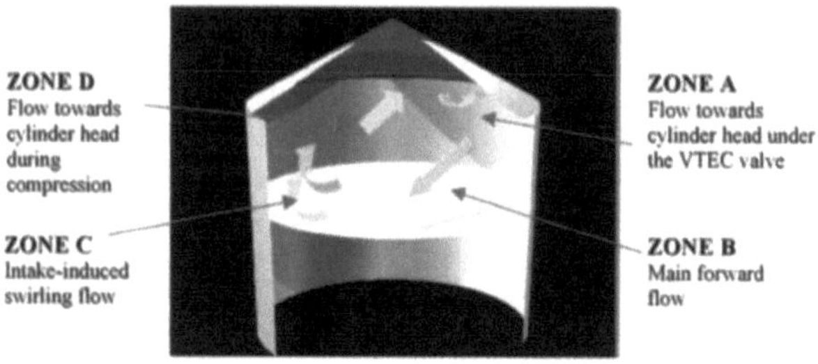

FIG. 4. Four zones of the in-cylinder flow pattern

the outset, in the case described by Hardalupas et al. [5]. The gas motion at the start of the intake stroke is complicated, which no doubt accounts for the great sensitivity of the CBCV of IMEP to the design of the inlet port and placement of the injector. Zone C shows the characteristic clockwise swirling motion that is found in this engine. Zone D is similar to Zone B and indicates that the gas is moving towards the cylinder head and the spark plug during compression.

It is worth noting here that it is conventional to analyse the motion into the separate components of "tumble" and "swirl" and, in some sense, to think of them as features separate and independent from those of flow (rather like a gyroscope in gimbals). The frames from the animation show that there is, at least after about 110° CA after ITDC, a single recognisably organised gas "vortical" motion, which is at least as long as the stroke of the piston, with its axis of rotation (at least up to 270° CA after ITDC) at a large angle relative to the axis of the cylinder's bore. An incidental result is that it is probably misleading to think of tumble, as is sometimes reported, as being destroyed as the piston reaches its top dead centre of travel. A more accurate picture is that the axis of the vortical motion becomes inevitably ever more closely aligned and coincident with the axis of the cylinder. It is tempting to speculate that the organisation of the flow is so strong that, from cycle to cycle, the overall pattern remains identical and that at least some of the measured "turbulence" in velocity[1] (e.g., by an LDA) is due to "coherent" displacement or jitter, or procession of the axis of the organised "vortical" motion. Such speculation could be substantiated in the future by the use of *p*article *i*mage *v*elocimetry (PIV), presented elsewhere in this volume.

Figure 5 shows an image sequence from animation presenting the motion of hypothetical vapour particles released *continuously* at four locations near the liner, opposite the intake valve where the droplet stream strikes the liner. The initial locations for the release are shown and were chosen as two pairs with positions determined arbitrarily; other than that, the release of each pair took place respectively in the regions in which the experiments of Hardalupas et al. [5] and Carabateas et al. [6] showed that droplets splash on the cylinder wall and evaporate. The locations are identified by a filled square and triangle and by an open circle and square in Fig. 5 for the regions closest to the impingement identified by Hardalupas et al. [5] and Carabateas et al. [6], respectively. Each image of the animation consists of a triplet of subimages: elevation, pseudoisometric, and plan view. The first image occurs at 60° CA after ITDC and corresponds to about 5° CA after vapour generation. The last vapour particle was released at 100° CA

[1] The word "turbulence" is shorthand for the ensemble-averaged root mean square of the fluctuations at a given crank angle.

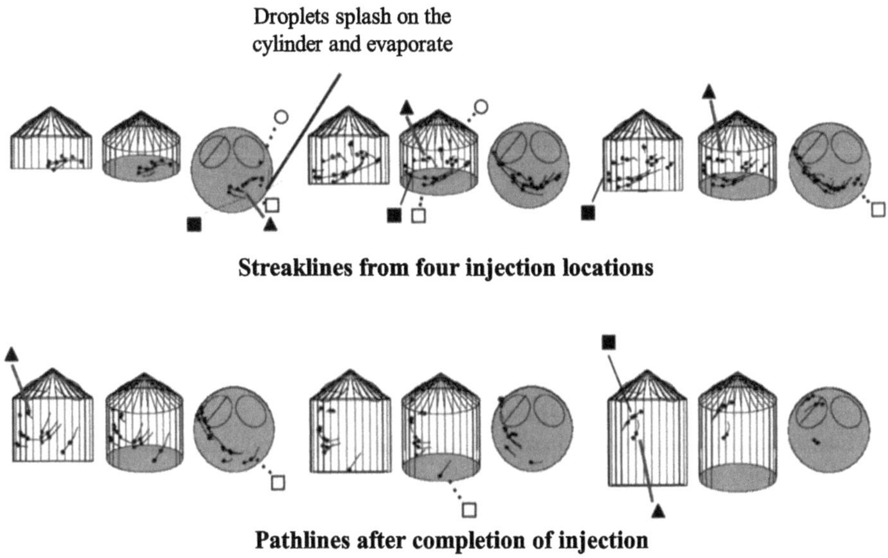

FIG. 5. Image sequence from animation. *Top left image triplet*: Symbols indicate a start-of-tracking position. *Filled symbols* indicate adjacent positions farther away from intake valves. *Open symbols* indicate adjacent positions closer to the main intake valve. Symbols at subsequent figures point to the location of vapour that originated from the respective designated positions in the first image triplet

after ITDC. The symbols in subsequent figures point to the location of the vapour that originated from the respective designated positions in the first image.

This animation sequence was produced with the method described above, although the visualisation graphics were taken directly from Mathematica, without the extra rendering performed in the sequence presented in Fig. 1. In addition, three subimages have been used here to facilitate depiction of the fluid particle trajectories. Also note that the engine cylinder displayed here is rotated 90° counterclockwise relative to Fig. 1 (as indicated by the location of the VTEC valve).

The trajectory corresponding to vapour release at the open circle in Fig. 5 rapidly moves towards the pentroof and is lost from the calculation domain. In contrast, vapour released nearby, from the open square (Fig. 5), moves down towards the piston crown and is last seen in the fifth image close to the piston crown. The vapour particles released from the two filled symbols (Fig. 5), corresponding to the case examined by Hardalupas et al. [5], follow a completely different set of trajectories: either from the open symbols or from each other. Those released from the square are shown, in the last image, under the VTEC (disabled) valve and near the cylinder liner, whereas those released from the filled triangle lie under an exhaust valve, close to the centre line of the cylinder. Both, however, have the general characteristics that would be expected from the pattern in Fig. 4, namely, a swirl and a tumble component. It is unlikely that the original measurements resolve in sufficient detail the complicated gas flow pattern in the cylinder for us to be able to claim great precision for the numerical results. What is beyond dispute is that the trajectories followed by vapour released from the open symbols in Fig. 5 (corresponding to the likely location of vapour release in the experiments of Carabateas et al. [6])

are different from those followed by vapour released from the filled symbols (corresponding to the likely location of vapour release in the experiments of Hardalupas et al. [5]). Indeed, the trajectories followed even from initially closely spaced locations (filled square and triangle) are quite different.

Two comments are relevant based on these observations. They relate to the likely extent of fuel stratification at the time of ignition. The first is, given that the vapour is generated by evaporation over a finite region of space, there is likely to be strong "mixing" of the fuel vapour due to the *mean* in-cylinder flow[2] even in the absence of considerations relating to turbulence, which have not been addressed at all in the calculations. Some mixing is essential because the process of evaporation presumably results in an initial cloud that is virtually pure vapour; that is, it is not premixed with air. On the other hand, excessive mixing results in the local A/F ratio being too weak for ignition at the spark plug, and it is arguable that extension of the current lean stability limit would entail arranging for evaporation to take place.

Thus, if we assume that the current stability limit of the engine is due to the fuel–air mixture being too lean at the instant of ignition (although other explanations are possible, such as excessive stretch), the excessive dilution is likely to be ascribable as much to the strong dispersion that the mean flow generates as to the turbulence. The second is that the difference between the lean stability limits of the two cylinder heads investigated by Hardalupas et al. [5] and Carabateas et al. [6] has been ascribed to differences in the mean trajectories of the fuel cloud based on the measured mean velocities. It will be of interest, in future, to measure, perhaps using PIV, the extent to which the CBCV in the velocity field in the region of droplet impingement on the liner at the time of such impingement is due to "precession" or jitter of the mean velocity pattern. Even minor changes in the flow structure are likely to have profound differences in the A/F ratio at the spark-plug location at ignition timing; and close to the lean stability limit it may be the poor repeatability in the in-cylinder velocity pattern that is the limiting factor.

Cyclic Variability of Initial Flame Development

Flame Images

The combustion chamber view through either the piston crown or the pentroof window is displayed in Fig. 6a, with the inlet and exhaust valves clearly indicated and the secondary inlet valve (VTEC) denoted with a large cross. Flame images were obtained soon after ignition timing to study the factors that influence early flame kernel growth. Figure 6b,c shows flames[3] from two cycles for stoichiometric conditions at 20° CA AIT (A/F = 15, $\phi = 1.00$). For these conditions, the flames have more-or-less round shapes in general

---

[2] The word "mean" is used in the sense of crank-angle ensemble-averaged. "Mixing" is here considered in the particular, specialised sense of *spatial dispersion* of the initially contiguous concentrated vapour cloud, generated by evaporation from the cylinder liner, by *mean* velocities with dispersion defined at any given crank angle as the magnitude of the root mean square of the distances of all vapour particles from their common centre of mass at that crank angle. We intuitively associate large dispersion with strong *molecular* mixing.

[3] These images have been coloured by transforming the 256-gray scale to an indexed 256-colour scale to give a temperature impression. The actual colour of the lean flames is towards blue scales instead of the red-orange-yellow pseudo ones shown in the images.

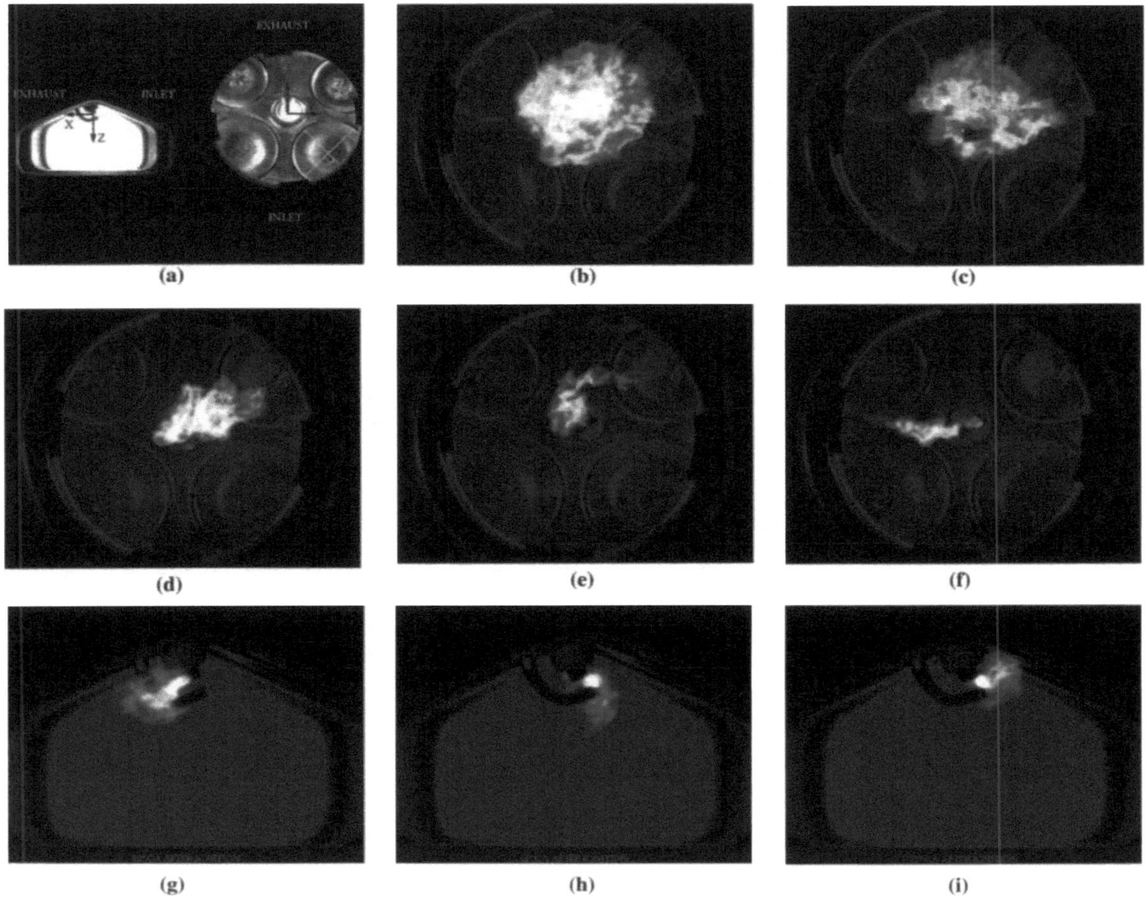

FIG. 6. **a** Combustion chamber geometry and the location of inlet and exhaust valves (north spark-plug orientation). **b,c** Stoichiometric flames (A/F = 15) as seen through the piston crown at 20° crank angle after ignition timing (east spark-plug orientation). **d–f** Lean flames (A/F = 22) as seen through the piston crown at 20° crank angle after ignition timing (east spark-plug orientation). **g–i** Lean flames (A/F = 22) as seen though the pentroof window at 12° crank angle after ignition timing (north spark-plug orientation)

and repeatable sizes from cycle to cycle, which leads to very low levels of COV_{IMEP} (1%–2%). For lean conditions (A/F = 22, $\phi = 0.68$), though, flame sizes and shapes are far from round, and there is no repeatability from cycle to cycle, as shown in Fig. 6d–f also for 20° CA AIT. This is unsurprisingly associated with COV_{IMEP} levels of the order of 15%. The visible local variations in the flame intensity indicate the dimensions of the flame along the line of sight and the level of turbulent convolutions. The swirl motion for the piston plane of view is in the clockwise direction, and observation has shown that most of the flames do indeed follow this flow motion. As might be expected, the flame area (and hence the volume and MFB) is substantially smaller for the leaner mixture.

Variations in flame sizes and shapes are also clear even earlier in the combustion process (e.g., at 12° CA AIT), as shown in the pentroof-window view images of Fig. 6g–i. The location of the spark is obvious, and it must be recalled here that the duration of the spark can be of the order of 20° CA [15]. Flame initiation occurs at about 5° CA AIT for the best lean cycles. Even from the first CA degrees after flame initiation shown in the images of 12° CA AIT, flame kernels have shapes far different from spherical. Some kernels are localised around the electrodes and others that, although anchored at the ground electrode, have moved out of the spark-plug gap. The anchoring is likely to be due to the duration of the spark rather than any aerodynamic bluff body effects in the near-wake of the J-electrode. It is clear from the images that the flame always develops in the direction towards which the spark appears to be stretched, suggesting that the dominant effects are those of the tumble motion, which is in the counterclockwise direction on the window plane of view. Figure 6i and to a lesser extent Fig. 6h are comparatively rare occurrences in that the flame is on the right of the spark plug, which is in the opposite direction to that of the mean tumble motion.

Two comments are needed here. The first is that it is difficult to envisage how "reverse" *tumble* can come about. The effect seen is more likely to be caused by the more mundane occurrence of the precession of the centre of the clockwise *swirling* component of velocity to the direct left of the spark plug in Fig. 6d–f. This would result in a swirl component of velocity of the correct direction to produce the convective transport of the flame as seen in Fig. 6i. We thus tend to subscribe to the view that the cause of the increasing departure in the flame shape from round, as the engine is made to operate progressively leaner than stoichiometric, is the decrease in the magnitude of the (laminar) flame front propagation velocity compared to the CBCV in the convective velocity vector of the gas phase at and immediately after ignition in the vicinity of the spark plug. This would explain the generation of flames with the high aspect ratios seen in Fig. 6f, for example. The second comment is that, whatever the true origin of the flame motion to the right, the speculation that such occurrences are associated with exceptional values of $\theta_{Xb5\%}$ or IMEP is not justified, as is discussed later.

It is generally believed that there are at least two main mechanisms that affect early flame development for a given equivalence ratio. The first is that low flow velocity, turbulence in the cylinder, or both cause the flame to stay close to the spark-plug gap and hence to be prone to quenching effects due to increased heat losses with the electrodes. On the other hand, high velocity, turbulence, or both cause large portions of the flame to be convected rapidly from the electrodes and the flame to quench during the resulting stretch. Thus, some intermediate flow velocity is presumably the best condition, as it convects the flame kernel away from the spark-plug gap (reducing the contact area of the flame surface with the electrodes) without quenching the flame kernel in the bulk flow. In general, most of the flame kernels appear "delicate" at this stage, and it is worth noting that individual islands of combusting regions have also been observed in the present study. Such islands are believed to be caused by local stretching and extinction but might also be related to insufficient local fuel concentration. That quenching is a possibility for these islands can be established as follows: The timescale of the larger energy-containing eddies at ignition timing has been estimated to correspond to about 6° CA for this engine using the velocity measurements of Carabateas [18] (5 m/s mean and 3 m/s RMS velocity) and the values of the integral length scale suggested by Fraser and Bracco [22] (10%–20% of the clearance height). Hence, the Karlovitz stretch factor (ratio of chemical to eddy life-times[4]) is of the order of 1.2 according to Bradley and Sheppard [24]. This magnitude indicates that the early flame kernel is indeed highly prone to quenching effects because an eddy dissipates before the reaction is completed. The measurements of Hall [25] made by a fibreoptic spark plug, although measured in a lower-speed lean-burn engine using a fuel different from that in the present study (methane), showed that the critical stage of flame initiation and development for the subsequent magnitude of P_{max} was the time when the early flame kernel had reached 5 mm in radius. This position occurred at about 20° CA for most of the cycles producing average levels of IMEP in this study.

According to Kalghatgi [26], the average flame speed increases under the influence of turbulence as the flame grows and reaches a fully developed value by the time the flame kernel has grown to a radius of 11 mm. This size of the flame kernel occurred at about 30° CA AIT for the average IMEP cycles of the present study. These two crank angles were also the position at which a strong correlation between enflamed volumes or projected areas and $\theta_{Xb5\%}$ began to be established (as is presented below), so there is as good a degree of accord between the two experiments reported in the literature and our own results as could be expected, given the differences in the experimental conditions.

Mass Fraction Burned and Enflamed Area

Typical mass fraction burned evolution curves obtained by the two-zone model from the pressure traces are shown in Fig. 7, and the strong correlation between IMEP and $\theta_{Xb5\%}$ is plotted in Fig. 8. The correlation coefficient for this relation was calculated to be −0.86, and the mean value of $\theta_{Xb5\%}$ was found to be at the FTDC position (40° CA AIT).

To investigate correlations with measures of IFKG closer to ignition, quantitative information from the flame images was obtained by calculating the projected enflamed area and centre of this luminous area. The flame-front locations were derived by assuming the flame to begin at a fixed intensity level in the image, and the flame areas were computed by summing the number of pixels that had intensity values above this threshold level and then multiplying this number by a scale factor derived by calibration. It was found that the optical resolution of one pixel corresponded to spatial dimensions of 0.09 mm for the piston images and 0.05 mm for the pentroof window images. Therefore, the spatial resolution of the flame images was of the order of the flame sheet thickness, normally 0.1 mm [10,27]. It should be noted here that because each image records the

[4] We acknowledge, but do not here examine, the fact that during IFKG the "self-stretch" of the rapidly growing flame is potentially more important than stretch by the turbulent flow field (e.g., see Gillespie et al. [23] for a review on this subject).

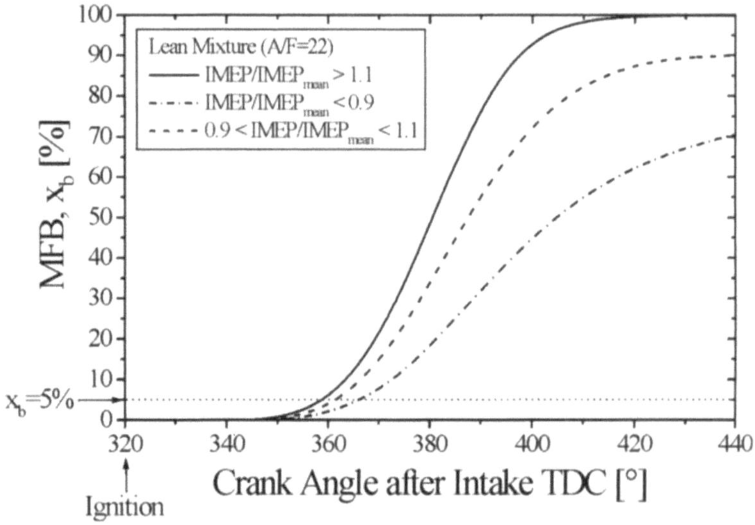

FIG. 7. Typical mass fraction burned (*MFB*, x_b) evolutions for A/F = 22 (crank angle measured from intake top dead centre, ITDC). *IMEP*, indicated mean effective pressure

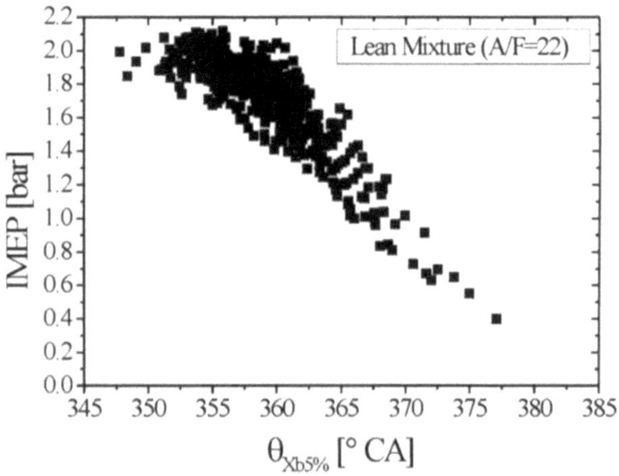

FIG. 8. Relation between indicated mean effective pressure (*IMEP*) and crank angle of 5% mass fraction burned ($\theta_{Xb5\%}$) for A/F = 22

spatial variation of the visible flame radiation integrated along the line of sight these are not necessarily contiguous flame front locations but the projected area of the flame front(s). The qualifying adjective "projected" is omitted in much of the text that follows. The position of the centre of the projected flame area was also calculated for each image to evaluate flame displacement away from the spark plug. We remind the reader that this information is new but should be read in conjunction with the measurements of flame volume provided by Aleiferis et al. [15].

The calculated enflamed area as seen through the piston crown for 40° CA AIT is plotted against the CA of 5% MFB ($\theta_{Xb5\%}$) in Fig. 9a. In general, the projected enflamed area as seen through the piston produced correlation coefficients for all CA degrees

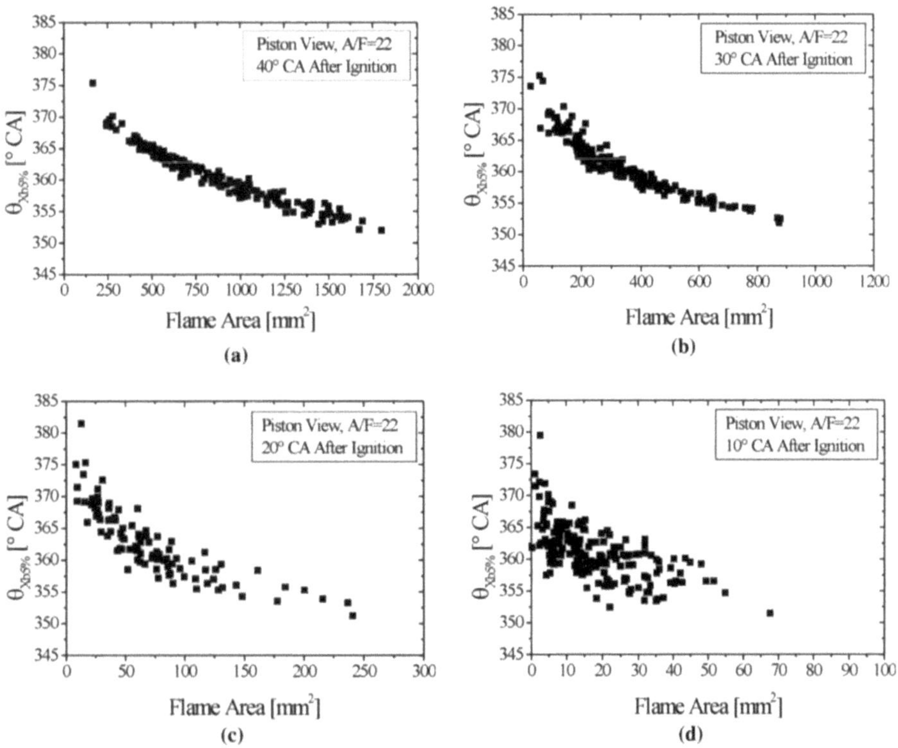

FIG. 9. Relation between crank angle of 5% mass fraction burned ($\theta_{Xb5\%}$) and projected enflamed area for A/F = 22. **a** 40°. **b** 30°. **c** 20°. **d** 10° Crank angle (*CA*) after ignition timing

higher than the corresponding coefficients calculated for the enflamed area as seen through the pentroof window. The correlation in Fig. 9a is high, producing a correlation coefficient of –0.96. We compare the correlation coefficient between the flame area and IMEP, which was found to be 0.85, with that between $\theta_{Xb5\%}$ and IMEP (measured by the pressure transducer), which was –0.86, as mentioned earlier. We can thus conclude that the optical measurement technique used in this study produces results of the same correlation level with the IMEP at this stage of combustion as the two-zone model. This result is reasonable, since 40° CA AIT corresponds to the mean value of $\theta_{Xb5\%}$ and demonstrates that the flame kernel *area* is a good measure of the MFB (i.e., a volume-based quantity) at that time.

However, the main task was to establish whether this correlation existed at earlier crank angles and if so to quantify it. Figure 9b–d presents scatter plots between the piston flame area and $\theta_{Xb5\%}$ for 30°, 20°, and 10° CA AIT, respectively (corresponding to MFBs of 1.0%, 0.2%, and 0.02%, respectively, based on estimates derived from the enflamed volume). The scatter increases as the CA position AIT decreases, but a correlation coefficient of –0.64 has still been calculated between $\theta_{Xb5\%}$ and the piston flame area at 10° CA AIT, corresponding to a flame radius of the order of 2 mm on average. The respective correlation coefficient with IMEP was found to be 0.48. This result is remarkable if one considers that the mean spark duration was about 18° CA; it demonstrates that the initial flame kernel growth speed and subsequent "quality" of a cycle are about half-determined by processes taking place within a short time of the spark event.

A factor that might be associated with the scatter between the projected flame areas and $\theta_{Xb5\%}$ or IMEP (i.e., that the correlation coefficient is less than unity, particularly for the earliest crank angles) is the effect of the flame front structure of each kernel on the subsequent performance of each cycle. This structure has large CBCV, as shown by laser sheet visualisation methods. Such images have revealed that, even for stoichiometric conditions, the laminar flame is distorted considerably by turbulence from its beginning [27] and that the size of the smallest flame-front corrugations are equivalent to the flame-front thickness, implying a combustion regime somewhere between the reaction sheet model and the distributed reaction model [28]. For lean-burn conditions, combustion appears to take place in the wrinkled flame regime and in a "pockets and peninsulas" regime; fingers of products have been widely observed [29]. The observed islands of unburned gas behind the flame front have been mainly associated with encirclement by large-scale structures rather than partial or total quenching due to flame stretch [30]. Such detailed information cannot be derived from the images presented in the present study due to integration of light along the line of sight. Nevertheless, the fact that the calculated flame areas throughout the early stage of combustion (10°–30° CA AIT) that correspond to MFB between 0.02% and 1.00% and flame radius between 2 and 10 mm) *do* correlate with parameters referring to later MFB positions, or cycle performance, suggests that whatever subsequent effects act on the flame they are of second order compared to the effects that cause the original variability in the IFKG. More to the point, it is remarkable that the flame size at 10° CA AIT exhibits correlation coefficients as high as –0.64 with $\theta_{Xb5\%}$, although flame initiation occurs at the earliest at about 4°–6° CA AIT for the fastest $\theta_{Xb5\%}$ cycles (meaning it burns for a time equal to or even shorter than the integral timescale of 6° CA).

This behavior suggests that the factors at work during the delay from spark onset to flame initiation play an important role and are perhaps associated with the fuel convection at the spark-plug gap at ignition timing. The latter might be directly related to the findings of Berckmüller et al. [31] that the A/F at the spark-plug location at ignition timing correlates with a coefficient of 0.6–0.7 with P_{max}. The present study has found that such high coefficients exist between the flame area viewed through the piston head and P_{max} at about 10°–15° CA AIT.

Other important factors for producing such high correlation coefficients with the flame size so early in the combustion process may also be related to subintegral length scales, as supported by Hill and Kapil [32] and measured by Johansson [33]. The latter author found that the most important in-cylinder eddies are in the region of 0.5–1.5 mm (which is smaller than the integral length scale) and that the increase in flame area when 1-mm eddies work on a 20-mm kernel is expected to be larger than the increase performed by 5-mm eddies (it should be recalled here that the size of the flame kernel at 10°–15° CA AIT was found in the present study to be about 5 mm).

Spark-plug Orientation and Flame Displacement

Because a conventional spark plug is not axisymmetrical, the spark-plug orientation relative to the in-cylinder flow-field influences the initial flame kernel growth. According to LDA measurements performed around a spark plug in a wind tunnel [34], the mean velocity and turbulence vary greatly within a spark-plug radius of the spark gap. Flame images were acquired for four spark-plug orientations, termed "north," "south," "east," and "west," through two views simultaneously, as described under *Engine and Experimental Techniques*, above, and depicted in Fig. 6a. For the north orientation, shown in the double-view image of Fig. 6a and the single-view images of Fig. 6g–i, the back of the spark-plug ground electrode faced the exhaust valves and its tip pointed towards the inlet valves. Exactly the opposite characterised the south orientation, as the latter referred to the spark plug being rotated 180° with regard to the north one. For the east and west orientations the spark plug was rotated 90° clockwise relative to the north and south ones, respectively (as seen through the piston crown), resulting in the ground electrode being parallel to the axis connecting the inlet valve centres. For the east orientation the ground electrode tip pointed towards the part of the cylinder containing the primary inlet valve (Fig. 6b–f), and for the west orientation the part of the cylinder containing the secondary inlet valve. LDA measurements near the spark plug of this engine [18,21] have shown that the north spark-plug orientation was, in the mean, actually a "downstream" ground-electrode orientation, the south orientation was an "upstream" one, and the east and west ones were "crossflow" orientations relative to the mean velocity vector at the spark-plug location at ignition timing.

The pressure measurements showed that the crossflow orientations exhibited about 25%–35% and 10%–20% lower COV_{IMEP} than the upstream and downstream orientations, respectively. To identify factors associated with these differences, the flame-centre locations on the window plane of view and the piston plane of view calculated from the two-view images at 20° CA AIT have been plotted in Fig. 10a,b for the north spark-plug orientation and in Fig. 10c,d for the east spark-plug orientation, with the scatter symbols parametrised according to $\theta_{Xb5\%}$. The trends concerning the positions of the flame centres in the combustion chamber were similar for all four spark-plug orientations. Detailed differences were as follows: the mean $(X^2 + Z^2)^{1/2}$ flame-centre displacement, in the window plane of view, of the west and east orientations were 50% and 30% greater, respectively, than for the south orientation. The corresponding mean displacement of the flame centres on the piston plane of view (X-Y plane) was also greater for the crossflow orientations than the one calculated for the south orientation (15%–20%). According to the LDA measurements of Carabateas [18], for the north orientation the mean velocity level around the spark-plug gap drops from 5 m/s at ignition timing to 3 m/s at 20° CA AIT; simple calculation suggests that these magnitudes would result in passive convection of the flame to a distance 6–8 mm away from the spark-plug gap within this time. This distance is, on average, equal to the one calculated from the images and suggests that the flame displacement is mainly proportional to the magnitude of the predominant velocity component in the spark-plug gap at the time of ignition (see below, however). This implies that the crossflow orientations allowed higher velocities to stretch the flame kernel downstream of the discharge channel, thereby increasing the surface area of the kernel and decreasing the proportion of this surface and the time in contact with the ground electrode (which leads to lower heat losses [12]). On the basis of these results, it is not clear which one of these two effects is dominant.

The preceding results and discussion might be taken to imply that the greater the flame convection away from the spark plug the faster is the flame growth, but this is not generally the case. For example, a high correlation coefficient was obtained only for the south orientation (–0.55), and it was between $\theta_{Xb5\%}$ and the flame-centre displacement of the window-view images $(X^2 + Z^2)^{1/2}$; the flame-centre displacement on the same plane for the other three orientations, as well as the piston-view $(X^2 + Y^2)^{1/2}$ flame-centre displace-

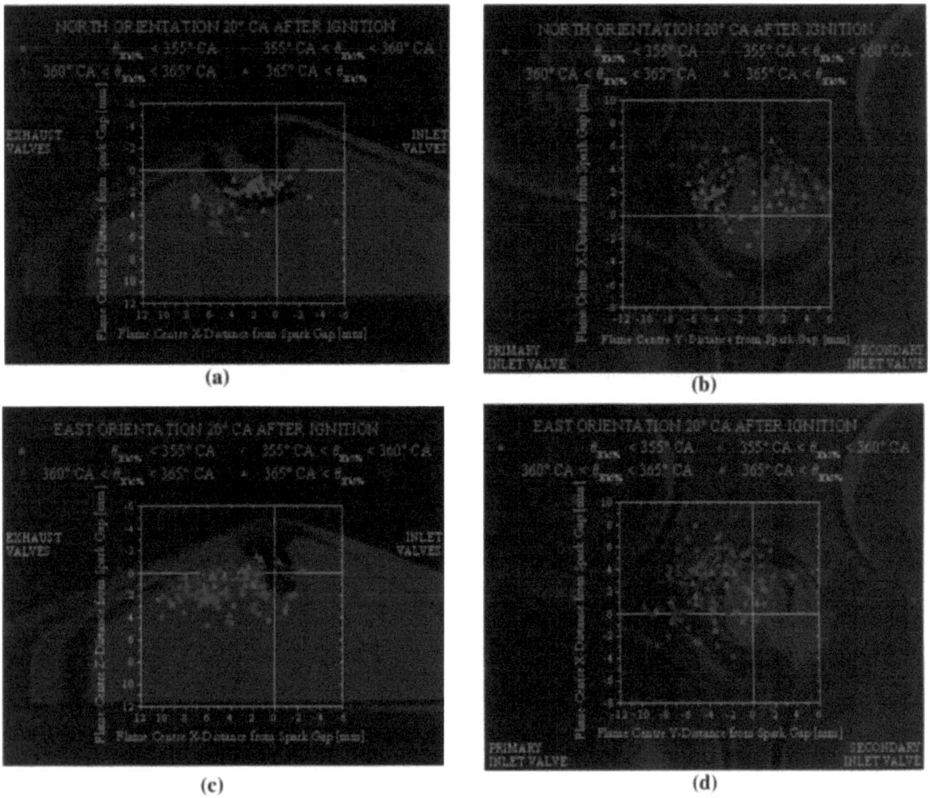

FIG. 10. Flame-centre locations for two spark-plug orientations at 20° crank angle (*CA*) after ignition timing for A/F = 22. **a** Pentroof view, north orientation. **b** Piston view, north orientation. **c** Pentroof view, east orientation. **d** Piston view, east orientation

ment and the three-dimensional flame displacement $(X^2 + Y^2 + Z^2)^{1/2}$ for all four spark-plug orientations, exhibited low correlation coefficients. Note that attempts to correlate the X or the Z component of velocity (measured by a single-channel LDA near but not in the spark-plug gap for the north orientation) with $\theta_{Xb5\%}$ also showed weak correlations [21].

It must be recalled here that the correlation coefficients between $\theta_{Xb5\%}$ and the individual flame areas of the two views showed that it is the flame area on the piston plane of view (swirl plane) that correlated better than the one on the window plane of view (tumble plane). These relations between $\theta_{Xb5\%}$ and the flame areas or displacements can be summarised by stating that the movement of the kernel out of the spark-plug gap by the tumble motion is probably more important for faster $\theta_{Xb5\%}$ than the one in the swirl plane; conversely, the increase in flame size is more important in the swirl plane than the one in the tumble plane.

Note also that there are a few measurements for which the flame centres lie to the right of the spark plug, and there are arguably too few from which to draw definitive conclusions. What evidence there is does not support the view that these displacements—apparently in the direction opposite to that of the mean tumble (as discussed earlier)—result in unusually good or poor cycles of combustion. It is also worth noting that because correlation coefficients for the window-view flame-centre displacement were highest for the upstream spark-plug orientation (south) and lower for the crossflow orientations (east and west), displacement of the flame centre is probably more important for the upstream orientation where the spark gap is shrouded by the ground electrode than for the crossflow orientations where the high velocities in the gap should be more repeatable on a cycle-by-cycle basis.

Summary and Conclusions

The CCD flame imaging and LDA velocity measurements have been performed in an optical SI engine with a Honda VTEC-E head to investigate the in-cylinder phenomena towards the extension of the lean operation limit. Images have been acquired from two views: one through the piston crown and the other through a pentroof window with direct access to the spark plug. Particle tracking based on the in-cylinder LDA measurements has shown that the in-cylinder airflow structures of swirl and tumble are visibly established at about 90° CA after intake TDC, and that these structures are actually an integral air motion that is obviously helical. Flame imaging has shown that under stratified in-cylinder conditions of a lean overall mixture cycle-by-cycle variations in combustion are already visible during the first 10° CA after ignition timing, referring to fractions of 0.1% MFB, and that cycle "quality" has already been established by 20° CA after ignition, referring to fractions of 1% MFB. Crossflow spark-plug orientations have been found to produce lower COV_{IMEP} values and to affect flame kernel behaviour by allowing higher velocities to carry the kernel out of the spark-plug gap.

Part II. Shadow Doppler Anemometry

Introduction

Scope and Background

The scope of this part of the chapter is to describe the work in progress on developing an optical (i.e., noninvasive or nonintrusive) instrument with four attributes: to make (1) in situ and (2) simultaneous measurements of the size and velocity of particles of arbitrary shapes and optical properties with high (3) temporal and (4) spatial precision contained in a flowing fluid stream. For reasons that become apparent below, the instrument is named the "shadow Doppler anemometer" (SDA). In contrast to Part I of this chapter, the focus of attention here is the instrument, not the application(s).

Sizing Anemometers Considered[5]

Modern attempts to develop sizing instruments can be dated from the widespread commercial availability of lasers, which generally permitted condition (1), above, to be realised (white light sources are generally too weak to permit this). One instrument, which was one of the first and has also been commercially successful, is based on

[5] This is not meant to be a comprehensive literature survey, many of which have appeared in recent years (e.g., see Taylor [35,36]).

Fraunhofer diffraction and is usually known by the trade name of one of the manufacturers, "Malvern" (e.g., see Jones [37] for a review). The advantages of the instrument include its ease of use and alignment, but it does not measure velocity [condition (2) above], nor does it have the ability to make spatially precise measurements of size distribution [condition (4), above].

There are many examples in industry and research in which the flow information to be extracted, with temporal and spatial resolution as high as possible, is as follows:

1. Particle concentration [m^{-3}]
2. Mass flux [$kg\,m^{-2}\,s^{-1}$]
3. Particle velocity moments of the *p*robability *d*istribution (PD) [$m\,s^{-1}$]
4. Size *e* probability (density) distribution
5. Continuous phase velocity moments of the PD [$m\,s^{-1}$]

The technical objective is usually to control (i.e., minimise or maximise) the particle dispersion. This depends on parameters such as the mass loading, or void fraction, of the dispersed phase and the velocity characteristics of the dispersed phase. This of itself justifies the need for developing sizing anemometers; and certainly until recently it implied instruments based overwhelmingly on the optical layout of the laser Doppler anemometer (LDA), which has excellent temporal and spatial resolution. An example of the technical importance of particulate flows (relevant to the theme of this book) is the combustion of *p*ulverized *f*uel (PF), which plays a crucial part in the raising of steam in boilers to generate electricity all over the world. It is achieved, however, at the expense of substantial amounts of "tail-pipe" emissions, particularly in the form of NO_X and particulates. The appropriate control of particle dispersion in the flame can limit the generation of NO_X.

For research purposes, we can add the need to measure items 1, 2, and 3 (in the list, above) conditionally on particle diameter (to distinguish between the behaviour of the "large" and "small" particles in the flow). This introduces the need to make simultaneous measurements of size and velocity: The continuous phase provides information on fluid flow turbulence (these flows being in many cases turbulent, although drop tube PF reactor measurements may be an exception) and information to make estimates of the Lagrangian and other timescales of the continuous phase flow. The all-important dimensionless parameter is the Stokes number, which is the ratio of particle timescale to flow timescale; it quantifies whether particles follow or are uninfluenced by the continuous phase. The particle timescale can be found analytically by solving the one-dimensional equation of particle motion in a uniform fluid stream. For homogeneous particles of *known* material density it is uniquely related to its "aerodynamic diameter," which is frequently dependent on the optical diameter.[6]

The LDA relies on the Doppler shift to the light scattered by the particles under consideration. Early attempts to size particles made use of the amplitude of the pedestal component of the Doppler signal (i.e., the low-pass frequency) or the "visibility," or modulation depth, of the signal observed by the LDA photodetector. Simple considerations show that, provided that diffracted light is the dominant mode of scattering,[7] (1) the

[6] Important exceptions to the latter statement include some PF "char" particles (i.e., PF after the "devolatilisation" stage) and many spray-dried products.

[7] Which renders the amplitude of the scattered electric vector insensitive to particle shape, which is never true of reflected or refracted light.

amplitude is proportional to the diameter squared and (2) the visibility should vary with particle diameter as a Bessel function (i.e., decrease from unity for small particles: compare the fringe spacing in the LDA measurement volume) to become small (and multivalued) for particles comparable to or greater than the fringe spacing. This simple picture is complicated by the fact that the underlying principles of these two approaches are intensity-based (resulting in "amplitude-modulated" signals); hence they are influenced by the fact that laser beams have Gaussian intensity distributions. The response curves (i.e., variation of the signal with particle diameter) mentioned above hold true only if the particles are measured in regions of known, uniform beam intensity. These conditions can be achieved at considerable optical complexity, but the instruments are prodigiously sensitive to misalignment. The electronics to make the measurements were complicated, and during the 1980s and early 1990s they had slow data rates. Worse, the instruments were sensitive to interruption of the beams by other particles outside the measurement volume and to fouling of any observation windows through which the beams had to pass. It is true to say that the number of extensive measurements reported in the literature using these instruments amounts, perhaps, to a few-score papers over almost three decades of research. The situation should be contrasted with the widespread adoption of the phase Doppler anemometer (PDA) for spherical particles (i.e., mainly droplets at low Weber numbers), which is fundamentally an interference-based measurement [giving spatial-*f*requency-*m*odulated (FM) signals], which are, to a first-order approximation, not affected by variations in the beam intensity or Gaussian distribution although second-order effects do exist. The PDA was, more or less as a consequence of being an FM instrument, optically robust, and the number of extensive investigations in the open literature must run into many thousands.

There remain many applications in which the "dispersed phase" is either not spherical or not transparent. Examples we have investigated or are investigating at the time of this writing, apart from the combustion of the PF, are the spraying of paint for coating car bodies, the spray drying of foods and medicines, particle dispersion in electrostatic filters, and the use of urea sprays in NO_X reduction techniques for diesel exhausts. For this class of flow, the SDA is an "FM" instrument based on the interference between the light scattered by the particle and the incident light from the original laser beam.

The purpose here is to present the SDA research and development tasks for the immediate future. In doing so, we start by briefly explaining some of the principles and practice behind the current version of the SDA.

Description of SDA

SDA is an imaging technique for sizing particles of arbitrary shape and arbitrary refractive index, and was first proposed by Hardalupas et al. [38]. The technique has been described in the literature several times (most recently by Morikita and Taylor [39]), so only a brief description is provided here.

In contrast to existing sizing methods based on the collection of diffracted laser light scattered by a particle, such as that of Orfanoudakis [40], SDA is not amplitude-based, as explained below. The independence of the method from the amplitude of diffractively scattered light means that no elaborate calibration is required before measurements, and reliable results can be obtained in *confined* two-phase *reacting* flows, where it is necessary to access the flow through optical windows that are susceptible to particle accretion and random movements due to thermal stresses developed on the surrounding metal parts confining the flow.

Generic Optical Arrangement of SDA

The generic optical setup of the SDA instrument is presented in Fig. 11. The transmitting optics are those of a conventional laser Doppler anemometer and consist, typically, of an Ar^+ laser operating at 488 nm with a power output of 200–700 mW and a beam-splitter and optical shifter unit provided by a single Bragg cell. The beams are focused by a f/600 singlet and form the LDA probe volume. The collection optics consists of 80 mm diameter f/500 and 82 mm f/300 achromats, which collect and focus the laser beams, respectively, on a 40× microscope objective lens, which is interchangeable with other lenses of smaller magnification ratios. The focal plane of the f/500 lens is on the intersection region of the two laser beams. A 25 × 35 mm aluminium-coated elliptical mirror is placed behind the f/500 lens and reflects part of the collected light, which is subsequently focused on a photomultiplier to measure particle velocity. The purpose of the microscope objective lens is to magnify the image of the LDA probe volume and to project it onto a linear photodiode array (model S4114-35; Hamamatsu, Japan). A linear photodiode array was selected as detector for the present implementation of the instrument because it offered the advantages over CCD cameras of high sampling rates and a smaller amount of information for processing, which were required for measurements in the velocity ranges encountered in the flow configurations of this work.

The two achromats relay the image of the sample space formed by the intersection of the two beams, and the microscope objective magnifies this image to a size convenient for detection by the linear array. When the particle is at the centre of the sampling space, the image projected at the detector plane is that of the outline of the particle at the centre of the space cast against the bright background of the incident laser beams. If the parti-

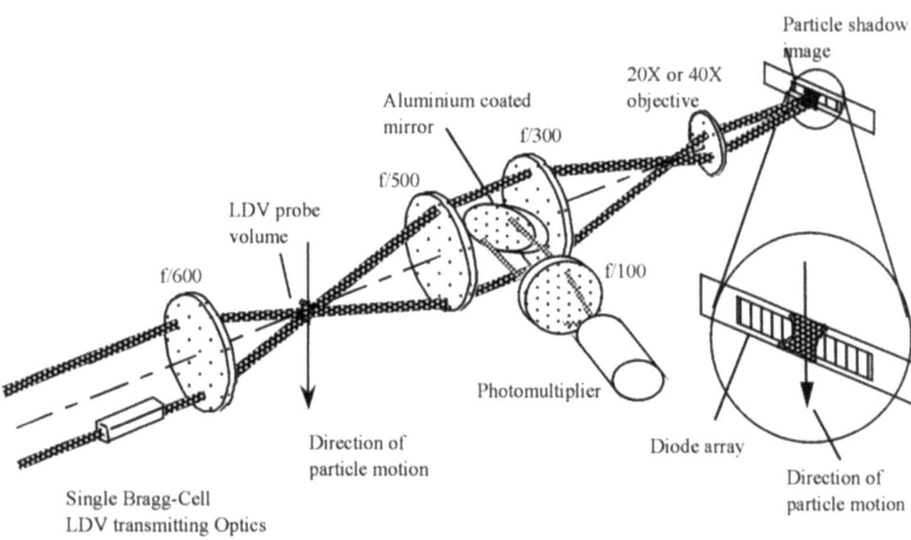

FIG. 11. Optical setup of the shadow Doppler anemometer (SDA). *LDV*, Laser Doppler Velocimetry

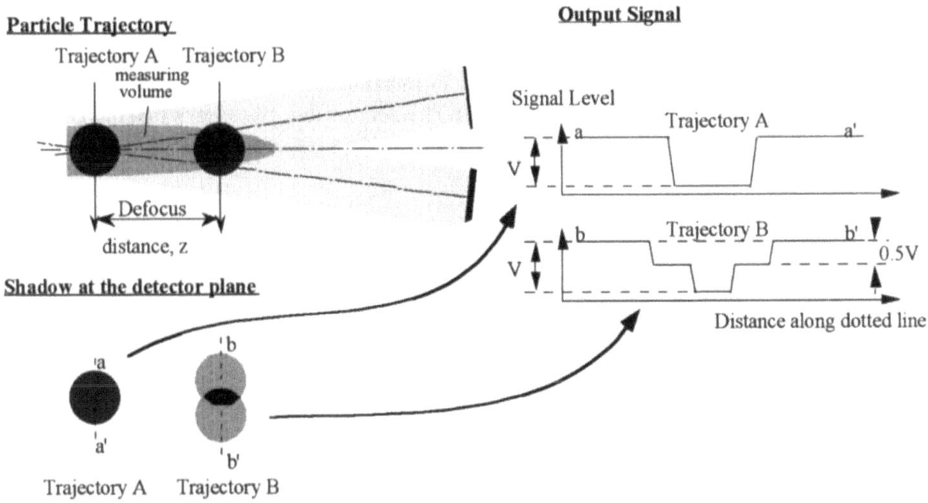

FIG. 12. Appearance of shadows of spherical particles, moving across the centre of the probe volume, on the detector plane as a function of the particle trajectory in the laser Doppler anemometer (LDA) probe volume and their respective output signals. The signals are proportional to irradiance detected from an array diode along the dotted lines *a–a′* and *b–b′*, respectively

cle traverses the central plane of the sample space (trajectory "A" in Fig. 12), the light intensity (or, equivalently, output signal level) at any point in the detector plane falls from the bright level corresponding to the absence of a particle to the dark level corresponding to the shadow of the particle. The rising and falling edges of the output signal (a–a′ in Fig. 12) are rapid but not instantaneous, even on the basis of simple theory. If a particle traverses a plane away from the central one (trajectory B in Fig. 12), the image at the detector plane is of two overlapping images of the projected area of the particle. Where these two images overlap, the "shadow" region is as dark as for trajectory A; where the images do not overlap, the shadow is of an intermediate brightness. The reader should note that the distance between the central plane and trajectory B is termed the defocus distance, for obvious reasons, and is a measurable quantity. The slope between the shadow and bright portions of the signal becomes shallower as the defocus distance increases and the particle size decreases. The detection electronics seeks signals at three or four user-specified levels at most. The description of the principles for operating the instrument shows that it is not fundamentally based on intensity but on the patterns of the shadows.

SDA: Research Perspectives

This section considers four subtasks, which can be grouped as follows: The first has to do with fundamental questions concerning the sizing accuracy and resolution of the SDA, both to be answered by calculating the scattering patterns from each of the beams. The second has to do with exploiting the excellent delimitation of the measurement area of the test volume of the SDA to evaluate the uncertainties that bedevil the delimiting of the measurement volume of the PDA. The third has to do with the trajectory bias effects causing current limitations of the detector used in current designs and the means to remove these limitations. The fourth concerns speculation on changes in the detector layout of the SDA, some of which may lead to relaxing the constraints from a single-particle counter to a multiple-particle counter. This section describes work that has either just started or is projected to take place in the future. As a consequence, it is deliberately brief.

Sizing Accuracy and Lower Limit of Measurable Size

It is interesting to establish the accuracy with which one can size particles and the lower limit of size measurable from a theoretical point of view—if only because the experiments necessary are difficult to perform. This work is currently being addressed in two ways.

In the first, the *g*eneralised *L*orenz-*M*ie *t*heory (GLMT) is being used to predict the characteristics of the shadow formed in the detector plane as a function of the position of the particle in the sampling space and, in particular, as a function of the defocus distance (G. Gréhan, G. Gouesbet, K.-F. Ren, A. M. K. P. Taylor, personal communication, 2000). The outcome is a function of the choice of the threshold levels chosen to process the signals, but it is clear that there are increasing inaccuracies for the current SDA configurations as the size decreases below about 20 μm. The interested reader is directed to this forthcoming publication for full details.

In the second, the classical Mie theory is being used to predict the accuracy of the instrument's response with emphasis on particles smaller than about 20 μm (A. R. Jones, N. T. Parasram, A. M. K. P. Taylor, personal communication, 2000). The Mie theory is used in preference to GLMT because the latter is so time-consuming. Moreover, the important advantage of GLMT—its ability to describe the scattering from particles under nonuniform illumination—is of great advantage for large, rather than small, size ranges.

It is worth establishing, in a qualitative way at least, why there is a limit to the measurable size. The preceding description of the SDA seems to imply that the formation of the shadow is due to the imaging on the detector of the "geometric" shadow created by the particle. Hence, there is no obvious lower limit other than that set by the spatial resolution of the detector or, perhaps, the fringe spacing. This is a misleadingly simple explanation of the formation of the shadow. It is, in fact, formed by the *interference* of the scattered light with the unscattered incident beam, as indicated in Fig. 13. At first approximation, the light scattered at small angles to the forward scatter direction is dominated by the diffractive component and is essentially described by the Airy pattern for a circular aperture. The angle of the first minimum in the intensity of the Airy pattern (i.e., the Airy disk) is inversely proportional to the particle diameter. If a sharp shadow is to be formed, it is necessary to capture all of the intensity contained in the Airy disk. It thus becomes apparent that the collection lens's aperture must be large enough to collect this component and that diffraction effects ultimately limit the shadow. The

FIG. 13. Details of the SDA sampling space through which the particle volume flux was measured, indicating the coordinate system and the dimensions defined in the text. In addition, examples of shadow images on the detector plane are also shown as a function of particle position in the LDA probe volume

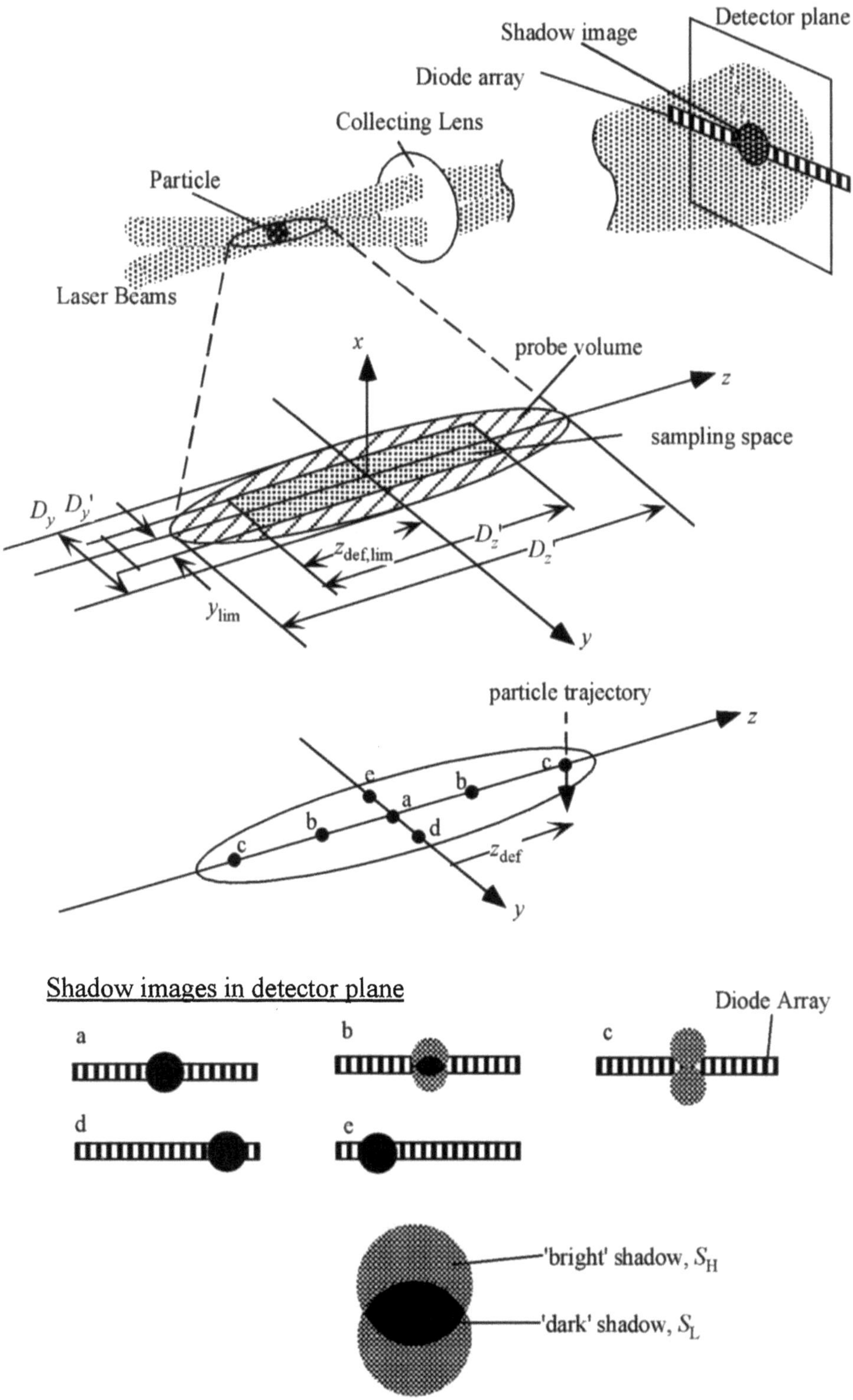
Detector plane
Shadow image
Diode array
Collecting Lens
Particle
Laser Beams
probe volume
sampling space
D_y D_y'
$z_{def,lim}$
D_z'
D_z'
y_{lim}
x
z
y
particle trajectory
z_{def}
Shadow images in detector plane
Diode Array
a
b
c
d
e
'bright' shadow, S_H
'dark' shadow, S_L

reason for the loss in accuracy as the particle size decreases below 20 μm is initially because increasingly less of the Airy disk pattern is being captured and, as a consequence, the edge of the shadow is no longer sharp. Fixed thresholds are no longer optimised for shadows, and hence output signals, with "fuzzy" edges. As the particle size decreases still further (i.e., to less than about 10 μm), this process continues; and increasingly the Airy pattern from each disk is so diffuse (the angle of the Airy disk is so large) that defocused particles no longer cast two easily identifiable, "separate" shadows. This is, at least, the case with collection lenses with focal lengths longer than a couple of decimetres, which are convenient for conventional measurements in a laboratory. Current work is directed to establishing the extent to which the lower limit of sizing (currently between 5 and 10 μm) can be reduced with practical lens configurations.

Flux Accuracy of the SDA

For an optical anemometer, the mean mass flux of the particulate phase, $G_\perp$ [$\mathrm{kg\,m^{-2}\,s^{-1}}$], is vector-defined in the direction normal to the plane containing the interference fringes in the LDA probe volume, given by the amount of particle mass crossing unit area per unit time, as follows [41]:

$$G_\perp = \frac{\rho_P}{T_S}\sum_i\left[\frac{V(i)}{A_\perp(i)}\sum_{j=1}^{n(i)}\frac{u_\perp(i,j)}{|u_\perp(i,j)|}\right] \tag{4}$$

where T_S is the total sampling time of the measurement; ρ_P is the particle density; $V(i)$ is the volume of the particle; $u(i,j)$ is the velocity of the particle sample j in size class i; $n(i)$ is the total number of particles in size class i; and $A_\perp(i)$ is the area of sampling space through which particles flow with unit normal perpendicular to the plane containing the fringes. The cross-sectional area $A_\perp(i)$ is a function of the particle size, and the method for its calculation is described elsewhere [42–44]. In the case of irregular particles, the particle volume $V(i)$ must be estimated from the projected area equivalent diameter, d_P.

Measuring the flux of a dispersed phase represents a greater or lesser difficulty for all optical single-particle counters. For both spherical and irregular particles, which are typically measured by PDA and diffracted light amplitude [e.g., 44–48], respectively, determination of the local particle flux is associated with substantially more uncertainty than is the case for the size and velocity. This is because flux is not measured directly but calculated from the particle velocity and size and the corresponding cross-sectional area $A_\perp(i)$ of the optical probe volume. It is the uncertainty in the latter quantity, and particularly its dependence on particle size due to the Gaussian irradiance of the incident laser beams, that are presumed to be the major contributor to the uncertainties. (See Taylor [36], for example, for a review of the definition of flux and concentration of a dispersed phase measured by optical methods.)

With PDA, the size (i.e., area) of the optical probe volume is preferably determined in situ from each measurement [e.g., 49–51], but the formulas rest heavily on the presumed Gaussian nature of the intensity distribution in the control volume. This, however, is a questionable assumption in many practical configurations. Extensive experience with the use of PDA has proven that the uncertainties are usually of the order of several tens of percent [e.g., 52,53]. Theoretical investigations show, in addition, that the "trajectory ambiguity effect" [54], also due to Gaussian irradiance, and the "slit effect" [55] are likely to result in an effective probe volume different from that estimated by the formulas of, for example, Saffman [49]; this error source is presumably a large potential contributor to the observed discrepancies. In addition, determination of the probe volume dimensions can be erroneous when measuring a two-dimensional flow using single-channel PDA because in the latter case the usual formula applied for the size of the probe volume assumes that the flow is always normal to the fringes [49]. Much of the basic reasoning behind the conjectures in this paragraph is likely to be true, but the magnitude of the effects is greatly complicated by the details of the signal processing techniques used and the particular settings chosen by the researcher. It is unlikely that there can be much useful progress in identifying the source of the error when measuring flux until direct measurements of $A_\perp(\mathrm{i})$ with more accuracy than those obtained with the PDA and taken simultaneously with it become available. It is this prospect which the SDA has opened, and the next few paragraphs outline how this is so.

The SDA obtains direct information on $A_\perp(\mathrm{i})$ in a different way than the PDA, essentially by making use of its quasi-imaging property. In principle the SDA has much higher accuracy.[8] The measurement relies on the work of Morikita et al. [44], who showed that it was possible to determine the location of the particle during its trajectory through the probe volume.[9] As a consequence, for particle mass flux measurements using the SDA, the size of the effective probe volume could be accurately defined. (Hereafter this parameter is termed the "sampling space" to indicate that it is explicitly user-delimited rather than set by a combination of the Gaussian irradiance profile and electronic trigger levels with the resulting sampling space size dependence estimated from on-line measurement, as in the case of PDA.) The sampling space is defined as that region of space over which a particle generates two superimposed shadows, the overlapping region of which, over some part of its trajectory, passes over "trigger" diodes and is not vignetted by the finite width of the linear array. Morikita et al. [44] showed that the area $A_\perp(\mathrm{i})$ of the sampling space for particle size d_P with defocus limitation at $z_{\mathrm{def,lim}}$ is a nonmonotonic function of particle diameter:

$$A_\perp(d_\mathrm{P}) = \int_{-z_{\mathrm{def,lim}}}^{z_{\mathrm{def,lim}}} d_\mathrm{W}(dP, z_{\mathrm{def}})\,dz_{\mathrm{def}} \tag{6}$$

where the so-called [41] net value of the width of the sampling space, d_W, for any particle size is given by

$$d_\mathrm{w}(d_\mathrm{p}, z_{\mathrm{def}}) = \min\left[d'_\mathrm{w}(d_\mathrm{p}, z_{\mathrm{def}}), d''_\mathrm{w}(d_\mathrm{p})\right] \tag{7}$$

where

$$d'_\mathrm{W}(d_\mathrm{P}, z_{\mathrm{def}}) = \sqrt{d_\mathrm{P}^2 - 4z_{\mathrm{def}}^2 \tan\gamma} + d_\mathrm{t} \tag{8}$$

and

$$d'_\mathrm{w}(d_\mathrm{p}) = W - d_\mathrm{p} \tag{9}$$

[8] Note that the sizing accuracy of the SDA instrument has been evaluated by comparins it with microscope measurements. It was found that the sizing accuracy was better than 10% with a further ±5% inaccuracy caused by the "defocusing" of the particle from the local plane of the receiving optics by up to ±500 μm [56].

[9] One can also measure the angles of particle trajectories relative to the axis of the photodiode array. This fact is important when measuring, among other things, nonunidirectional flow.

where z_{def} is the defocus distance, d_P is the particle diameter, γ is the laser beam half-angle of intersection, d_t is the width of the image of the triggering area in the LDA probe volume (given by the width of eight trigger elements of the diode array divided by magnification of the collection optics, d_t), and W is the width of the image of the detector (corresponding to the 30 segments of the diode array) in the sampling space of the anemometer; W depends on the magnification of the collection optics. The major advantage of this formulation is that all of the quantities are directly measured by the SDA, and there are no obvious inferences based on any presumed intensity distribution in the sample space.

The above description shows that the SDA can be used to locate the trajectory of a particle passing through an SDA sampling space. This is useful because the transmitting optics of a PDA are identical to those of an SDA, which (in principle at least) allows the SDA to make measurements simultaneously with a PDA. Hence with appropriate data communication between the two instruments it becomes possible to map *empirically* the PDA's sampling space against the more accurate measurements of the SDA. This procedure is straightforward for the case of droplets flowing unidirectionally through the sampling space. There are theoretical and practical difficulties, however, with the SDA measurements for the *equally important* case where the trajectories through the sampling space of the PDA/SDA are random, or omnidirectional. This is due to the dependence of the sampling space size on the particle size, shape, orientation, and trajectory, which could cause a systematic bias in the validation rate, especially for trajectories near the edge of the sampling space.

The reasons for this are briefly outlined here. In general, the validatable sampling space is smaller than predicted by the equations above if there are particles of random shapes and with random orientations. Figure 14 shows a pseudoimage (i.e., the output of the linear detector array as a function of time) that illustrates these cases. In Fig. 14a,b a spherical particle passed through the sampling space at trajectory angles relative to the axis of the array of 90° (Fig. 14a) and other than 90° (Fig. 14b). The trajectory in Fig. 14a is accepted, but that of Fig. 14b is rejected by software because during the course of the trajectory part of the shadow image was vignetted and not recorded, although the particle, at the time of triggering, lies within the area defined by the above equations. Figure 14c,d illustrates the first of two possible effects of particle shape and orientation for the particular case of identical ellipsoids with an aspect ratio of about 2.0 and a trajectory normal to the diode array axis, passing through the centre of the probe volume. The choice of ellipsoids is for the sake of example only and does not limit the generality of application of the conclusions drawn here. Inspection shows that the width of the vali-

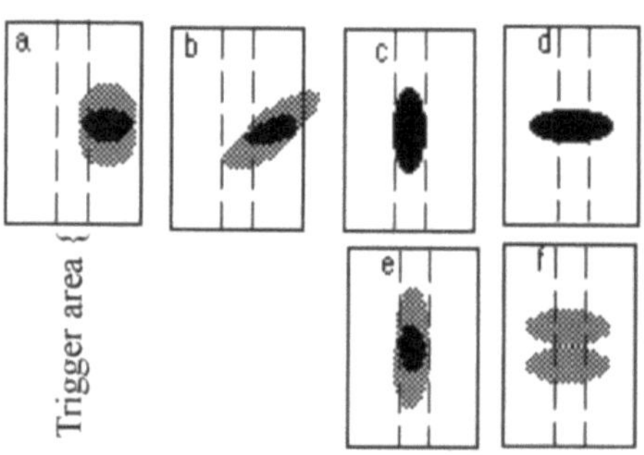

FIG. 14. SDA image in the image buffer for different cases: spherical particle at 90° (**a**) and other than 90° (**b**); ellipsoid in-focus (**c,d**) and defocused (**e,f**). The width of the frame surrounding each image represents the width of the active diodes of the array. The height represents samples of the array as a function of time

datable cross-sectional area of the sampling space in Fig. 14c is larger than in Fig. 14d. Figure 14e demonstrates the second effect, which is to limit the maximum defocus distance up to the point at which the two shadows just overlap. Particles with the orientation of Fig. 14e can have larger maximum defocus distances than with the orientation of Fig. 14f before they fail to trigger the instrument, and the length of the validatable area of the sampling space is therefore larger. The magnitude of these two effects depends on the details of particle shape and orientation. Only some of these effects are important when evaluating the area of the PDA's sample space because the particles under investigation should be spherical. It is clear that caution is required when performing this experiment.

Trajectory Bias Effects

Because the sizing technique is based on a linear photodiode array, size-validated measurements can be obtained only when the particles' velocity component perpendicular to the axis of the array ($V_\perp$) exceeds about 0.5 m/s, a limit that has been determined experimentally. This limit is due to the fact that the shadow image of a particle *must* travel over the array, so that a finite number of one-dimensional slices is obtained before the particle's projected image can be reconstructed.

This condition can be violated in one of two distinct ways. First, if the shadow image of a particle were travelling with velocity $V_\perp$ (practically zero), as could happen in a recirculation zone, the one-dimensional slices obtained during sampling would correspond to the same position of the particle, and no particle image could be reconstructed. Second, and more generally, a flow with a nonzero local mean velocity vector that is closely aligned with the axis of the array will have a $V_\perp$ value of practically zero, but the finite mean velocity parallel to the axis of the detector will result in the covering, at some point in time, of the extreme diodes of the array. This again results in an invalid measurement.

For a similar reason, a measurement is validated, provided the trajectory of the particle is not parallel, or nearly parallel, to the axis of the photodiode array, which in practice means that only particles with trajectories at angles ϕ smaller than about 70° are measured (Fig. 15b). A consequence of these two effects is that there is a combination of particle trajectories and velocities that do not contribute to the particle velocity statistics, and it is therefore necessary to investigate theoretically the extent of velocity biasing caused by these two effects. The magnitude of the likely effects is to be presented in a forthcoming technical note (I. Prassas, A. M. K. P. Taylor, in preparation, 2000).

Detector Developments

The linear diode array detector described in this work has provided the basis for a useful start for the SDA. It is clear that it has limitations in terms of both readout speed (and hence particle velocity) and completeness of information. A new version of the instrument is under construction based on a larger number of elements; it is capable of detecting faster-moving particles. The improvements to be made to the detector geometry are clear: Linear arrays may well not be the best for omnidirectional flows, and the limitation is likely to be the speed with which the potentially vast amounts of information are presented to the memory of the logging computer. An exciting possibility is that the size measurement principle, rather than the velocity measurement, is not obviously limited to single-particle measurement in the sampling space. At present, multiple

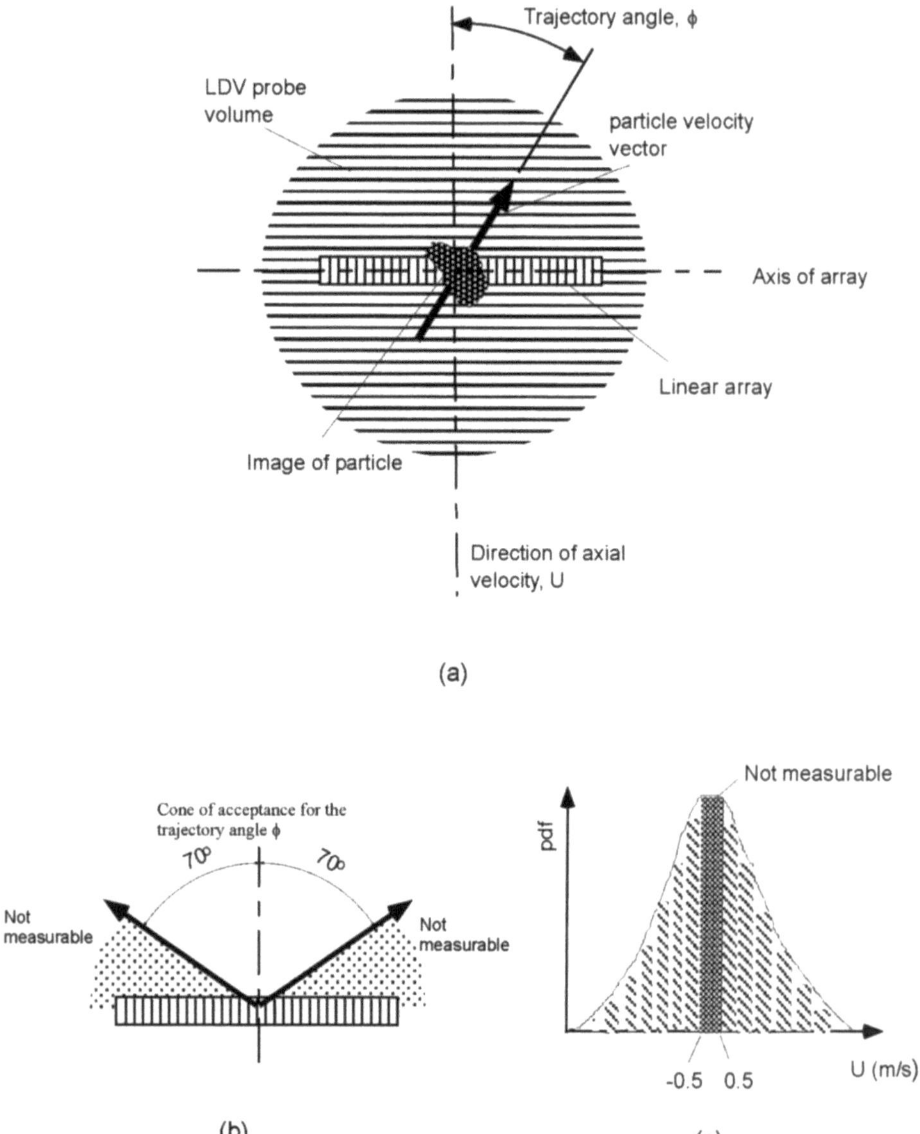

FIG. 15. Velocity bias effects due to SDA. **a** Image of the LDA probe volume on the diode array and a particle trajectory through the probe volume at angle ϕ relative to normal for the array. **b** Range of experimentally determined measurable trajectory angles ϕ. **c** Range of nonmeasurable velocities normal the array that are missing from the velocity PDF

particle sizing can be envisaged using a CCD camera for very slow moving flows. With improvements in detector technology, it is possible that the speed limitation can be relaxed in the future.

Summary and Conclusions

The above brief description of the latest developments show that the SDA, although in existence for more than half a decade and with several diverse applications to its credit, remains a fertile field of research. Fundamental aspects relating to its accuracy, the optimum detector configuration, and its use for answering important questions about the accuracy of the flux measurements made by the PDA are important topics to cover during the next few years.

Acknowledgments. The authors gratefully acknowledge Dr. Y. Hardalupas for his helpful advice and fruitful discussions during the preparation of the present work. The financial support of Honda R & D Co., Ltd., Japan and the technical advice of Dr. K. Ishii and Mr. Y. Urata are gratefully acknowledged. The contributions of Dr. H. Morikita and Kanomax, Inc., Osaka towards the development of the SDA are also much appreciated.

References

1. Germane GJ, Wood CG, Hess CC (1983) Lean combustion in spark-ignited internal combustion engines: a review. SAE paper 831694
2. Ozdor N, Dulger M, Sher E (1994) Cyclic variability in spark-ignition engines: a literature survey. SAE paper 940987
3. Quader AA (1982) The axially-stratified-charge engine. SAE paper 820131
4. Berckmüller M, Tait NP, Greenhalgh DA (1996) The time history of the mixture formation process in a lean-burn stratified-charge engine. SAE paper 961929
5. Hardalupas Y, Taylor AMKP, Whitelaw JH, Ishii K, Miyano H, Urata Y (1995) Influence of injection timing on in-cylinder fuel distribution in a Honda VTEC-E engine. SAE paper 950507
6. Carabateas NE, Taylor AMKP, Whitelaw JH, Ishii K, Yoshida K, Matsuki M (1996) The effect of injector and intake port design on in-cylinder fuel droplet distribution, airflow and lean burn performance for a Honda VTEC-E engine. SAE paper 961923
7. Ault JR, Carabateas NE, Taylor AMKP, Whitelaw JH, Ishii K, Sasaki T, Urata Y (1997) Cyclic variability of lean combustion and gas motion at the spark plug of a VTEC-E Honda engine. In: Proceedings of the 10th international symposium on transport phenomena in thermal science and process engineering, Kyoto, Japan, pp 743–748
8. Ishii K, Sasaki T, Urata Y, Yoshida K, Ohno T (1997) Investigation of cyclic variation of IMEP under lean-burn operation in a spark-ignition engine. SAE paper 972830
9. Nakai H, Ichi S, Ikeda Y, Nakajima T (1998) Flame propagation variation due to insufficient HC concentration. SAE paper 982565
10. Gatowski JA, Heywood JB, Deleplace C (1984) Flame photographs in a spark-ignition engine. Combustion and Flame 56:71–81
11. Baritaud TA (1987) High-speed Schlieren visualisation of flame initiation in a lean operating SI engine. SAE paper 872152
12. Pischinger S, Heywood JB (1990) How heat losses to the spark-plug electrodes affect flame kernel development in an SI engine. SAE paper 900021
13. Bates SC (1989) Flame imaging studies of cycle-by-cycle combustion variations in a SI four-stroke engine. SAE paper 892086

14. Arcoumanis C, Bae CS, Whitelaw JH, Xu HM (1994) Imaging of lean premixed flames in spark-ignition engines. SAE paper 942052
15. Aleiferis PG, Taylor AMKP, Whitelaw JH, Ishii K, Urata Y (2000) Cyclic variations of initial flame kernel growth in a Honda VTEC-E lean-burn spark-ignition engine. SAE paper 2000-01-1207
16. Horie K, Nishizawa K, Ogawa T, Akazaki S, Miura K (1992) The development of a high performance four-valve lean-burn engine. SAE paper 920455
17. Lavoie GA, Heywood JB, Keck JC (1970) Experimental and theoretical study of nitric oxide formation in internal combustion engines. Combustion Science and Technology 1:313–326
18. Carabateas NE (1997) Two-phase flow and combustion in SI engines. PhD thesis, University of London
19. Farwig R (1986) Rate of convergence of Shepard's global interpolation formula. Mathematics of Computation 46:577–590
20. Franke R (1982) Scattered data interpolation: tests of some methods. Mathematics of Computation 38:181–201
21. Ault JR (1999) Measurements and imaging of flow and combustion in spark-ignition engines. MPhil thesis, University of London, UK
22. Fraser RA, Bracco FV (1989) Cycle-resolved LDA integral length scale measurements investigating clearance height scaling isotropy and homogeneity in an IC engine. SAE paper 890615
23. Gillespie L, Lawes M, Sheppard CGW, Woolley R (2000) Aspects of laminar and turbulent burning velocity relevant to SI engines. SAE paper 2000-01-0192
24. Bradley D, Sheppard CGW (1988) Limitations to turbulence-enhanced burning rates in lean-burn engines. paper C46/88 IMechE
25. Hall MJ (1989) The influence of fluid motion on flame kernel development and cyclic variation in a spark-ignition engine. SAE paper 890991
26. Kalghatgi GT (1985) Early flame development in a spark-ignition engine. Combustion and Flame 60:299–308
27. Zur Loye AO, Bracco FV (1987) Two-dimensional visualisation of premixed charge structure in an IC engine. SAE paper 870454
28. Baritaud TA, Green RM (1986) A 2-D flame visualisation technique applied to the IC engine. SAE paper 860025
29. Mantzaras J, Felton PG, Bracco FV (1988) Three-dimensional visualisation of premixed-charge engine flames: islands of reactants and products; fractal dimensions and homogeneity. SAE paper 881635
30. Hicks RA, Lawes M, Sheppard CGW, Whitaker BJ (1994) Multiple laser sheet imaging investigation of turbulent flame structure in a spark-ignition engine. SAE paper 941992
31. Berckmüller M, Tait NP, Greenhalgh DA (1997) The influence of local fuel concentration on cyclic variability of a lean-burn stratified-charge engine. SAE paper 970826
32. Hill PG, Kapil A (1989) The relationship between cyclic variations in spark-ignition engines and the small structure of turbulence. Combustion and Flame 78:237–247
33. Johansson B (1994) The influence of different frequencies in the turbulence on early flame development in a spark-ignition engine. SAE paper 940990
34. Halldin C (1992) Velocity and turbulence measurements close to a spark plug. SAE paper 920154
35. Taylor, AMKP (1992) Measurements techniques for two-phase flows. In: Durão DFG, Heitor MV, Whitelaw JH, Witze PO (eds) Combusting-flow diagnostics. Kluwer Academic, Dordrecht, pp 233–289
36. Taylor AMKP (1994) Two-phase flow measurements. In: Buchhave P, Lading L, Wigley G (eds) Optical diagnostics for flow processes. Plenum, New York, pp 205–228
37. Jones AR (1993) Light scattering for particle characterisation. In: Taylor AMKP (ed) Instrumentation for flows with combustion. Academic, London

38. Hardalupas Y, Hishida K, Maeda M, Morikita H, Taylor AMKP, Whitelaw JH (1994) Shadow Doppler technique for sizing particles of arbitrary shape. Applied Optics 33:8417–8426
39. Morikita H, Taylor AMKP (1997) Application of shadow Doppler velocimetry to paint spray: potential and limitations in sizing optically inhomogeneous droplets. [Special issue on non-spherical/non-homogeneous particles.] Measurement Science and Technology 9:221–231
40. Orfanoudakis NG (1994) Measurements of size and velocity of burning coal. PhD thesis, University of London
41. Prassas I (1998) The combustion of pulverized coal in swirl burners. PhD thesis, University of London
42. Morikita H (1996) PhD thesis (in Japanese), Keio University
43. Morikita H, Hishida K, Maeda M (1994) Simultaneous measurement of velocity and equivalent diameter of non-spherical particles. Particle and Particle Systems Characterization 11:227–234
44. Morikita H, Hishida K, Maeda M (1995) Measurement of size and velocity of arbitrary shaped particles by LDA-based image technique. In: developments in laser techniques and applications to fluid mechanics. Springer-Verlag, Berlin Heidelbery New York pp 354–375
45. Hirleman ED, Moon HK (1982) Response characteristics of the multiple-ratio single-particle counter. Journal of Colloid and Interface Science 87:124–139
46. Orfanoudakis NG, Taylor AMKP (1992) The effect of particle shape on the amplitude of scattered light for a sizing instrument. Particle and Particle Systems Characterization 9:223–230
47. Orfanoudakis NG, Taylor AMKP (1995) Evaluation of an amplitude sizing anemometer and application to a pulverised coal burner. Combustion Science and Technology 108:255–277
48. Yeoman ML, Azzopardi BJ, White HJ, Bates CJ, Roberts PJ (1982) Optical development and application of a two-colour LDA system for the simultaneous measurement of particle size and particle velocity. Presented at the Engineering applications of laser velocimetry, ASME winter annual meeting, Phoenix, Arizona
49. Saffman M (1987) Automatic calibration of LDA measurement volume size. Applied Optics 26:2592–2597
50. Hardalupas Y, Taylor AMKP (1989) On the measurement of particle concentration near a stagnation point. Experiments in Fluids 8:113–118
51. Schöne F, Bauckhage K, Wriedt T (1994) Size of the detection area of a phase-Doppler anemometer for reflecting and refracting particles. Particle and Particle Systems Characterization 11:327–338
52. Hardalupas Y, Taylor AMKP, Whitelaw JH (1994) Mass flux, mass fraction and concentration measurement of liquid fuel in a swirl-stabilised flame. International Journal of Multiphase Flow 20:233–259
53. Maeda T, Morikita H, Hishida K, Maeda M (1996) Determination of effective measurement area in a conventional phase-Doppler anemometry. Presented at the 8th international symposium on applications of laser techniques to fluid mechanics, Lisbon, Portugal
54. Gréhan G, Gouesbet G, Naqwi A, Durst F (1992) Trajectory ambiguities in phase Doppler systems: use of polarisers and additional detectors to suppress the effects. Presented at the 6th international symposium on applications of laser techniques to fluid mechanics, Lisbon, Portugal
55. Xu T-H, Tropea C (1994) Improving the performance of two-component phase Doppler anemometers. Measurement Science and Technology 5:969–975
56. Hardalupas Y, Morikita H, Taylor AMKP (1995) Theoretical evaluation of sizing non-spherical particles using diffracted light. Measurement Science Technology 6:717–726

Industrial Drying Technologies: Current Status and Future Trends

Arun S. Mujumdar

Summary. Industrial drying accounts for 12% to 25% of a developed nation's total industrial energy consumption. Because most of the energy required is obtained by combustion of fossil fuels, this consumption is associated with significant emissions of carbon dioxide as well as other noxious gases. This chapter starts with a tutorial overview of the diverse dryer types, selection criteria, need for innovation, and features of novel dryers, and concludes with a discussion of recent trends as well as the drying technologies of the future.

Key words. Energy—Environmental impact—Classification of dryers—Selection of dryers

Introduction

Drying involves removal of a liquid (generally water) from a solid, semisolid, or liquid feedstock to produce a solid product by thermally induced phase change. It is perhaps the oldest, the most common, and yet, at the fundamental level, a very complex operation. It is common to almost all industries. Indeed, there is a direct correlation between the amount of water thermally removed in industrial operations and the GDP of the country. Thus, as the developing economies of the world raise their GDPs there will be an increase in the energy needs for drying and the resulting increase in environmental emissions of the greenhouse gas, CO_2, and NO_x, because of fossil fuel combustion as well as fine particulate emissions and in some cases emissions of volatile organic carbon (VOCs) and odorous compounds.

Estimates of energy consumption for industrial drying in the developed countries range from 11% (e.g., Canada, France, Sweden, USA) to more than 20% (e.g., Denmark, Germany). Most of this energy comes from combustion of fossil fuels. Most industrial dryers operate at thermal efficiencies ranging from a low of 20% to 80% with a median value of about 50%. The efficiency of a dryer depends to some extent on the type (e.g., convective, contact, radiant) as well as the product being dried (e.g., whether heat sensitive). In most cases, however, there is potential for improvement of energy efficiency.

Mechanical Engineering Department, National University of Singapore, 10 Kent Ridge Crescent, Singapore 119260

Improved efficiency is directly correlated to environmental impact except when renewal energy sources (e.g., solar, wind, tidal power) are used rather than fossil fuels. Overall, almost all industrial drying is carried out using nonrenewable energy sources.

This chapter starts with a coarse classification of industrial dryers, guidelines for selection of dryers, some innovative drying technologies with significant advantages from the energy and environmental standpoints, recent trends in industrial drying, and some R&D needs and opportunities in the field of drying. Readers interested in a more detailed exposure to fundamentals of drying as well as discussion of various dryer types and industrial practice are referred to the Handbook of Industrial Drying [1].

More than 400 dryer types are reported in the literature; some 100 are available commercially from different vendors. In view of the tremendous physical and chemical diversity of products that must be dried industrially, the variety of ways one could supply energy to the material for drying, the wide ranges of operating temperature and pressure, etc., it is not surprising that so many different classes of dryers are possible. Indeed, even for each type of dryer, for example, spray dryer or fluid bed dryer, one can identify scores of subclasses each with some definite distinct features that make them particularly suited for some niche applications. New drying concepts and techniques are being reported at an increasing rate over the past two decades. Indeed, some 250 U.S. patents per year are awarded in the area of drying and dryers, a number that is at least an order of magnitude larger than patents awarded in other areas such as adsorption, crystallization, and membrane separations. While applied R&D in drying is proceeding at a healthy clip, fundamental research in drying usually carried out in academia is not proceeding at a comparable rate at this time. It is noteworthy that drying is an interdisciplinary and multidisciplinary area that is best advanced via active industry–academia interaction and collaboration.

Figure 1 is a conceptual schematic of how a small wet particle may be dried. Note the variety of heat transfer modes that may be employed as well as the potentially diverse mechanisms of moisture transfer within the material. The energy input may be continuous or intermittent. Different modes of energy input may be deployed to advantage in many cases, although purely convective dryers using hot air or combustion gases constitute more than 80% of all industrial dryers. The difficulty of recovering energy from the humid exhaust from a convective (direct) dryer cost effectively makes such dryers energy inefficient.

Table 1 summarizes a coarse classification of dryers. Although extensive, it does not include all possible criteria one could use for such a classification.

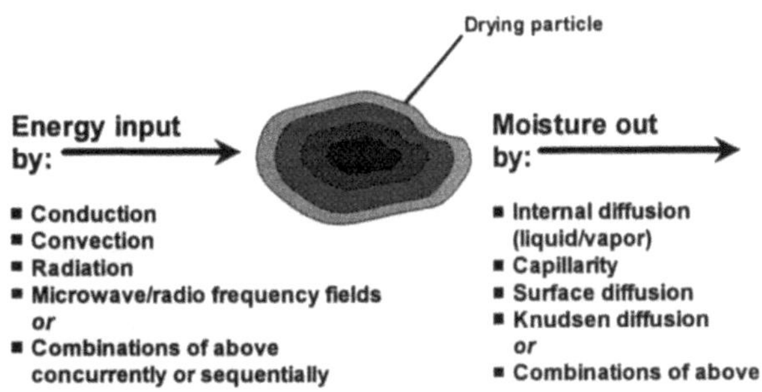

FIG. 1. Schematic representation of the possible modes of heat input and moisture transport in drying of a particle

Table 1. Classification of dryers

Criterion	Types
Mode of operation	Batch Continuous[a]
Heat input type	Convection[a], conduction, radiation, electromagnetic fields, combination of heat transfer modes Intermittent or continuous[a] Adiabatic or nonadiabatic
State of material in dryer	Stationary, moving, agitated, dispersed, etc.
Operating pressure	Vacuum Atmospheric[a]
Drying medium (convection)	Air[a] Superheated steam Flue gases
Drying temperature	Below boiling temperature[a] Above boiling temperature Below freezing point
Relative motion between drying medium and drying solids	Cocurrent Countercurrent Mixed flow
Number of stages	Single[a] Multistage
Residence time	Short (<1 min) Medium (1–60 min) Long (>60 min)

[a] Most common in practice

Dryer Selection

According to a recent survey of several major European chemical companies, more than 80% of them had experienced problems due to incorrect selection of the dryer. Costs of incorrect selection can be very significant in terms of higher capital costs, increased running costs, cost of downtime, lost production, higher maintenance, longer startup times for new installations, production of off-spec product, and in extreme cases the cost of replacement with the right dryer. The number of criteria to be considered at the selection stage is such that an expert system may be needed if the companies do not have an in-house expert or an outside consultant to help with this key step. Mujumdar [1] provides a comprehensive list of criteria to be applied in the selection of dryers. Indeed, it is important to select a drying system rather than a dryer; the former includes selection of the predrying stages (e.g., filtration, centrifugation, backmixing with dry solids if needed) as well as postdrying stages (e.g., gas cleaning, cooling of product, agglomeration, granulation). Finally, it is also necessary to operate the dryer at optimal operating conditions so as to obtain the desired product quality at highest efficiency and lowest costs. Often several different dryers or drying systems may yield the same product, in which case other criteria, such as unit cost of drying, footprint of equipment, environmental and energy aspects, and delivery times dictate the final selection.

The following sequence is suggested for selection of dryers on the basis of their energy efficiency and ecofriendliness. Ability of the dryer to physically handle the feed to be dried and produce a dry product of specified characteristics (e.g., quality) is the key issue in selection of a dryer. Often several dryer types can meet this criterion. If this is the case, one can then evaluate the most energy-efficient dryer or the most ecofriendly one. Table 2 is a coarse rating scheme for the ecofriendliness of various commonly used dryers. Here, L represents a low rating due to its higher energy consumption, leading to increased pollution, and M and H denote medium and high ratings, respectively. The indirect-type dryers and convective dryers with superheated steam as the drying medium or with heat pumps have the highest rating. Of course, these dryers tend to have higher capital costs. Over the lifetime of the dryer the energy savings may offset the initial high cost, however. Also, the cost of auxiliary equipment needed to control emissions of vapors, odorous compounds or fine particulates, especially for convective dryers, may well make the cost of the overall drying system lower for the indirect dryers. A technoeconomic evaluation over the lifetime of the drying system is the only way to arrive at the correct decision on dryer selection. Interestingly, the choice of optimal drying system also depends on the production rates and turndown ratios desired. Typical steps in dryer selection are outlined in Fig. 2.

Table 3 lists the competing criteria in dryer selection when quality of the product is of paramount or of little significance. Often quality of the product overrides the energy efficiency as the controlling parameter affecting dryer selection.

Further, whatever is the dryer choice it must be operated at optimal conditions, preferably with automatic control to attain best efficiency and minimal or no emissions. It is often necessary to develop a reliable mathematical model to select the optimal drying conditions. It can also be used with a model-based control system to ensure optimal operation even when there are process upsets. Use of fuzzy logic and artificial neural nets

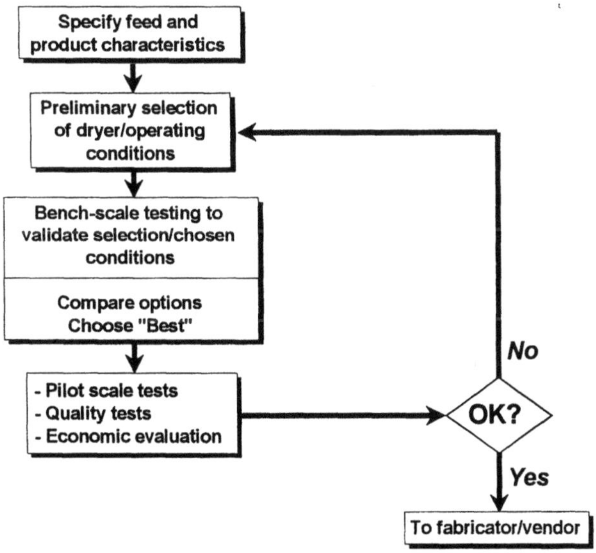

FIG. 2. Basic steps in selection of industrial dryers

TABLE 2. Ranking of some dryer types according to their environmental friendliness

Type	Rating	Remarks
Convective (direct)		
No recirculation	L	Low efficiency; emissions with exhaust difficult to treat because of large volumes
With partial recycle of exhaust	M	Partial energy recovery
With superheated steam	H	High potential for recovery; also for containment of emissions
With heat pump	H	Capital intensive; CFCs may leak out, emissions may concentrate
Conduction (contact/indirect)		
Vacuum	H	No exhaust gases; no emission of toxic vapors or particulates; higher capital costs
Atmospheric	M	Emissions with exhaust; cleaning equipment needed
Radiant/MW/RF		
Vacuum	H	Capital intensive; energy intensive
With convection	M	Smaller quantities of gas means easier emission control

L, low; M, medium; H, high

TABLE 3. Competition between energy and quality criteria in selection of drying systems depending on value of product

Value of product	Criteria (descending order)	Comments/examples
High	1. Quality 2. Quality 3. Quality	Pharmaceuticals, advanced materials, some food products, biotech products
Medium	1. Quality 2. Energy 3. Capital cost	Foods, agricultural products, paper, wood, ceramics, dyes, pigments
Low	1. Overall cost 2. Energy 3. Environmental aspects	Commodity chemicals, sludges, waste products

to develop control strategies is expected to become prevalent in years to come when reliable mathematical models cannot be developed.

Innovation in Drying Equipment Design

In the past three decades there has been a nearly exponential rise in R&D in the inter- and multidisciplinary field of drying. Drying of solids is a complex operation involving transient transport phenomena coupled with physical/chemical/biochemical transformations, which, in turn, may lead to changes in heat and mass transfer mechanisms. Our understanding of drying at the microscopic level is still rudimentary while our knowledge base with regard to dryers has continued to expand over the years. Scale-up of most dryer types continues to be complex and empirical, and is often equipment- and product specific because of the highly nonlinear nature of the governing conservation equations of transport processes.

Many commonly used drying technologies have matured and perhaps reached their respective inherent asymptotic limits of performance. New products, new processes, higher production rates, more stringent environmental regulations, increased safety concerns, etc. often demand better performance levels at lower costs than is possible with traditional dryers. This need has led to some innovation in drying technologies. Many of the innovative concepts are still at the pilot stage while several have reached successful commercialization. Due to the lower risk involved and their shorter gestation periods, many of the successful innovations are evolutionary in nature; revolutionary innovations involve greater risks, are more difficult to scale-up because of lack of experience, and hence are less readily embraced by industry.

Conventional Versus New Drying Technologies

Kudra and Mujumdar [2] have classified and discussed various novel dryers, ranging from laboratory-scale curiosities (e.g., acoustic drying, drying of slurries by impinging sprays over a hot surface) to pilot-scale demonstrations (e.g., pulse combustion dryers, ultrasonic spray dryers, impinging stream dryers) to full-scale commercial dryers (e.g., pulsed fluid beds, superheated steam fluid bed/flash dryers, rotary dryers with drying air injected into the rolling bed). A full discussion of the truly bewildering variety of nonconventional dryers is beyond the scope of this chapter. The interested reader is referred to the new book by Kudra and Mujumdar [3] for a comprehensive coverage of new drying concepts and technologies.

Table 4 summarizes the key features of the newer dryers as compared to those of conventional ones for drying of various physical forms of the wet feed material. Note that the new designs are not necessarily better than the traditional ones for all products, but they do offer some advantages that may make them a better choice in some applications.

Table 5 compares some key features of the newer or emerging drying technologies with those of the more commonly used conventional techniques. In terms of the sources of energy there is no difference. However, in terms of how this energy is delivered and transferred to the wet solid there are some significant differences.

In the chemical industry the most common drying application involves production of dry particulates from pumpable liquids (solutions, suspensions, or slurries), thin or thick pastes (including sludges), or granular solids. Spray and drum dryers are used most commonly for such applications. Spray dryers today no longer just convert a pumpable liquid to a powder but can be used to produce "engineered" powders with specific particulate size, as well as structure (e.g., agglomerates, granules, or large mono-sized spherical particles). Personnel safety on and around dryers, prevention of environmental pollution, and emphasis on production of a high-quality product at minimum cost are paramount considerations in the design of spray dryers today. With the help of computer simulations, better designs of the dryer chamber and air flows within the dryer have led to minimal wall deposit problems in spray dryers.

According to Masters [4], there is a movement toward full-containment spray drying, which implies these concerns:

TABLE 4. Conventional versus innovative drying techniques

Feed type	Dryer type	New techniques[a]
Liquid suspension	Drum Spray	Fluid/spouted beds of inert particles Spray/fluid bed combination Vacuum belt dryer Pulse combustion dryers
Paste/sludge	Spray Drum Paddle	Spouted bed of inerts Fluid bed (FB) (with solid backmixing) Superheated steam dryers
Particles	Rotary Flash Fluidized bed (hot air or combustion gas)	Superheated steam FBD Vibrated bed Ring dryer Pulsated fluid bed Jet-zone dryer Yamato rotary dryer
Continuous sheets (coated paper, paper, textiles)	Multicylinder contact dryers Impingement (air)	Combined impingement/radiation dryers Combined impingement and through dryers (textiles, low basis weight paper) Impingement and MW or RF

MW, microwave; RF, radiofrequency
[a] New dryers do not necessarily offer better technoeconomic performance for all products

TABLE 5. Comparison of conventional versus emerging drying technologies

	Conventional	Emerging trends
Energy (heat source)	Natural gas Oil Biomass Solar/wind Electricity (MW/RF) Waste heat	
Fossil fuel combustion	Conventional	Pulse combustion
Mode of heat transfer	Convection (>85%) Conduction Radiation (<1%) Microwave/radio frequency (<0.1%)	Hybrid modes Nonadiabatic dryers
Drying medium (convective dryers)	Hot air Flue gases	Superheated steam Hot air + superheated steam mixture
Number of stages	One—most common Two or three—same dryer type	Multistaging with different dryer types
Dryer control	Manual Automatic	Fuzzy logic Model based Artificial neural nets

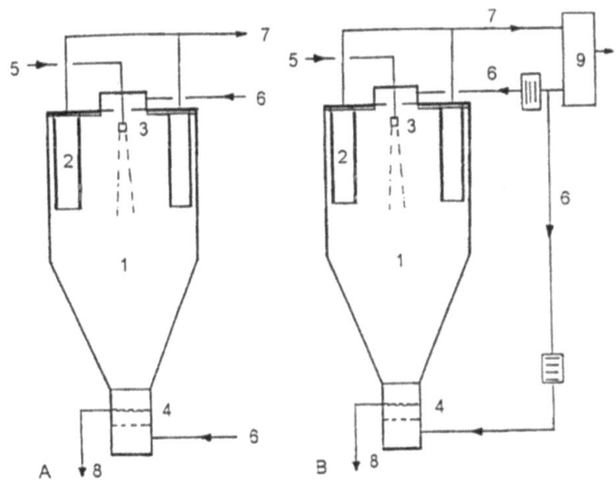

FIG. 3. Spray drying chamber with integrated fluid bed and particulate filters. *1*, drying chamber; *2*, integrated particulate filter; *3*, atomizer; *4*, integrated fluid bed dryer; *5*, liquid feed; *6*, drying air; *7*, exhaust air; *8*, powder discharge (single point); *9*, condenser (evaporated volatiles recovery)

TABLE 6. Spray drying of emulsion—PVC: effect of atomizer type on dryer performance (a comparison between different atomizers)

Parameter	Rotary disk	Two-fluid (sonic)	Two-fluid (standard)
Dryer geometry	Conical/cylindrical, $H/D \approx 1.2$–1.5	Tall-form cylindrical, $H/D \approx 4$	Tall-form cylindrical, $H/D \approx 5$
Evaporation capacity (water)	1600 kg/h	1600 kg/h	1600 kg/h
Size ($D \times H$)	6.5 m × 8 m	3.5 m × 15 m	3 m × 18 m
Number of nozzles	1, 175-mm disk, 15 000 rpm	16 nozzles, 4 bar pressure	18 nozzles, 4 bar pressure
Power for atomizer	25 W/kg slurry	20 W/kg slurry	80 W/kg slurry
Capital cost	High	Medium	Medium
Operating cost	Medium	Low	High

H, height; D, diameter

- Need to reduce large volumes of drying gas exhausting to atmosphere using recycle of exhaust (partially or fully closed dryer layout) to minimize exhaust air plumes with associated thermal, odor, and particulate emissions
- Limiting the presence of product to dryer itself, thus simplifying product handling and simplifying operational safety and cleaning procedures
- Need to make more compact dryer layouts

The development of in-place cleanable bag filters makes it possible to retain the particulates within the dryer chamber; this is achieved by mounting the filter elements in the roof of the spray dryer chamber. No external cyclones are needed in this case (Fig. 3). This technology, coupled with the popular fluidized-spray dryer featuring a fluid bed dryer integrated into the base of the spray dryer chamber, allows efficient production of dustfree agglomerated or granulated products at substantially lower product temperatures than those found in conventional spray dryers.

Another significant development in spray drying is the two-fluid sonic atomizer, which produces a nearly monodisperse spray for fluid of viscosities up to 250 cp. Table 6, adapted from Shah and Arora [5], compares the chamber size, power consumption, and relative capital and operating costs for spray-drying 1600 kg/h of emulsion PVC. The advantages of the sonic nozzle are obvious. More recently, ultrasonic nozzle atomizers have also been successfully tested for small-scale spray drying applications. Note that the conventional disk and two-fluid nozzles give a polydisperse product.

Another common dryer used to produce flaky (as opposed to the powdery product from spray dryers) product is the classical steam-heated drum dryer. Its application is severely constrained by the very low production capacity for a given size of equipment. Attempts have been made to enhance the drying capacities of drum dryers using arrays of impinging jets, radiant heaters, and also by utilizing inductive heating of the drum wall carrying the thin film of wet product being dried. The latter also permits online computer control of the drying process by adjusting the inductive heat input. Note that increased complexity of the dryer involves higher capital costs but the payback time may still justify it by means of the reduced energy costs and premium price for the higher-quality product.

For drying granular or particulate solids, the most common dryers in use today are cascading rotary dryers with or without internal steam tubes, conveyor dryers, and continuous tray dryers (e.g., turbo or plate dryers), which must compete with fluidized-bed dryers (with or without internal heat exchangers) and vibrated bed dryers, among others. At least 20 variants exist of the fluidized-bed dryer alone. For larger particles, a spouted bed dryer is preferable to the conventional fluidized-bed dryer. For difficult-to-fluidize, sticky particles or feedstocks with a wide particle size distribution, the vibrated bed dryer offers advantages over the conventional fluid bed because it allows use of low drying air velocities while mechanical vibration assists in pseudofluidizing the solids.

Fluidized-bed dryers have been operated successfully using superheated steam as the drying medium both at near atmospheric pressures (e.g., drying of pulverized lignite) and at elevated pressures (3–5 bars for drying of beet pulp). In addition to eliminating potential fire and explosion hazards, use of superheated steam permits utilization of the exhaust steam by condensation, reheating, or compression. Of course, such steam is often contaminated and must be cleaned before reutilization depending on the application. Net energy consumption in superheated steam dryers may be as low as 700 to 1000 kJ/kg water evaporated, which is five to ten times lower than many conventional convective air drying systems consume. Mujumdar [1] has discussed the principles, practice, and potential of the rapidly emerging superheated steam drying technologies in various industrial sectors, principally for drying paper, wood, some processed foods, sludges, etc.

Use of volumetric heating by microwave (MW) and radiofrequency (RF) fields has yet to make major inroads in the chemical industry. It is well established that MW/RF-assisted drying is faster, but the energy costs are also significantly higher and scale-up for large production capacities is much more difficult. Such dryers are expected to find some niche applications. RF drying under vacuum has been applied successfully on a commercial scale by a Canadian company for drying of timber and thick veneer with future applications anticipated for drying of chemicals, polymers, and foods. An RF dryer, in conjunction with impingement with hot air jets, is already a commercial process for drying of coated paper.

Recent Trends in Drying Technologies

The following trends have emerged over the past decade and are expected to continue over the next decade.

- Use of combined heat transfer modes, for example, convection coupled with conduction, radiation, or microwave/RF fields to reduce the amount of drying gas that is usually exhausted without heat recovery for economic reasons. Such combinations often involve application of known technologies, reduced risk, smaller equipment size, simpler gas cleaning, and smaller footprint. More than 85% of dryers in use today are of the convective (direct) type; this percentage is expected to decrease over the next decade as a result of energy and environmental concerns. As the energy costs rise, indirect dryers will become increasingly more competitive.
- Use of superheated steam as the drying medium rather than air or combustion gases offers many advantages (e.g., low net energy consumption, low containment of volatile emissions, safe operation, better quality of products that are susceptible to oxidative damage) but also several limitations. At atmospheric pressure operation, the product temperature is at least 100°C when removing surface moisture and higher when subcritical water is being removed. Thus, heat-sensitive materials must be dried at lower pressures in superheated steam, which increases the cost of equipment because it must be leakproof.
- Flash, fluidized-bed, and impingement dryers using superheated steam as the drying medium are already commercialized for certain products. Successful pilot tests have been reported with spray dryers as well. This technology is particularly suited for feeds containing large amounts of water to be removed at high production rates. Clearly the initial investment costs are high but the energy costs over the lifetime of the dryer are significantly low. It is noteworthy that up to 70% of the life cycle cost of a convective dryer is for energy whereas only 12% is attributed to the capital investment. Thus, energy-efficient dryers will become increasingly common in the coming decade.
- Batch dryers with time-varying energy input such as variable gas flow rate, variable gas temperature, variable operating pressure, etc. A cursory examination of the batch drying curve and the mechanism of moisture transfer in the falling rate period shows that improved energy efficiency accrues from supplying energy for drying intermittently or at a rate dictated by the drying kinetics of the product [6]. Numerous experimental studies as well as numerical simulations have demonstrated the benefits of intermittent drying in terms of energy savings and better quality of product, especially if it is heat sensitive or fragile. As expected, a marginal increase in drying time is needed. This idea of intermittency has been applied successfully on a commercial scale in the pulsed fluid bed dryer where the bed is fluidized only periodically. Pilot tests have verified the advantages of periodic spouting in a novel rotating jet spouted bed when drying in the falling rate period proposed by Jumah et al. [7].
- Scheduling the drying conditions in an optimal manner has been the conventional practice in such major batch drying equipment as wood-drying kilns and freeze dryers, which require extremely long drying times. In fact, some of the traditional continuous dryers operate in an inherently intermittent mode, for example, rotary dryers in which convective heat transfer occurs when the particles cascade down under gravity from the top of the rotating shell while internal diffusion of moisture from the core of the particle to the surface continues as the particles travel from the bottom of the shell to the top without contact with drying air before cascading down again. In the case of the rotary dryer, such intermittency cannot be controlled independently of the other operating parameters, however.
- It is expected that novel gas-particle contactors (e.g., pulsed beds, rotating jet spouted beds, impinging streams) will find new applications as dryers for particulates

as well as slurries. In the latter case, the particles are inert (e.g., Teflon, glass balloting, sand). The liquid to be dried is sprayed into the dryer and coats the particles with a thin film that dries rapidly and peels off as a powder as the result of collisions between the particles. For smaller-scale production, such a dryer can be more cost-effective than a spray dryer although the quality of the powder obtained is very different.

- Repeated attempts have been made in the past decade to utilize the relatively old technology of pulse combustion to dry liquid slurries directly in the tailpipe exhaust of a pulse combustor. It is well known that combustion efficiencies of up to 99.9% with low NO_x emissions are possible in a pulse combustor. Also, by simply injecting the feed liquid jet into the highly turbulent and strong acoustic field (SPL ≈ 160 dB) in the tailpipe, the liquid atomizes and dries within a matter of milliseconds. Unlike the spray dryer, atomizer or fan power is not needed. Even highly heat-sensitive materials can be dried in such a dryer. Recently, Toho Gas Company of Osaka, Japan, has introduced a new line of valveless pulse combustion dryers for drying of a variety of liquid form or sludgelike feeds to produce fine powders. Materials ranging from contaminated sludges to iron oxide, pigments, vinyl chloride resin, toners for printers, and surface-active agents are dried successfully in units with evaporation capacities up to 2000 kg/h. Earlier designs of pulse combustion dryers used rotating mechanical valves and were designed mainly to dry highly heat-sensitive materials such as yeast and vitamins. These units are no longer available commercially, however. Attempts have been made to couple pulse combustion technology to other dryers, for example, for flash-drying of fine particulate solids and even for drying of wood. However, it is unlikely that pulse combustion drying will be suitable for other drying applications than drying slurries to produce fine powders in very short drying times, typically fractions of a second when only surface moisture can be removed. This technology can potentially offer serious competitive edge over spray and flash drying. Further R&D is needed to improve scale-up capabilities and reduce noise levels to acceptable levels, however.
- A trend toward multistage drying to enhance energy efficiency, improve quality, and to reduce costs will be increasingly observed in the coming decade. This has already occurred in the spray drying area. For very large scale operations, spray drying from a pumpable feed consistency to the final dry product is often not economically feasible because of the size of the chamber needed to provide the required residence time in the dryer. If the spray chamber is designed only to remove surface moisture, the partially dried product still containing internal moisture can be removed in a small fluidized bed (integral to the spray chamber or external to it) or a vibrated bed or a conveyor-type through dryer located at the base of the spray dryer. In fact, the use of a fluid bed or vibrated bed permits agglomeration or "instantizing" of products such as milk, coffee, cocoa, and detergents. In drying of polyester chips to very low final moisture content, it is desirable to follow a two-stage approach because this operation involves a crystallization stage concurrently with the initial drying phase. A wide variety of two-stage operations involving different dryer types for the first and the second stage have already been commercialized for various products. In general, if a product has significant amounts of surface as well as internal moisture it is reasonable to evaluate a two-stage drying process, for example, a flash dryer to remove surface moisture (which also eliminates possible stickiness of certain materials), followed by a fluid bed or a conveyor dryer.

- Control of dryers using model-based control strategies, fuzzy logic, or artificial neural nets will become increasingly popular. Advances in sensor technologies are expected to permit the control of dryers by directly sensing on line the quality of the product, which is currently done indirectly.
- Finally, for energy-efficient drying of high-value products that are heat sensitive, convective dryers using mechanical heat pumps to dehumidify drying air and also recover the latent heat from dryer exhaust are expected to find a niche market. Such dryers can be operated in the batch or continuous mode. Economics of such dryers could be improved by utilizing infrared or microwave assists as well as by operating batch dryers with intermittent energy input by convection and/or other modes.

Some Suggestions for Future Drying Technologies

The following ideas are suggested for those interested in developing the next generation of dryers.

Miniaturization of Dryers

One of the characteristic features of the conventional industrial drying equipment is that it is usually very large in volume as well as footprint because of the extremely low holdup of the material being dried within the dryer volume. It ranges from being negligible (e.g., thin film on a drum dryer, wet sheet on a multicylinder or impingement dryer for paper) to a more respectable 10% to 20% in a fluid bed, spouted bed, or rotary dryer. Even simple conveyor dryers usually have low holdup of the material being dried. Large size of dryer implies larger duct sizes for air handling, larger volumes of drying air, larger product collection systems for particulate materials, and, of course, higher capital and running costs. For convective dryers, about 60% to 70% of the life cycle cost of operation is due to the energy consumed; thus, smaller dryer volumes are clearly highly desirable.

Unfortunately, because the dryer, first and foremost, must be able to physically transfer the material from the dryer entrance to the exit, this places some severe restrictions on the geometry of the dryer and the volume useful for actually carrying the product within it, such as a drum dryer or paper machine dryer. The cascading rotary dryer cannot optimally process holdups higher than about 15% by volume. The fluidized-bed dryer volume cannot be reduced much as it needs a significant volume for the plenum chamber to distribute the gas uniformly and another to allow the entrained product to disengage from the exhausted gas due to the ploydisperse nature of most feeds. Furthermore, the drying time needed directly affects the dryer size for a given throughput. Enhancement of drying rate using techniques listed in Table 7 can reduce the dryer volume somewhat; dramatic reductions are generally not cost-effective, however.

It may be impossible in the near future to reduce dryer volumes for several dryer types. For particulate solids, there is some potential to enhance drying rates as well as increase product holdups by using modified rotary or fluid bed dryers. In rotary dryers the dryer volume can be halved, as shown by Yamato [8] for their patented TACO Rotary Dryer, by injecting the hot drying air directly into the rolling bed in the rotating cylinder rather than using the conventional axial flow of drying air. The resulting intimate contact between the gas and the particles more than doubles the volumetric heat and mass trans-

TABLE 7. Techniques for enhancement of drying rates in the constant rate and falling rate periods

Drying regime	Enhancement techniques	Comments
Constant rate period; surface moisture removal	Increase temperature, velocity of drying medium; IR heating	Limited by heat sensitivity of material
	Decrease humidity of drying medium	Higher cost
	Increase turbulence	Higher running cost
	Use pulsing or vibration to enhance external heat/mass transfer rates	Higher initial cost
	Apply vacuum	Higher cost; justified for heat-sensitive, high-value products
	Use superheated steam above inversion temperature as the drying medium	High cost, high efficiency
Constant rate period; surface moisture removal	Apply acoustic field	High cost, noise problem, low efficiency
Falling rate period	Above techniques	Only second-order effect due to internally controlled heat/mass transfer rates
	Use microwave/radiofrequency fields	Expensive, affect product quality

fer coefficients; this allows increase of the holdup to about 30% of the shell volume. This simple modification is a true innovation in rotary drying, which make it a competitor to fluid bed drying for some large-scale operations.

The need to have large disengagement volumes as well as plenum volumes can be obviated by using mechanical vibration to pseudofluidize beds of polydisperse particles instead of achieving true fluidization with gas alone. There may be significant cost savings in energy consumption, although the noise problem may need to be addressed in this case. In principle, it is possible to reduce the overall dryer volume this way although the "footprint" may not be reduced significantly. Use of low gas velocities needed in this type of dryer also reduces attrition and reduces the size of the ducts and gas cleaning equipment.

Very often the dryer volumes used in industry are large for no obvious reason. If the designer carefully questions the need for "ineffective volumes" in their design of the dryer or drying systems, it should be possible to make substantial savings with existing technologies.

Development of "Smart" Dryers

The revolutionary concept for industrial dryers of "smart" dryers requires online sensors to measure solids properties such as moisture content, temperature, and possibly even some quality parameters (e.g., color, shape) as the wet solid travels through the dryer. The dryer design and operation will also become correspondingly more complex, requiring real-time computer control. The idea here is to monitor the solids properties and adjust the local drying conditions according to a preprogrammed control scheme using a mathematical model, fuzzy logic or artificial neural nets. Such a dryer will clearly be difficult to design and operate. It will be cost-effective mainly for very high value product initially.

Closing Remarks

Although drying technologies are considered mature they are not sufficiently well understood to ensure reliable design and off-design analysis. There is potential to improve existing technologies and to design intelligent combinations of current technologies that will lead to better quality product, smaller equipment size, greater reliability, safer operation, lower energy consumption, and reduced environmental impact while reducing the overall cost of drying. Further R&D is needed coupled with close interaction between industry and university researchers to enhance our understanding of the fundamental drying processes so that they may be applied to better design, optimize, and operate the wide assortment of dryers in use today.

In view of the direct relationship between the massive energy needs of drying operations in all industrial sectors and emissions of greenhouse as well as noxious combustion products, it is obvious that there is a need to develop and operate environmentally friendly drying processes. As use of renewable energy sources for industrial drying will remain limited, there is need to devise more efficient combustors as well as drying equipment to obtain high-quality products with the least consumption of resources. Active collaboration between industry and researchers is needed to meet this goal in a reasonable time. With the rapid advances made in computer technology, material science, and our understanding of the underlying transport phenomena in drying of solids, there is scope for rapid development of more efficient drying technologies that could be deployed in both the developed and developing parts of the world.

References

1. Mujumdar AS (ed) (1995) Handbook of industrial drying, 2nd edn. Dekker, New York
2. Kudra T, Mujumdar AS (1995) Specialized drying techniques and novel dryers. In: Mujumdar AS (ed) Handbook of industrial drying, 2nd edn. Dekker, New York, pp 1087–1149
3. Kudra T, Mujumdar AS (2000) Advanced drying technologies. Dekker, New York
4. Masters K (1999) Full containment spray drying. Drying Technology—An International Journal 17:2341–2348
5. Shah RM, Arora PK (1996) Two-fluid nozzles and their application in spray drying of E-PVC. In: Strumillo C, Pakowski Z, Mujumdar AS (eds) Drying '96: Proceedings of the 10th international drying symposium, Lodz, Poland, pp 1361–1366
6. Devahastin S, Mujumdar AS (1999) Batch drying of grains in a well-mixed dryer—effect of continuous and stepwise change in drying air temperature. Transactions of the American Society of Agricultural Engineers 42:421–425
7. Jumah RY, Mujumdar AS, Raghavan GSV (1996) Batch drying kinetics of corn in a novel rotating jet spouted bed. Canadian Journal of Chemical Engineering 74:479–486
8. Yamato Y (1996) A novel rotary dryer with through air combination. In: Strumillo C, Pakowski Z, Mujumdar AS (eds) Drying '96: Proceedings of the 10th international drying symposium, Lodz, Poland, pp 627–630

Critical Heat Flux in Subcooled Flow Boiling

GIAN PIERO CELATA

Summary. This chapter reports a survey of recent experimental work performed to obtain parametric trends of critical heat flux (CHF) in water-subcooled flow boiling, with special concern for high liquid velocity and subcooling, small-diameter channels, and low–intermediate pressure, together with a review of current modeling analysis. Under these thermal hydraulic and geometric conditions, the CHF is proved to be very high (up to many tens of MW/m^2) such as to match the requirement of fusion reactor high heat flux components for thermal hydraulic design. From the experimental aspect, experiments carried out so far show parametric trends of CHF in terms of subcooling, mass flux, pressure, channel diameter and length, channel orientation, tube wall material and thickness, and peripheral nonuniform heating. Passive techniques for CHF enhancement, such as twisted tapes, helically coiled wires, and the hypervapotron, are also reviewed. Apart from the specific difficulty of properly modeling the CHF phenomenon for a large range of conditions, a major drawback of many existing theories is the use of empirical constants to justify the model with the data, making complex correlations that distort the original idea of the model. Among the many existing models, the superheated layer vapor replenishment and liquid sublayer dryout theories proved capable of predicting CHF under these conditions, extending to uniform and nonuniform heating of the channel and swirl flow as well.

Key words. Critical heat flux—Subcooled flow boiling—Experimentation—Modeling—High heat fluxes

Introduction

Water subcooled flow boiling, under conditions of high mass flux, high liquid subcooling, and small-to-intermediate channel diameter, can accommodate very high heat fluxes. As is well known, this forced convective boiling involves a locally boiling liquid, whose bulk temperature is below the saturation temperature, flowing over a surface exposed to a high heat flux. However, successful use of subcooled flow boiling requires the critical heat flux (CHF), which is described as a sharp reduction in the energy

---

Energy Department, National Institute of Thermal-Fluid Dynamics, ENEA, CR Casaccia, Via Anguillarese, 301, I-00060 S.M. Galeria, Rome, Italy transfer from a heated surface, not to be reached. The occurrence of CHF, in a heat flux controlled system, results in a significant increase of the wall temperature, which is usually well above that at which serious damage or 'burnout' of the heating surface occurs.

Subcooled flow boiling has been widely investigated in the past [1–3], with particular reference to the thermal hydraulic design of light water reactors (LWRs), for which the order of magnitude of heat fluxes to be removed is about 1 MW/m^2. Boyd [4,5] has reported a thorough review of CHF in subcooled flow boiling.

Some components of fusion thermonuclear reactors, such as divertors, plasma limiters, or first-wall armor, are believed to be subjected to operating conditions characterized by extremely high thermal loads. It is therefore necessary to remove from the surface of these components very high heat fluxes, ranging from 2 to about 20 MW/m^2. This need has called, over the past 10 years, for a new investigation of the CHF under fusion reactor-relevant conditions, that is, low-to-intermediate pressure (up to 5 MPa), high liquid velocity (up to 40 m/s), high liquid subcooling (up to 250 K), and small-to-intermediate channel diameter (1–15 mm). Such studies clearly showed that the CHF mechanisms under these extreme conditions may be largely different from those observed in the past for LWRs.

The aim of the present chapter is to provide an updated state-of-the-art review of the CHF in subcooled flow boiling with special regard to the high heat flux region. Information provided spans from the experimental point of view, showing the basic parametric trends observed and the most recent visualization studies, to the point of view of mechanistic modeling, presenting the most widely used models.

Experimentation

Many experimental studies have been conducted to characterize the CHF in subcooled flow boiling under fusion reactor-relevant conditions. Part of these have been devoted to ascertain the parametric trends under these extreme conditions, mainly using pipes of different geometry under different thermal hydraulic conditions. Some research has been performed to specifically analyze possible enhancement techniques for the CHF. Finally, some studies have been devoted to the visualization of the CHF to obtain direct information about the physical mechanisms leading to the CHF and possibly to develop a more realistic mechanistic model of the thermal crisis.

Experiments for Parametric Trends

A significant number of experimental data have recently been collected for a total of about 2000 data points [6] under the following operating conditions: $0.1 \leq p \leq 8.4$ MPa; $0.3 \leq D \leq 25.4$ mm; $0.0025 \leq L \leq 0.61$ m; $1 \leq G \leq 90$ Mg/m^2 s; $25 \leq \Delta T_{sub,in} \leq 255$ K (see Appendix for definitions of symbols). Such a database refers to peripheral and axial uniform heating in smooth channels. Critical heat flux values ranging from 3.3 to 227.9 MW/m^2 have been obtained, and basic parameters trends have been outlined.

Influence of Subcooling

As reported by Boyd [5], most of the early experimental studies showed that the relationship between subcooling and CHF is almost linear, even though Bergles [7] indicated

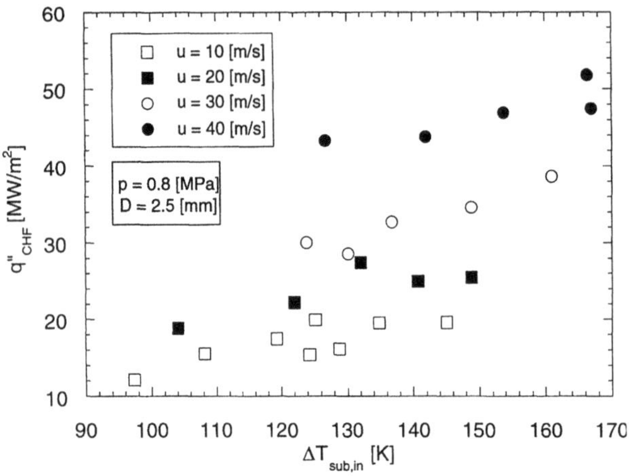

FIG. 1. Critical heat flux versus inlet subcooling [10]

that this relationship may deviate from linearity in several ways. Recent experiments under conditions of high liquid subcooling [8–10] confirmed the almost linear relationship between CHF and subcooling. Figure 1 shows the CHF versus inlet subcooling for data determined by Celata et al. [10] in 2.5-mm-i.d. stainless steel tubes, 0.25-mm wall thickness, 10 cm long, uniformly heated by the Joule effect, with vertical upflow of water. The functional dependence of the CHF on subcooling is practically linear, up to very high subcooling and very high liquid velocity. The CHF versus $\Delta T_{sub,in}$ curves, plotted at different liquid velocities, are parallel with each other, and no interrelation between u and $\Delta T_{sub,in}$ seems to exist. The same conclusions have been drawn by other investigators. The influence of subcooling on the CHF is quite relevant, and the absolute increase looks almost constant (CHF versus ΔT_{sub} curves at different liquid velocity are almost parallel).

Influence of Mass Flux

CHF is an increasing function of the mass flux (or fluid velocity) in the subcooled region, as has been observed up to very high values of mass flux (90 Mg/m^2 s) [9–13]. Figure 2 shows the results of experiments carried out by Boyd [11,13] using water as a fluid in horizontal test sections of amzirc (copper-zirconium alloy) with an inner diameter (i.d.) of 3.0 mm, wall thickness about 0.5 mm, and a heated length of 0.29 m [11,12], or 10.2-mm i.d., 0.125-mm wall thickness, 0.5 m long, and copper as a material [13]. Tests were performed at a constant inlet temperature of 20°C. The CHF increases with the mass flux G, in less than linear fashion, and it is possible to observe that passing from 10 to 40 Mg/m^2 s the CHF can be increased up to a factor of about 3 or more.

Although from a physical point of view an increase of the fluid velocity is always positive for the enhancement of CHF, nonetheless a design upper limit is given by the pressure drop increase (with the square of the velocity) and by flow-induced vibrations.

Influence of Pressure

In the range of 0.1 to 5.0 MPa, direct influence of pressure on CHF was observed to be weak, other conditions being equal (i.d. for the same subcooling and liquid velocity).

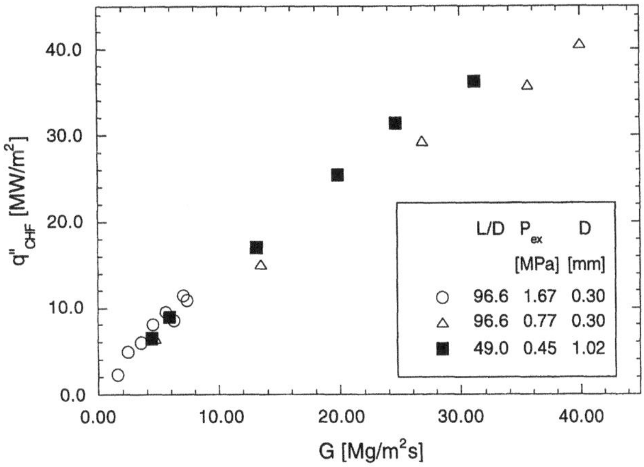

FIG. 2. Critical heat flux versus mass flux [11–13]

This is demonstrated in Fig. 3, where the CHF is plotted versus exit pressure p [9], obtained with stainless steel tubes of 1.07 mm i.d. and 26.75 mm long. Virtually no pressure effect was noted; in fact, there seemed to be a very slight decrease of CHF with increasing pressure. Figure 4 shows the results of Celata et al. [14] obtained with stainless steel tubes of 8.0 mm i.d., 10 cm long, with uniform heating. At a lower liquid velocity than in the Vandervort et al. data [9] (10 m/s instead of 25 m/s), the CHF versus subcooling data lie on a unique curve independent of the pressure, indicating the negligible effect of this parameter. The same conclusions on the weak influence of pressure on the CHF were drawn by Nariai et al. [8] and Celata et al. [10]. Although pressure was observed to have a very slight influence on the CHF, nonetheless it enables the enhancement of the CHF because of the higher inlet subcooling possible, other conditions being equal.

Influence of Channel Diameter

Recent experiments to ascertain the influence of channel diameter on the CHF were carried out by Vandervort et al. [9] in the range 0.3 to 2.5 mm, by Celata et al. [15] in the range 2.5 to 8.0 mm, and by Celata et al. [16]. Figure 5 shows the CHF versus the channel diameter D [9]. As also observed by previous researchers, for given values of exit thermal hydraulic conditions, heated length, and liquid velocity, the CHF increases with the decrease of the tube inside diameter, but the effect was less significant for decreased mass flux. A threshold is observed beyond which the effect of the tube inside diameter may be considered negligible, that is, a function of the channel geometry and thermal hydraulic conditions. Celata et al. [17], on the basis of a purely theoretical approach, in a range of diameter from 0.5 to 32 mm, found the following relationship between CHF and channel diameter:

$$\frac{(CHF)_D}{(CHF)_D = 8\,\mathrm{mm}} = \left(\frac{D}{8}\right)^n \tag{1}$$

with $n = -0.3$, where $(CHF)_D$ is the CHF for a diameter of interest, $(CHF)_{D=8mm}$ is the CHF for an 8-mm tube (typical reference value for look-up tables [18]), and D is the tube diameter value in millimeters. Equation 1 was proposed by Doroshchuk et al. [19,20] with

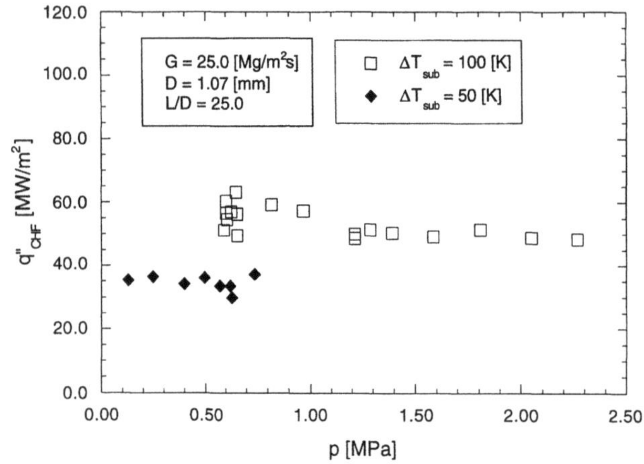

FIG. 3. Critical heat flux versus pressure [9]

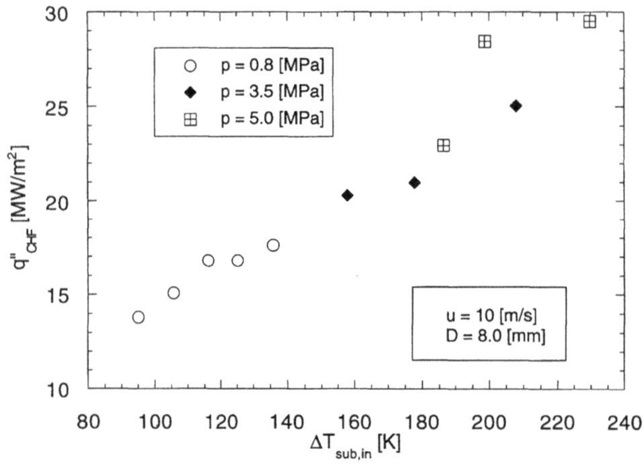

FIG. 4. Critical heat flux versus inlet subcooling [14]

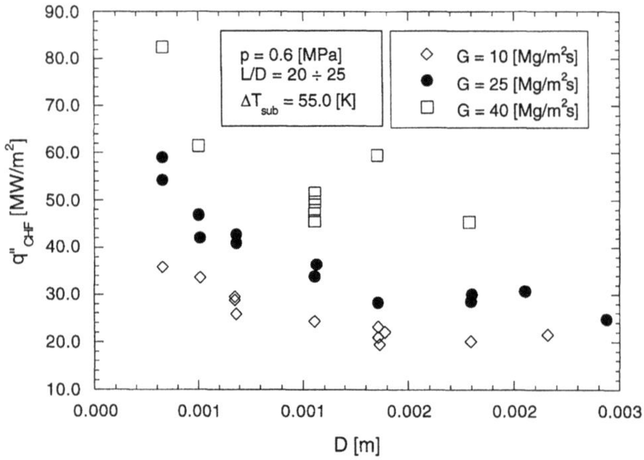

FIG. 5. Critical heat flux versus channel diameter [9]

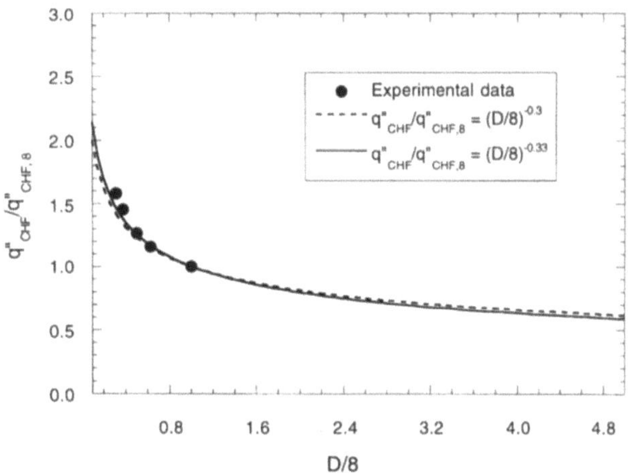

FIG. 6. Functional dependence of the critical heat flux on the channel diameter [17]

a value of −1/2 for the exponent n, for diameter values between 4 and 16 mm. Groeneveld et al. [21] found better agreement for the value $n = -1/3$ with their database of diameter values between 4 and 20 mm. Smith [22] extended the value of −1/3 to a tube diameter of 32 mm. Therefore, Eq. 1, with $n = -0.3$, would seem to provide an extension of the Groeneveld exponent, widening its validity in the range of tube diameters from 32 mm down to 0.5 mm. Figure 6 reports the comparison between predictions given by Eq. 1 and the data of Celata et al. [15].

To possibly explain the observed dependence of CHF on tube inside diameter, Bergles [7] argued that for a tube with a smaller inside diameter we have (1) a small bubble diameter, (2) an increased velocity of the bubbles with respect to the liquid, and (3) the fluid subcooled bulk closer to the growing bubbles (collapsing in the bulk). From the analysis of experimental data of void fraction in narrow tubes, Nariai and Inasaka [23] concluded that, as tube inside diameter decreases and mass velocity increases, the diameter of generated bubbles or, better, the thickness of the two-phase boundary layer becomes smaller because of the intense condensation effect by subcooled water at the core region, and the void fraction becomes smaller, making the CHF higher. The decrease in diameter gives rise to an increase in the slope of the velocity profile in the two-phase boundary layer, making the detachment of growing bubbles and the consequent condensation in the core region easier. The higher the mass flux, the more consistent the effect.

Celata et al. [16], using test pipes with an inner diameter ranging from 0.25 to 1.5 mm, found that the CHF increases as the channel diameter decreases until a value of 0.7 mm is reached. As the channel diameter is further decreased, the CHF is found to be practically constant and independent of the channel diameter; this may be the result of a change in the boiling crisis mechanism, passing from liquid sublayer dryout model (large diameters) to "flooding" of the channel by the vapor phase. This latter effect results from the size of the bubbles, which may be of the same order of magnitude for channel diameter less than 0.7 mm.

At this stage of the research, available experimental data do not allow us to draw any systematic and quantitative conclusions, but only to have generic qualitative information on the feature of the dependence.

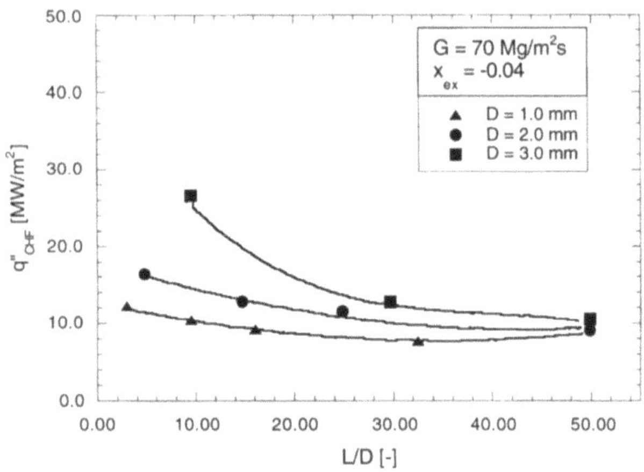

FIG. 7. Critical heat flux versus channel length [8]

Influence of Channel Length

The heated length of the channel seems to be inversely related to the CHF. Generally, investigators use the ratio of the heated length to the inside (or equivalent) diameter of the channel L/D, as the characteristic nondimensional length, but this still needs to be established. Recent experiments have been reported [8,9,16]. Figure 7 reports results [8] showing the CHF versus L/D. The CHF increases as L/D decreases, and the effect is more significant for smaller channel diameter. As the effect seems to be greatest for $L/D < 20$ (depending on the diameter), this would indicate that the CHF is related to the state of development of the bubble boundary layer, as also claimed by Vandervort et al. [9] and Celata et al. [16], who verified that the functional dependence between CHF and L/D is independent of mass flux. As for channel diameter, experiments showed the presence of a threshold beyond which the CHF is practically independent of L/D ($L/D > 40$).

Influence of Channel Orientation

The effect of flow orientation (e.g., horizontal versus vertical upflow) may be significant if the buoyancy force is a nonnegligible percentage of the axial inertial force in flow boiling. Quantitatively, this can be evaluated by the modified Froude number, defined as

$$Fr = \frac{G\cos\alpha}{\rho_l\left[gD\left(\frac{\rho_l-\rho_g}{\rho_l}\right)\right]^{1/2}} \tag{2}$$

where $\alpha = 0$ represents the horizontal case. For modified Froude numbers greater than 5 to 7, effects of stratification and orientation may disappear. Wherever flow orientation plays a relevant role, the CHF for horizontal flow is always less than the value for vertical flow [24,25].

Recent experiments were carried out by Celata et al. [14], under conditions relevant to the NET/ITER divertor: D = 8.0 mm, u up to 8.0 m/s, p = 3.5 MPa, T_{in} = 30°C. Figure 8,

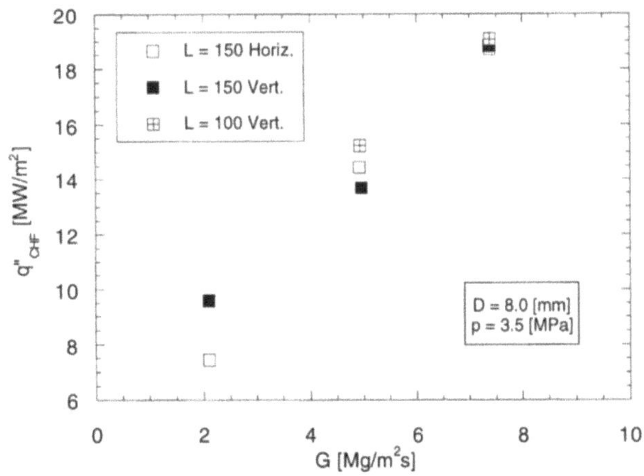

FIG. 8. Critical heat flux versus mass flux for different channel orientation and length [14]

where the CHF is plotted versus mass flux G, shows that for liquid velocity greater than 5.0 m/s (for which the modified Froude number is greater than 20), there is no significant difference between horizontal and vertical data.

Influence of Tube Wall Material and Thickness

A number of researchers have questioned the effect of tube wall material or wall thermal properties on CHF. Vandervort et al. [9] used five different materials in their experiments on CHF: stainless steel 304 (wall thickness, 0.203 mm), stainless steel 316 (wall thickness, 0.203 mm), nickel 200 (wall thickness, 0.254 mm), brass (70/30) (wall thickness, 0.356 mm), and inconel 600 (wall thickness, 0.254 mm). Experimental data carried out to this purpose under very similar geometric and thermal hydraulic conditions indicated that no significant difference was noted.

Vandervort et al. [9] also tested the influence of possible changes in the metal crystalline structure. Indeed, it is known that stainless steel 304 is susceptible to carbide precipitation (movement along a temperature gradient) at temperatures near 1000°C. This temperature may be reached approaching the CHF at very high heat fluxes. Yet, changes in the structure might affect the electrical resistivity, potentially causing nonuniformities in the power generation and resulting in local "hot spots." Vandervort et al. [9], using pure nickel (nickel 200) test sections where no change of this type can occur, did not note any changes in the CHF level.

The tube wall thickness is also a matter of discussion among researchers. Celata et al. [16] tested a number of stainless steel 304 tubes having almost the same inner diameter and different wall thickness to ascertain its possible influence on the CHF. Results are plotted in Fig. 9, where the ratio between the calculated and the experimental CHF is reported versus the wall thickness, *t*. The calculated CHF is obtained using the Celata et al. [26] model, described hereafter. Considering the extended range of the wall thickness, up to a factor of 6, its effect on the CHF may be regarded as very small.

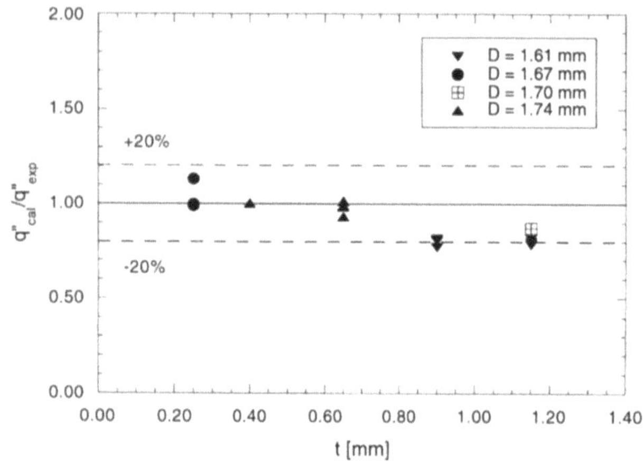

FIG. 9. Influence of tube wall thickness [16]

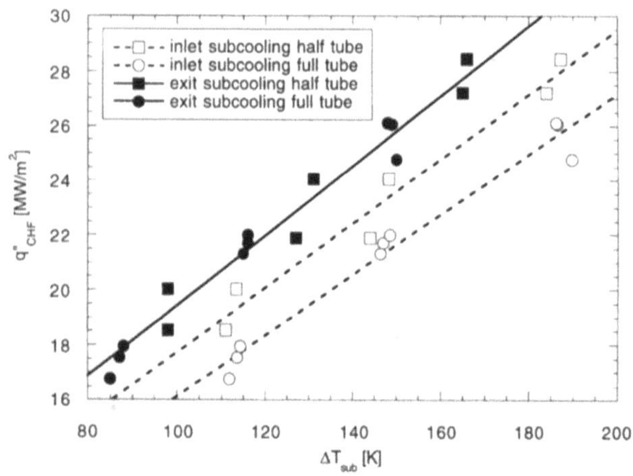

FIG. 10. Influence of nonuniform heating on critical heat flux [28]

Influence on Nonuniform Heating

The effect of nonuniform heating along the tube circumference on the CHF is of relevant importance in the thermal hydraulic design of fusion reactors high heat flux components, as the divertor is thermally loaded only on one side. Ad hoc experiments were therefore recently performed by Nariai et al. [27] and Gaspari [28]. In particular, Gaspari made a comparison between peripherally full and half-heated tubes, straight flow, analyzing the CHF at both inlet and exit thermal hydraulic conditions. Using a 10-mm-i.d. channel, 0.15 m long, Gaspari observed that, under constant inlet liquid subcooling, higher CHF values were observed for half-heated tubes. When the CHF versus exit liquid subcooling is plotted, such differences tend to disappear, as reported in Fig. 10, where the CHF is plotted versus inlet/exit subcooling. Results plotted in Fig. 10 show that the influence of the channel heating on the subcooled boiling crisis (circumferentially uniform or nonuniform heating) may be neglected if we refer the CHF to local conditions.

Experiments for the Augmentation of the CHF

Under conditions typical of the divertor design (u around 10 m/s, D = 10–15 mm, T_{in} = 30°–60°C, p = 3.5–5.0 MPa), the order of magnitude of the attainable CHF with water is about 30 MW/m^2. If we consider safety factors, this is still too low if compared with that requested by fusion reactor thermal hydraulics designers [29]. On the other hand, it is interesting to observe that according to Gambill and Lienhard [30] the above-mentioned value of the CHF is only a small fraction (about 1%) of the maximum heat flux that can conceivably be achieved in a phase transition process. In fact, according to Gambill and Lienhard [30], "if one could contrive to collect every vapour molecule that leaves a liquid vapour interface without permitting any vapour molecules to return to the liquid," the highest heat flux attainable can be estimated as

$$q''_{\max} = \rho_g \lambda \sqrt{\frac{RT}{2\pi}} \tag{3}$$

where R is the ideal gas constant on a unit mass basis. Under the above-mentioned conditions, Eq. 3 provides a maximum theoretical heat flux of about 3 to 4000 MW/m^2. According to Gambill [30], the most serious restriction that prevents reaching this limit in practice "is that many vapour molecules will inevitably be returned to the interface by molecular collisions." The return flow of vapor molecules can only be slowed, not eliminated. "Another problem lies in the premise that all the heat ultimately passes through a liquid–vapour interface. The problem is to get the heat to flow through the liquid, up to an interface, and away from the interface on the vapour side." Researchers contrived to do this with the help of turbulence promoters, or swirl inserts, such as twisted tapes and helically coiled wires. A very interesting method is represented by the hypervapotron technique.

Use of Twisted Tape Inserts

Twisted tape inserts inside circular tubes provide a means of increasing the surface heat transfer coefficient and, in the case of flow boiling, the CHF, by inducing secondary radial and circumferential velocity components in the coolant. A considerable increase in the pressure drop with respect to smooth tubes is observed with the use of twisted tapes. Among the many studies conducted in the past, two of the earlier ones in subcooled flow boiling were carried out by Gambill et al. [31,32]. The most recent and systematic studies were performed by Celata et al. [14], Nariai et al. [33], and Achilli et al. [34]; Nariai et al. [33] used a 6.0-mm-i.d. tube, 0.1 m long, and a zirconia twisted tape with the twisted tape ratio (γ) ranging from 2.7 to infinite. Thermal hydraulic test conditions were pressure from 0.1 to 1.47 MPa, inlet temperature between 32° and 45°C, and liquid velocity from 5.3 to 7.4 m/s. A gap of 0.2 mm existed between the twisted tape and the heated wall. A graph of the experimental results, together with the Gambill et al. data [32], is presented in Fig. 11, where the ratio between the CHF with the twisted tape and the value of the

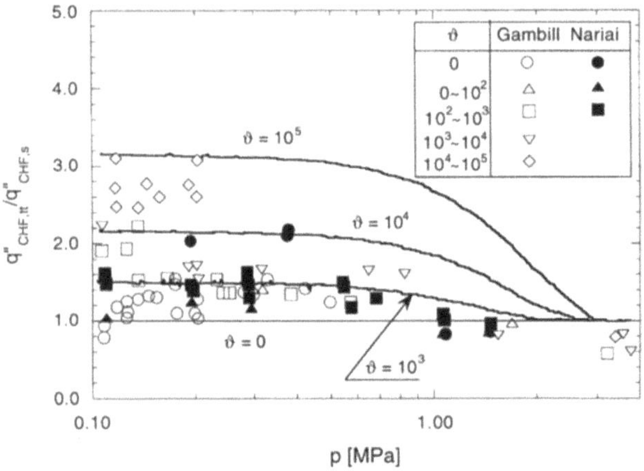

FIG. 11. Critical heat flux enhancement using twisted tapes [33]

straight tube is plotted versus pressure for different values of the nondimensional centrifugal acceleration term, ϑ:

$$\vartheta = \frac{a_t}{g} = \frac{\pi^2 u^2}{2g\gamma D} \tag{4}$$

which is defined as the ratio between the tangential centrifugal acceleration and the standard gravitational acceleration. Augmentation of CHF in subcooled flow boiling seems to be particularly effective at a pressure below 0.5 MPa (up to a factor of 3 with very low twisted tape ratio, γ). The thermal efficiency of the twisted tape decreases as pressure increases and becomes insignificant when pressure is above 2.0 MPa. This effect is probably due to the presence of a gap between the wall and the twisted tape. In fact, the clearance allows steam trapping (which is an increasing function of pressure) in the tube–tape gap, which may result in premature CHF.

Experiments [34,35] were carried out using GLIDCOP tubes 8.0, 10.0, and 15.0 mm i.d., 150 and 300 mm long, at pressure from 1.0 to 6.0 MPa, velocity from 5 to 20 m/s, and inlet temperature from 25° to 155°C. The authors obtained increases in the CHF, with respect to smooth tubes, ranging from 1.4 to 2.0, with γ equal to 1.0. The pressure drop increase was observed to range from 5 to 11, respectively. These authors did not find any effect of the system pressure on the thermal efficiency of the turbulence promoters such as that observed by Nariai et al. [33].

Use of Helically Coiled Wire Inserts

Helically coiled wires were used in the past as turbulence promoters to enhance the heat transfer in single-phase flow (air, water, water-glycerol solution, oil) [36–38]. Enhancement of heat transfer was found (and expected) to be coupled with much larger increase in frictional power loss. Wires were recently used by Celata et al. [14] to enhance the CHF in highly subcooled flow boiling using 8.0-mm-i.d. tubes, 0.1 m long, uniformly heated, at pressure of 3.5 and 5.0 MPa, inlet temperature of 60°C, and liquid velocity

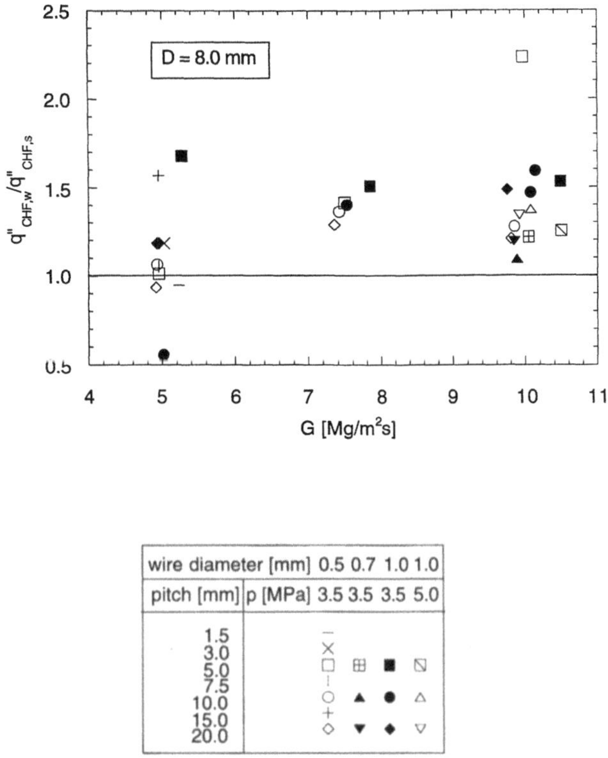

FIG. 12. Critical heat flux enhancement using helically coiled wires [14]

from 2 to 10 m/s. Wires were of spring steel, having a diameter of 0.5, 0.7, and 1.0 mm, and a relative spacing of promoter (pitch) from 1.5 to 20.0 mm. Experimental results (Fig. 12) show the ratio between the CHF with the wire and the value with the straight tube plotted versus mass flux. The authors have observed an increase of the CHF up to 50% using a 1.0-mm wire (p = 3.5 MPa). Low velocity tests reveal very scattered and, as an average, a low efficiency in the CHF enhancement, probably because the relative roughness (wire diameter/hydraulic diameter of the channel) of the turbulence promoted determines the Reynolds number and then the velocity at which the promoters become effective, as stated by Sutherland [36], who established the heat transfer performance of boundary-layer turbulence promoters. Contrary to the performance of twisted tapes, where the increase of the thermal efficiency and the associated increase of the pressure drop are strictly interrelated, the thermal efficiency of wires is independent of the wire pitch, whereas pressure drop is inversely related to it. This latter can therefore be properly reduced decreasing the relative spacing of the promoter, without affecting the thermal performance of the turbulence promoter itself. The effect of the pressure on wire heat transfer performance is observed to be negative, in the sense that tests carried out at 5.0 MPa reduce to only 30% the increase of the CHF produced by the wire with respect to 3.5 MPa data, as already observed by Nariai et al. [33] using twisted tapes.

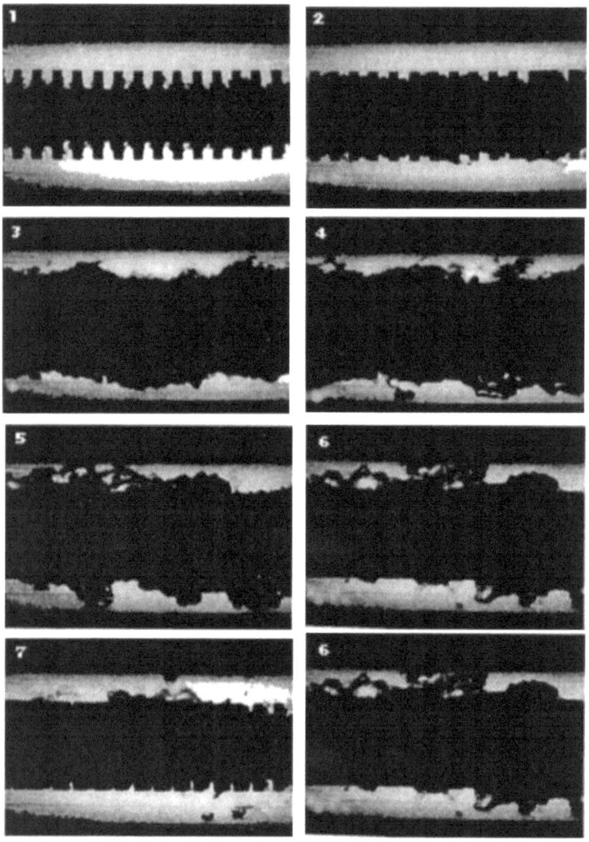

FIG. 13. Typical video sequences of hypervapotron effect [39]

Use of the Hypervapotron Technique

A very intriguing technique using fins, but placed perpendicular to the fluid flow (in subcooled flow boiling), is the so-called hypervapotron technique. From the physical aspect, the hypervapotron effect consists of the following succession of events. The liquid inside two adjacent fins of high-conductivity material and in contact with the heated wall starts boiling while the fluid bulk outside the fins is under subcooled conditions. Once the slot is full of steam, this latter undergoes a quick condensation in the subcooled liquid bulk, emptying the slots and making their replenishment with cold liquid easier. The heated wall is rewetted until the wall temperature during the uncovered phase is below the Leidenfrost temperature. The base of the fin is allowed to operate at a temperature greater than the CHF temperature while the remaining portion operates near the temperature for the onset of stable nucleate boiling. This continuous boiling and condensation sequence (frequency between 10 and 40 Hz) allows a high CHF, essentially on the basis of the transport of the latent heat extracted from the heated wall during boiling and transferred to the coolant outside the fins during condensation. Cattadori et al. [39] obtained a maximum CHF of 29.4 MW/m^2. A typical picture from visualized tests is reported in Fig. 13.

Experiments of Visualization of the CHF

Detailed knowledge of the phenomena causing the boiling crisis on the heat transfer surface is scarce, and, in particular, there are still relevant disagreements among researchers about the mechanisms triggering the boiling crisis. Disagreements are caused by the fact that the direct observation of the heater surface during the boiling crisis, by means of flow visualization, is very difficult because of small vapor bubble dimensions, bubble crowding, and high liquid velocity. Therefore, relatively few works on burnout visualization were carried out in the past, and, in particular very few of them reported direct observations of the boiling phenomena at the CHF.

Fiori and Bergles [40] described the physical phenomena occurring before and at the CHF in subcooled flow boiling of water. The flow pattern during the boiling crisis is slug or froth flow. They also measured the wall temperature and found an oscillatory behavior for it at CHF. The authors proposed the following boiling crisis mechanism in the slug flow regime: the CHF is caused by a hot spot formation in the liquid film during the passage of a slug along the channel.

Gunther [41] carried out experiments in subcooled flow boiling of water making direct observations of flow pattern at the CHF using a high-speed movie. At the boiling crisis he noted the formation of local vapor film caused by bubble coalescence. The passage of these long vapor films causes a reduction in the heat transfer on the heater surface, with associated wall temperature increase. At the CHF, with sufficiently high wall temperature increments, vapor films could spread, thus causing further temperature increases up to the heater burnout.

Kirby et al. [42–44] studied the CHF, visualizing the subcooled flow boiling. They found that the flow pattern in the channel at burnout depends on liquid subcooling, and vapor bubbles tend to keep to the heater surface. These authors suggested a CHF mechanism based on small dry spot formation under some nucleation bubbles.

Celata et al. [45] presented the experimental results of a visual study of the CHF in subcooled flow boiling of water. In their tests they used a pseudoannular geometry, formed by a central heater rod contained in a duct characterized by a square cross section. From the video sequences of the flow boiling phenomena during the boiling crisis the authors obtained useful observations on flow patterns during the boiling crisis and quantitative information on vapor bubbles.

More recently, Celata et al. [46] reported the experimental results of a visualized study on the CHF in subcooled flow boiling of water performed with a pseudoannular test section similar to the previous one, although different in the flow channel configuration. The new test channel allowed the measurement of the average temperature of the heater rod. The authors recorded high-speed movies of flow pattern in subcooled flow boiling, from the onset of nucleate boiling up to the physical burnout of the heater. From the analysis of video images, the authors obtained a qualitative description of the near-wall flow pattern for increasing values of heat flux up to the CHF. Besides, they obtained bubble dimensions from images and measured the heater wall temperature during each experimental test. As a result, they provided, in graphs, a detailed description of the phenomena from the boiling crisis up to the heater burnout, joining the results of bubble parameters and wall temperature measurements, as well as direct observations of flow patterns, for all flow regimes. These graphs provide a useful global point of view of boiling phenomena from ONB (Onset of Nucleate Boiling) to boiling crisis up to heater burnout (Fig. 14). Based on their experimental observations, the authors divided the flow

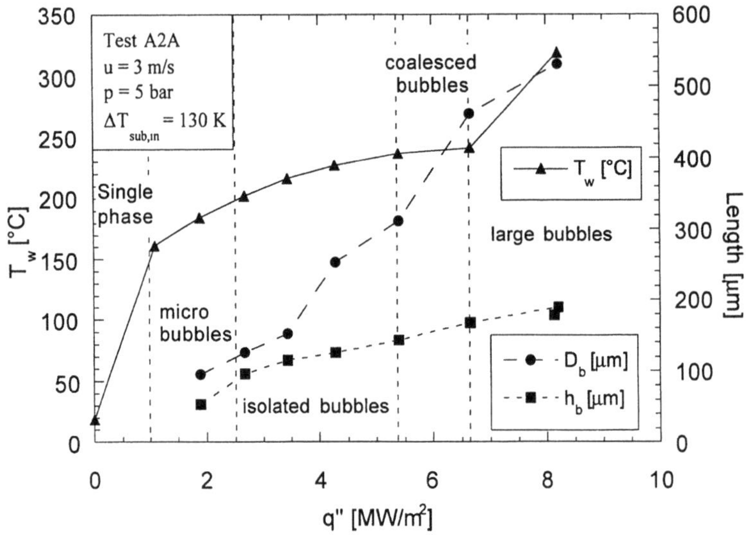

FIG. 14. Measured wall temperature and vapor bubble parameters versus heat flux [46]

pattern, for increasing value of heat flux, into the following groups (Fig. 14): microbubbles, isolated bubbles, coalesced bubbles, and large bubbles. The authors reported, in the large bubbles region during the heat flux increments, the appearance of a bright red hot spot on the heated wall, typical of white-hot metals, and after a period of time depending on test conditions, burnout of the heater occurred.

On the basis of their observations, Celata and coworkers proposed a sequence of mechanisms of the boiling crisis phenomena in subcooled flow boiling. The proposed CHF mechanism sequence was based on the following points: (1) at high heat fluxes, the formation of dry patches under large vapor bubbles caused a significant wall temperature increase; (2) when the heater temperature reached the Leidenfrost value, dry patches were no longer rewetted and the heater surface became covered with a vapor film; (3) on the surface covered with a vapor film, heat transfer was reduced by both thermal insulation and nonboiling; and (4) on the insulated surface, the wall temperature rose up to its melting point, resulting in the burnout. Typical pictures taken during the experiments are shown in Figs. 15 and 16, where images refer to 93.7% of the CHF and to the occurrence of CHF, respectively.

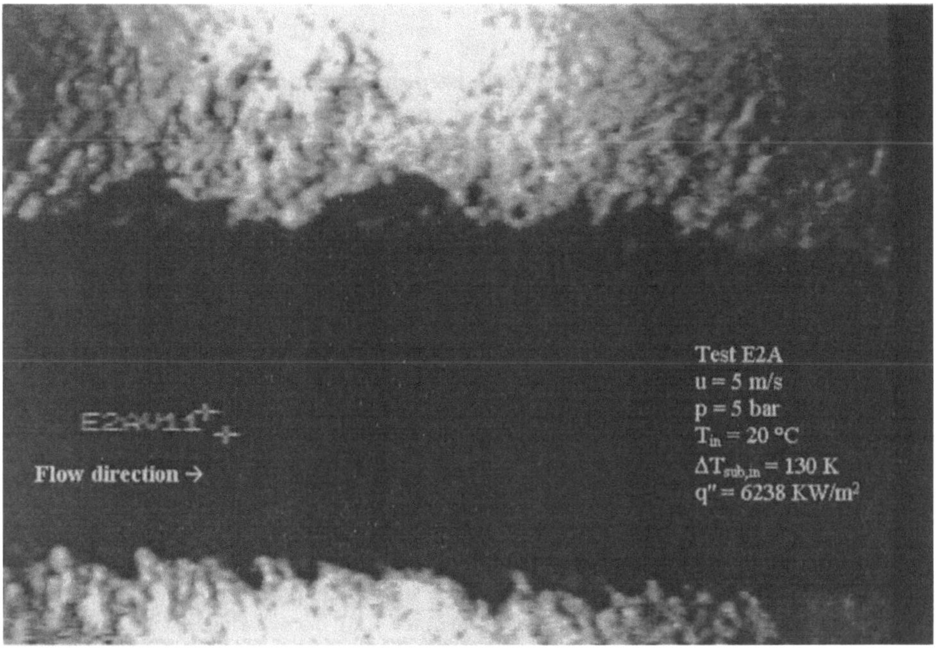

FIG. 15. Typical picture of the flow pattern before critical heat flux (q″ = 93.7% of the critical heat flux)

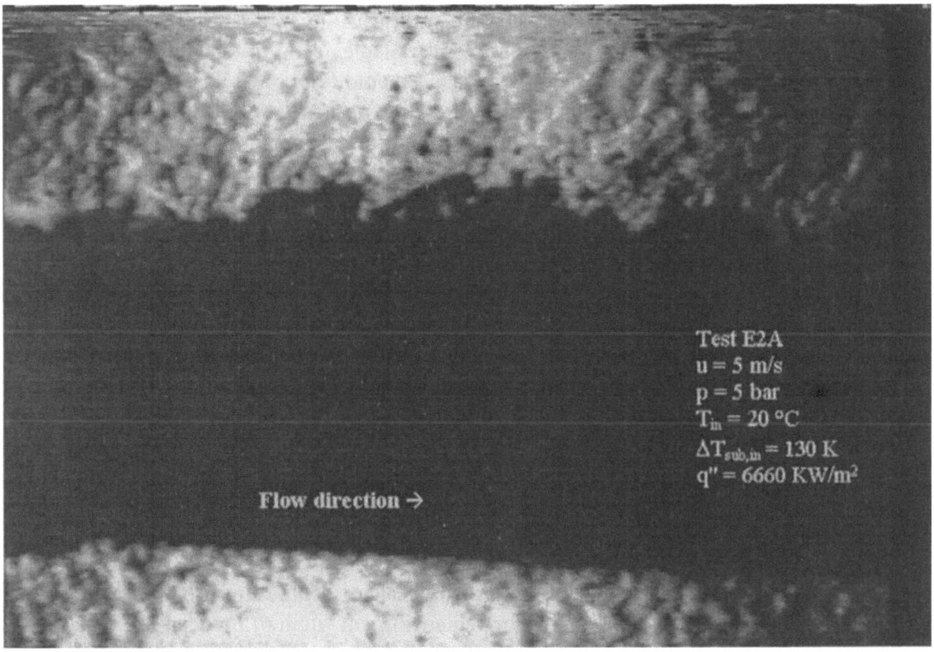

FIG. 16. Typical picture of the flow pattern at critical heat flux

Modeling

Recent reviews of modeling of CHF have been published [47–50]. Major theoretical approaches to CHF can be categorized into six groups, according to the basic mechanism assumed by the respective authors as the main cause of the CHF occurrence.

1. Liquid layer superheat limit model: the difficulty of heat transport through the bubbly layer causes a critical superheating in the liquid layer adjacent to the wall, giving rise to the occurrence of the CHF [51].

2. Boundary-layer separation model: this model is based on the assumption that an 'injection' of vapor from the heated wall into the liquid stream causes a reduction of the velocity gradient close to the wall. Once the vapor effusion increases beyond a critical value, the consequent flow stagnation is assumed to originate CHF [52–57]. The weak physical basis of this model has been demonstrated by the studies reported previously [40,58–60].

3. Liquid flow blockage model: it is assumed that CHF occurs when the liquid flow normal to the wall is blocked by the vapor flow. Bergelson [61] considers critical velocity is raised by the instability of the vapor–liquid interface.

4. Vapor removal limit and near-wall bubble crowding model: it is assumed that the turbulent interchange between the bubbly layer and the bulk of the liquid may be the limiting mechanism leading to the CHF occurrence. CHF occurs when bubble crowding near the heated wall prevents the bulk cold liquid from reaching the wall; it is postulated that [62–64] CHF occurs when the void fraction in the bubbly layer, calculated under the assumption of homogeneous two-phase flow in the bubbly layer, just exceeds the critical value of 0.82. The void fraction in the bubbly layer is determined through the balance between the outward flow of vapor bubbles and the inward liquid flow at the bubbly layer–bulk liquid flow interface. The Weisman and Ileslamlou model [65] is an improvement of the Weisman and Pei model [63] for subcooled exit conditions. Research carried out by Styrikovich et al. [66] showed that measured void fraction at CHF ranges from as low as 0.3 to as high as 0.95, making the validity of the near-wall bubble crowding models questionable. In addition, the models are quite empirical in the determination of the turbulent exchange in the bubbly layer.

5. Liquid sublayer dryout model: this model is based on the dryout of a thin liquid sublayer underneath an intermittent vapor blanket or elongated bubble, due to coalescent bubbles, flowing over the wall [26,67–70].

6. Superheated layer vapor replenishment model: because of bulk subcooled conditions and considering the temperature distribution in the main stream, there will be a distance from the wall at which the fluid temperature is equal to the saturation value. This distance is called the superheated layer and is the only region where a bubble may exist. Because of the accumulation and condensation of the vapor generated from the heated wall, a thin elongated bubble, called a vapor blanket, is formed, rising along the near-wall region as distorted vertical vapor cylinders. The CHF is postulated to occur when the vapor blanket replenishes the superheated layer, coming into contact with the heated wall [71].

Currently, only the near-wall bubble crowding (item 4), the liquid sublayer dryout (item 5), and the superheated layer vapor replenishment (item 6) theories are receiving significant attention for the prediction of the CHF in the highly subcooled region.

The Vapor Removal Limit and Near-Wall Bubbles Crowding Theory

The near-wall bubble crowding theory preceded the liquid sublayer dryout one. Although the general idea was first suggested by Hebel et al. [62], these investigators were only able to obtain a correlation that provided a lower envelope for the data examined.

The Weisman and Pei model [63] assumes the following conditions:

a. During subcooled and low-quality boiling, the bubbly layer builds up along the channel until it fills the region near the wall where the turbulent eddies are too small to transport bubbles radially. At the CHF site, the bubbly layer is assumed to be at this maximum thickness.
b. CHF occurs when the volume fraction of steam in the bubbly layer just exceeds the volume fraction (critical void fraction) at which an array of slightly flattened ellipsoidal bubbles can be maintained without significant contact between the bubbles.
c. The volume fraction of steam in the bubbly layer is determined by a balance between the outward flow of vapor and the inward flow of liquid at the bubbly layer–core interface.

Considering a bubbly layer control volume, they can write the total mass balance on the bubbly layer taking into account the total flow rate from core to bubbly layer, which must be equal to the total flow rate from bubbly layer to core plus the axial flow in and out of the bubbly layer control volume. From a simple mass balance over the bubbly layer they obtain:

$$\frac{q''_{CHF}}{\lambda G'} = \frac{x_2 - x_1}{F} \tag{5}$$

where G' represents the mass flow rate into the bubbly layer. This mass flow rate is determined by the turbulent velocity fluctuations at the bubbly layer edge. The distance from the edge of the bubbly layer to the wall is taken as the distance at which the size of the turbulent eddies is k times the average bubble diameter. Only a fraction of the turbulent velocity fluctuations produced are assumed to be effective in reaching the wall. The effective velocity fluctuations are those in which the velocity exceeds the average vapor velocity away from the wall produced by the vapor being generated at the wall. The quantities x_1 and x_2 represent the vapor qualities in the core region and bubbly layer, respectively, at the CHF (these are actual values and not thermodynamic equilibrium qualities). The factor F represents the fraction of the heat flux-producing vapor that enters the core region, given by the ratio between the difference of the enthalpy of saturated liquid and that at bubble detachment point, and the difference between the enthalpy of liquid at a given axial location and that at the bubble detachment point.

The occurrence of the CHF is for that quality in the bubbly layer that corresponds to the maximum void fraction which is possible in a bubbly layer of independent bubbles just before agglomeration. For slightly flattened elliptically shaped bubbles with a length-to-diameter ratio of 3 : 1, this void fraction is estimated as 0.82.

Weisman [49] rebutted the criticism raised on basis of the measurements of bubbly layer void fractions at CHF [66], which showed that the void fraction varies from 0.3 to 0.95, making incorrect the assumption of void fraction equal to 0.82 at the CHF. Weisman claims that a closer examination of the Styrikovich et al. [66] data indicates that the mea-

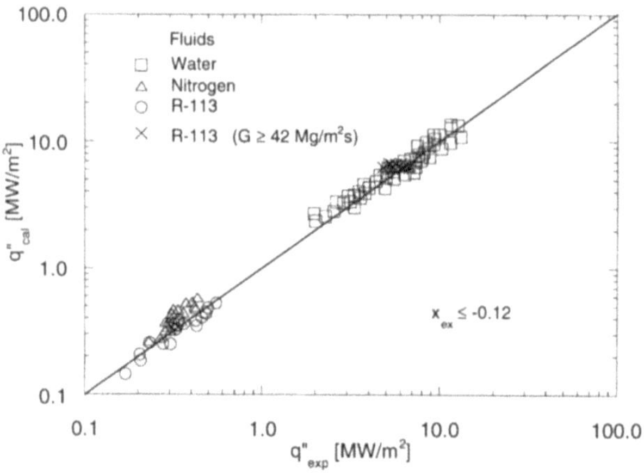

FIG. 17. Comparison of predicted and measured critical heat flux values for water, nitrogen, and refrigerant 113 using the Weisman and Ileslamlou model [65]

sured void fractions actually were fairly constant in the range 0.8 to 0.92 for all heat fluxes below 3.8 MW/m^2 (water at 3.0 MPa); only at very high subcooling and very high heat fluxes does the wall void fraction at CHF drop significantly.

To overcome the problems shown by the model in the region of high subcooling (region where the liquid enthalpy is close to the enthalpy of bubble detachment) Weisman and Ileslamlou [65] extended the Weisman and Pei model to the high subcooling region using the basic energy balance for the bubbly layer to compute the CHF. The applicability of the proposed approach to water data [72], liquid nitrogen data [73], and freon-113 data [74,75] is shown in Fig. 17, where the predicted versus calculated CHF is plotted for an exit quality greater than –0.12. The model is capable of predicting CHF data for all three fluids, showing reasonable predictions also for very high mass flux data [75]. Currently, the approach of Weisman and coworkers is the only theoretically based CHF prediction procedure that has been shown to give good accuracy with fluids other than water, especially with refrigerants. This is an important feature for the testing new designs inexpensively in a refrigerant system and readily scaling the results to steam–water conditions.

Predictions of water data at high liquid velocity and subcooling [50] are shown in Fig. 18, where the predicted versus calculated CHF is plotted. Although many data points are predicted within ±30%, a clear systematic error is present, the model tending to overpredict the CHF at relatively low heat fluxes (<20 MW/m^2), while underpredicting CHF data at very high heat fluxes (>40 MW/m^2). As a result, the model cannot be recommended for application to ranges typical of the fusion reactor high heat flux components, although it has the great merit to be suitable to predict CHF with different kind of fluids. Also, its application, in terms of mathematical description, is not straightforward.

The Liquid Sublayer Dryout Theory

At present, the liquid sublayer dryout theory (item 5) is receiving significant attention, is well developed, and is able to provide good predictions over a wide range of conditions, especially those of interest for fusion reactor high heat flux component thermal hydraulic design. For subcooled flow boiling in a uniformly heated tube, approaching the CHF, the situation at the tube exit end may be presumably that sketched in Fig. 19.

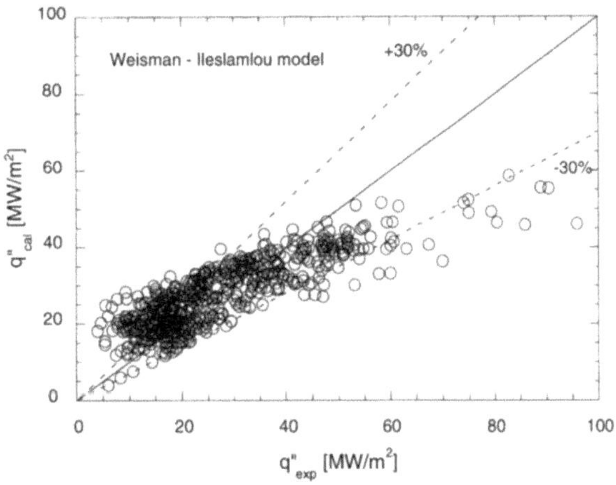

FIG. 18. Critical heat flux predictions using the Weisman and Ileslamlou model [65] for the database reported by Celata and Mariani [6]

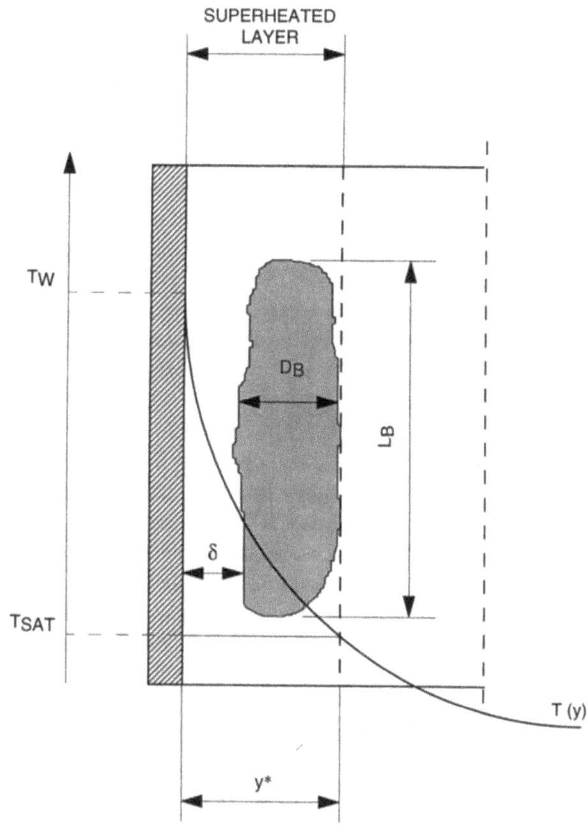

FIG. 19. Schematic of the liquid sublayer dryout model [26]

Because of accumulation and condensation of the vapor furnished from the heated wall, a thin elongated bubble, called a vapor blanket, is formed as a consequence of the coalescence of small bubbles rising along the near-wall region as distorted vertical vapor cylinders. The vapor blanket is overlying a very thin liquid sublayer adjacent to the wall, and CHF is assumed to occur when the liquid sublayer initial thickness is extinguished by evaporation during the passage time of the vapor blanket. The minimum heat flux necessary for the evaporation of the liquid sublayer is given by

$$q''_{CHF} = \frac{\delta \rho_l \lambda}{L_B} U_B \tag{6}$$

where δ is the initial liquid layer thickness, ρ_l is the liquid layer density, λ is the latent heat of evaporation, and L_B and U_B are blanket length and velocity, respectively. Lee and Mudawar [67] were the first to develop and propose a mechanistic model based on the liquid sublayer dryout theory. Following the same principle as Lee and Mudawar, Katto [68,69] developed a generalized CHF model applicable to not only water but also nonaqueous fluids (R 11, R 12, R 113, water, nitrogen, and helium). Katto [70] then extended this model so as to cover the CHF of water boiling at low pressures also.

The Lee and Mudawar and the Katto models make use of empirical constants determined through experimental data. The Lee and Mudawar model makes use of three empirical constants. The first empirical constant is used in the analysis for evaluating the temperature of the liquid entering the sublayer. The other two empirical constants are used in the analysis for evaluating the lateral force on the vapor blanket necessary to calculate the sublayer thickness. The three empirical constants employed are evaluated through comparison with the experimental data. In his model, Katto, for evaluating the vapor blanket velocity, makes use of an empirical parameter called the velocity coefficient obtained by an empirically based relationship as a function of Reynolds number, liquid and vapor density, and void fraction deduced from experimental data. It is necessary to say here that the choice of introducing the velocity coefficient was made to avoid applying a simplistic mechanistic assumption to the complicated state of two-phase flow near the vapor–liquid interface.

However, the use of empirical constants or correlations evaluated through comparison with the experimental data somewhat limits the use of these models within the database on which they have been assessed. The Lee and Mudawar model closely predicts several well-known databases at high pressure and low mass flux ($4.9 < p < 17.6\,MPa$; $4 < D < 16\,mm$; subcooling $< 50\,K$; void fraction < 0.7; $1000 < G < 5000\,kg/m^2 s$). The Katto model, although yielding acceptable predictions of very high heat flux CHF data points in a wide range of pressure, and having the capability to apply to not only water but also various nonaqueous fluids (such as R-11, R-12, R-113, nitrogen, and helium), is not able to calculate the CHF in those cases where the local void fraction in the near-wall bubbly layer is higher than 70%. This is the limit considered by the author for the validity of the assumption of homogeneous flow in the bubbly layer. It happens in all cases where inlet thermal hydraulic conditions are such that the bulk liquid at the exit is slightly subcooled, and was verified for about 51% of high-flux CHF data reported so far in the literature [50].

More recently, Celata et al. [26] developed a new model which, starting from the foregoing ones, does not make use of empirical constants yet is capable of predicting the CHF of water boiling in the comparatively low-to medium-pressure region. Of course, existing models based on the liquid sublayer dryout theory differ in their physical structure, that is, the calculation of the three unknowns in Eq. 6, δ, L_B, and U_B.

Evaluation of Vapor Blanket Length and Velocity

In the foregoing three models, the vapor blanket length is assumed to be given by the Helmholtz instability wavelength at the interface to the liquid sublayer, which is calculated under the assumption that $U_B >> U_L$, U_L being the velocity of the liquid sublayer, with respect to the respective movements of the vapor blanket and liquid sublayer. The vapor blanket velocity U_B is evaluated by considering the velocity distribution of the main stream in the tube under the assumption of homogeneous flow. Roughly speaking, Lee and Mudawar and Celata et al. assume vapor blanket velocity in the turbulent flow stream as equal to the superposition of local liquid velocity (calculated at the centerline of the vapor clot) and relative vapor blanket velocity, determined by a balance between the buoyancy and drag forces exerted on the blanket. [Note: The effects of gravity on the vapor clot motion can be neglected at high mass velocities.] The two models use the same correlation to calculate the drag coefficient, that is, that recommended by Harmathy [76] and Ishii and Zuber [77], but make use of a different method for the calculation of vapor blanket thickness, which is required in the drag coefficient correlation. Both models assume that the circumferential growth of a vapor blanket is strongly limited by adjacent blankets and by the steep velocity gradient in case of high liquid velocity. It is therefore assumed that the equivalent diameter of each blanket, that is, its thickness, may be approximated by the diameter of a bubble at the departure from the wall. In other words, it is assumed that departing bubbles may coalesce into a distorted blanket that stretches along the fluid flow direction (due to vapor generation by sublayer evaporation) and keeps almost a constant equivalent diameter (thickness). Lee and Mudawar use a correlation among those suggested by Levy [78] (on the basis of a balance between surface tension and shear forces created by liquid motion on a bubble in contact with the wall) [79,80].

Celata et al. use the Staub model [81], based on a balance of forces to growing bubbles attached to the heated surface, to approximate the diameter of bubbles at departure. In the Staub model, the bubble is considered to detach from the surface when dislodging forces overcome adhesive forces. Among the several forces acting on the bubble (surface tension force, dynamic force due to the momentum change of the liquid resulting from the growing bubble, drag force, buoyancy force, dynamic forces due to the liquid inertia and to the evaporating vapor thrust), Staub considered surface tension force (adhesive) and drag force (dislodging) to be predominant. To calculate the friction factor, as required in the Staub model, Celata et al. use the classical Colebrook–White equation combined with Levy's rough surface model [78], recommended for highly subcooled nucleate boiling.

Katto assumes the vapor clot velocity U_B as proportional to the local velocity of the main stream at the level of the interface facing to the liquid sublayer. The proportionality factor, called the velocity coefficient, is correlated empirically, through comparison with the experimental data, to the Reynolds number (assuming homogeneous two-phase flow in a tube), the density ratio, and the void fraction of the main stream.

Evaluation of the Initial Liquid Sublayer Thickness

The three models are quite different from each other in the procedure to evaluate δ. Lee and Mudawar determined δ on the basis of the mechanistic analysis of the balance of forces exerting on the vapor blanket in the radial direction (introducing two empirical constants). The rate of momentum produced by sublayer evaporation into the vapor blanket, which pushes the vapor blanket away from the wall, is balanced by a lateral force caused by the rotation of the blanket due to the relative velocity between the two phases and the velocity gradient associated with the liquid boundary layer. Katto [82] makes use of a correlation proposed for the evaluation of the macrolayer thickness in pool boiling. As the liquid sublayer underneath the vapor blanket is generally very thin, its initial thickness may be determined as the critical length of tiny vapor jets anchored to active sites on the heated wall, using only the fraction of the heat flux that contributes to evaporation, and not considering the fraction of the heat flux transferred by forced convection of subcooled liquid. Meanwhile, the Celata et al. model considers the temperature distribution of the main stream in the tube under the assumption of homogeneous flow [83], determining the distance from the heated wall beyond which a vapor blanket cannot develop or exist because of subcooling. A vapor blanket can develop and exist only in the near-wall region where the local liquid temperature is above the saturation value.

Considering the temperature distribution from the heated wall to the center of the channel, the distance from the wall at which the temperature, decreasing as we proceed toward the center of the tube along the radius, is equal to the saturation value at the local pressure is defined as the superheated layer and indicated with y^*. For a distance from the wall greater than y^*, the blanket (and each single bubble) will collapse in the subcooled liquid bulk. Considering also that the vapor blanket is pushed toward the center of the tube by the velocity gradient, it is assumed that the vapor blanket location in the superheated layer is such that it occupies the region closer to the saturation limit, that is, as far as possible from the heated wall, but within the superheated layer, y^*. As the temperature distribution is linked to the inside tube wall temperature, the latter is obtained by equating the local cross-section average fluid temperature given by the coolant heat balance with that provided by the temperature profile. Then, δ can be finally determined as the difference between the superheated layer y^* and vapor blanket thickness.

Treatment of Conjugate Heat Flux

In subcooled flow boiling, heat flux is formally expressed by two contributions: the part used by latent heat of evaporation and that used by sensible energy heating. This aspect is taken into account by the models of Lee and Mudawar and Katto using, in Eq. 6, a latent heat of evaporation modified by considering sensible energy heating. On the other hand, in the Celata et al. model, the heat flux evaluated from Eq. 6 using latent heat is regarded as the CHF itself, because this model stands on the assumption that the existence of the vapor blanket and the underlying liquid sublayer are restricted in the superheated liquid layer.

Logical Limitations of Applicable Range

In all three models the main stream in the tube is assumed to be homogeneous flow, and accordingly Lee and Mudawar and Katto impose the limitation of void fraction lower than 0.7; this limits application to many available data [26,50]. The Celata et al. model does not impose this limit, but has a limit of exit subcooling perhaps greater than 25 K (but depending on mass flux and pressure), or exit quality greater than −0.15. The less the liquid is subcooled at the exit, the less the model assumptions are valid.

Performance of the Liquid Sublayer Dryout Model for Smooth Tubes, Uniform Heating

The Celata et al. model [26] provides a fairly good prediction of a high heat flux data set collected by Celata and Mariani [6], already described at the beginning of this chapter. Figure 20 reports predictions showing the calculated versus the experimental CHF. The Celata et al. model predicts about 91% of data points within ±30% with a RMS of 17.2%. The analysis of parametric trends also demonstrated the good accuracy of the Celata et al. model with regard to experimentally observed variations of mass flux, pressure, subcooling, and tube diameter.

To give an idea of the order of magnitude of the several parameters involved in the CHF calculation, Fig. 21 shows the normalized data of the predicted CHF, the vapor blanket velocity U_B, the wall temperature T_w, the initial liquid sublayer thickness δ, the length of the vapor blanket L_B, and its equivalent diameter D_B, as a function of mass flux (top figure), and of pressure (bottom figure), for data reported by Celata et al. [10].

The Superheated Layer Vapor Replenishment Theory

The superheated layer vapor replenishment theory has been recently proposed by Celata et al. [71], after an earlier idea developed by Katto [84]. The reference flow configuration is schematically illustrated in Fig. 22. As already stated in the discussion of the liquid sublayer dryout theory, because of the accumulation and condensation of the vapor furnished by the heated wall, a thin elongated bubble called the vapor blanket is formed, also as a consequence of the coalescence of small bubbles rising along the near-wall region as distorted vertical vapor cylinders. A continuous blanket may be formed along the inner wall of the tube as a consequence of blankets merging. The vapor blanket can

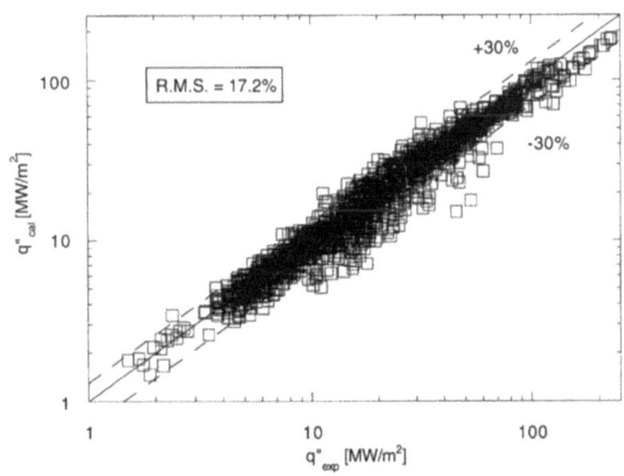

FIG. 20. Critical heat flux predictions using the Celata et al. model [26] for the database reported by Celata and Mariani [6]

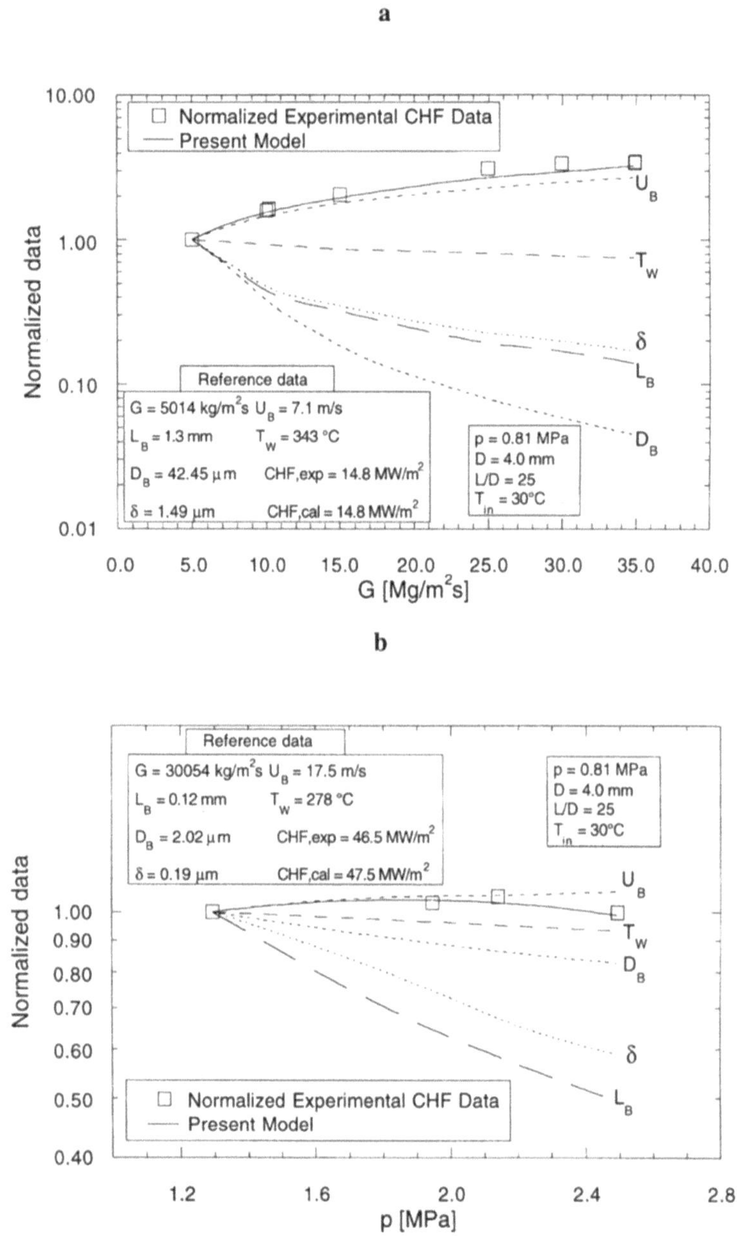

FIG. 21a,b. Typical values of predicted critical heat flux, blanket velocity, wall temperature, initial sublayer thickness, blanket length, and equivalent diameter using the Celata et al. model [26] (data from Celata et al. [10]). **a** Values of critical heat flux as a function of mass flux. **b** Values of critical heat flux as a function of pressure

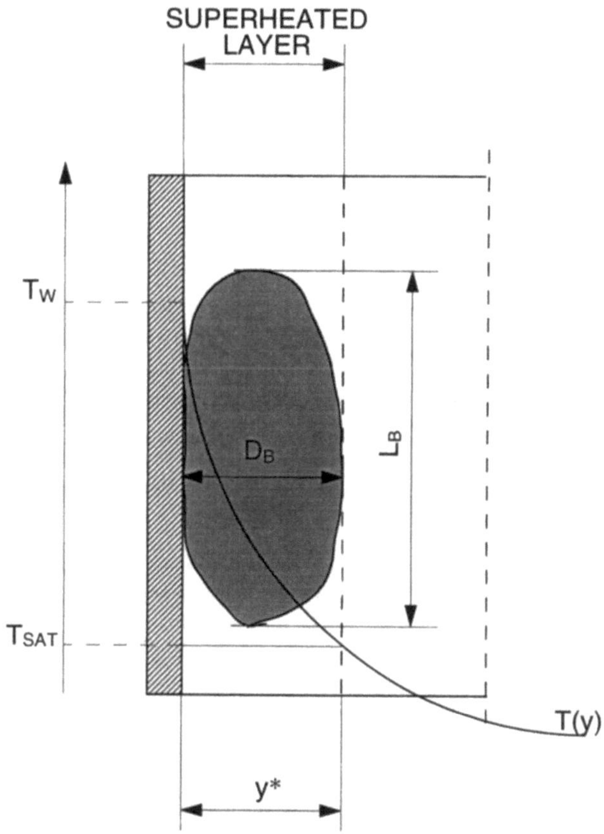

FIG. 22. Schematic of the superheated layer vapor replenishment model [71]

develop and exist only in the near-wall region where the local liquid temperature is above the saturation value.

Considering the temperature distribution from the heated wall to the center of the channel, a distance from the wall will exist at which the temperature, decreasing as we proceed toward the axis of the tube along the radius, is equal to the saturation value at the local pressure. This distance is defined as the superheated layer [26] and is indicated with y* (see Fig. 22). For a distance from the wall greater than y*, the blanket (and each single bubble) will collapse in the subcooled liquid bulk. As initially suggested [84], the CHF is postulated to occur when the vapor blanket replenishes the superheated layer, coming into contact with the heated wall, that is, the heat flux for which y* will be equal to D_B is assumed to be the CHF.

Evaluation of the Vapor Blanket Thickness

The vapor blanket thickness is evaluated as in the liquid sublayer dryout model [26]. As already assumed [67], the circumferential growth of the vapor blanket is supposed to be strongly limited by adjacent blankets and by the steep velocity gradient. It is therefore reasonable to assume the thickness of the vapor blanket, D_B, as approximately equal to the diameter of a bubble at the departure from the wall. The assumption is such that departing bubbles may coalesce into a distorted blanket that stretches along the fluid flow direction (due to vapor generation in the near-wall region) and keeps almost a constant thickness. To approximate the diameter of bubbles at the departure authors use the model proposed by Staub [81], based on a balance of forces acting on growing bubbles attached to the heated surface, and already described in the "Evaluation of vapor blanket length and velocity" section of the liquid sublayer dryout model. The calculation of D_B is performed exactly in the same way.

Evaluation of the Superheated Layer

The superheated layer y^* is calculated in the same way as by Celata et al. [26], that is, considering the temperature distribution of the main stream in the tube under the assumption of homogeneous flow (according to Martinelli [83]), as already described in the "Evaluation of the initial liquid sublayer thickness" section of the liquid sublayer dryout model.

Performance of the Superheated Layer Vapor Replenishment Model for Smooth Tubes, Uniform Heating

The Celata et al. model [71] was tested against the same data set as the liquid sublayer dryout model [26]. Figure 23 shows a comparison of calculated versus experimental CHF using this data set. About 85% of data points are predicted within ±25%, while more than 90% of the data points used are predicted within ±30%; the standard deviation is ±16.6%. Statistics are very similar to those of the liquid sublayer dryout model [26], as reported earlier and in Fig. 20.

Although the basic mechanisms on which the liquid sublayer dryout and the superheated layer vapor replenishment models rely are significantly different, they use exactly the same equations to calculate the vapor blanket thickness and the superheated layer. Of course, the closure equation, connected with the different mechanism leading to CHF, is different in the two models. The superheated layer vapor replenishment model is definitely simpler than the liquid sublayer dryout model in its mathematical formulation and calculation procedure, nonetheless providing exactly the same performance in the

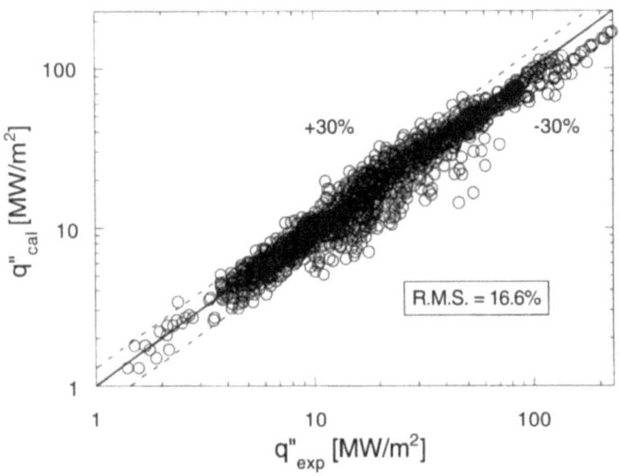

FIG. 23. Critical heat flux predictions using the Celata et al. model [71] for the data base reported by Celata and Mariani [6]

prediction of the same data set. Similar to the liquid sublayer dryout model, the present model shows a systematic underprediction of the CHF for those data characterized by exit conditions close to the saturation. It is obvious that, when approaching saturation conditions, assumptions made in the construction of the model may not hold any longer, and the model provides a systematic error in the prediction.

Model Prediction of the CHF for Peripheral Nonuniform Heating and Swirl Flow

The effect of nonuniform heating along the circumference of the tube can be relevant in the thermal hydraulic design of fusion reactors high heat flux components. In fact, as an example, the divertor is thermally loaded only on one side. Besides, to further increase the CHF, as mentioned earlier, swirl flow using twist tape inserts is often employed. Both the liquid sublayer dryout model [26] and the superheated layer vapor replenishment model [71] can be used to predict the CHF for peripheral nonuniform heating and swirl flow.

Peripheral Nonuniform Heating

Circumferential nonuniform heating can be accounted for in the two above-mentioned models [26,71] simply by considering the total thermal power delivered to the fluid in the coolant heat balance for the calculation of the local average coolant temperature. It is evident that exit bulk thermal hydraulic conditions, on which the CHF is strongly dependent, are only a function of the total thermal power delivered to the fluid, independent of its distribution. As boiling crisis in subcooled flow boiling is a strictly local phenomenon, all equations employed in the model mathematical description can also be used in the case of peripheral nonuniform heating. In fact, all calculations, except for the cross-section average temperature, are made using the maximum value of the heat flux, and all parameters used to calculate the CHF are local values. Possible distortion of such distributions is likely to happen in the bulk of the flow as the result of turbulent mixing. As velocity and temperature distributions are used locally in the calculation of the CHF (the superheated layer is a small portion of the channel flow diameter), it looks reasonable to continue making use of such distributions also in the case of peripheral nonuniform heating. The prediction of the data of Gaspari [28] is shown in Fig. 24, where the ratio between the calculated and the experimental critical heat flux is plotted versus the exit quality, x_{ex}, using the vapor replenishment model [26]. The liquid sublayer dryout model [26] provides quite similar results. The agreement is good, being most of 26 experimental data within ±20%.

Swirl Flow

The necessity of using twist tape inserts for the CHF enhancement has already been discussed. Both the liquid sublayer dryout and the vapor replenishment models can be extended to swirl flow CHF making use of the same corrections already used in empirical correlations. As the presence of a twisted tape inside a channel is associated with a relevant increase in the pressure drop, a friction factor correction for twisted tapes was suggested by Lopina and Bergles [85].

Based on experiments, Lopina and Bergles show that the friction factor, f_{tt}, for use in the twisted tape geometry varies as

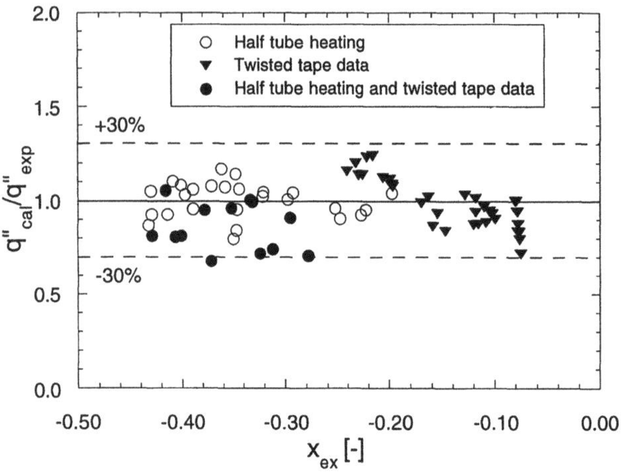

FIG. 24. Extension of the superheated layer vapor replenishment model [71] to peripheral nonuniform heating, straight flow critical heat flux data [28]; uniform heating, swirl flow critical heat flux data [33]; and swirl flow, peripheral nonuniform heating critical heat flux data [35]

$$f_{tt} = 2.75 f \gamma^{-0.406} \tag{7}$$

where γ is the twist ratio of the tape and f is the friction factor for straight tube. To extend the present model for use with twisted tapes, Eq. 7 was tried.

The prediction of the Nariai et al. data [33] is shown in Fig. 24, using the Celata et al. model [71]. The agreement is quite encouraging, as almost all the data are predicted within ±25%, showing that the procedure can be successful. Recent experiments [35] were carried out with twisted tapes inserted in peripherally half-heated tubes. Predictions obtained using the vapor replenishment model on these data are also shown in Fig. 24. The agreement can be considered satisfactory in view of the complexity of the situation in comparison with the original description of the model. Although a general slight underestimation of the CHF is observed, most of the experimental data are predicted within ±25%. Similar predictions are given by the liquid sublayer dryout model [26] in both cases.

Concluding Remarks

Forced convective, subcooled boiling at very high mass fluxes, high subcoolings, and with small-diameter tubes can accommodate high heat fluxes such as those required for some components of fusion thermonuclear reactors. Studies conducted so far showed that the CHF, the upper limit of the cooling system, is essentially a linear increasing function of coolant subcooling and an increasing function of mass flux. The effect of channel diameter and length is to decrease CHF as they increase, showing a threshold beyond which the influence of geometric parameters vanishes. The influence of pressure turned out to be negligible in the ranges of parameters relevant to fusion reactor thermal hydraulics. Effects on the CHF of secondary parameters such as tube wall material and thickness and channel orientation were found to be almost negligible. Influence of nonuniform heating does not affect the critical heat flux, local thermal hydraulic conditions being equal.

Although forced convective subcooled boiling allows accommodating very high heat fluxes, up to 276 MW/m^2 under extreme geometric and thermal hydraulic conditions

[86], engineering or design problems (increase of pressure drop, flow induced vibrations, etc.) must be faced. Therefore, with the aim of providing adequate safety margins, in the frame of fusion reactor thermal hydraulic operating conditions ($p = 5.0\,MPa$, $D = 15\,mm$, $u = 10\,m/s$, $T_{in} = 60°C$), a further increase of the CHF is necessary. Turbulence promoters such as twisted tapes or helically coiled wires were used, enabling an increase of the CHF ranging from 1.2 to 1.8, with a corresponding increase of the pressure drop to 11 times that observed in straight tube tests.

Among the many theories existing for the predictions of water subcooled flow boiling CHF, the superheated layer vapor replenishment [71] and the liquid sublayer dryout [26] theories are able to provide good predictions over a wide range of conditions, especially those of interest for the fusion reactor high heat flux components thermal hydraulic design, being also able to be easily extended to nonuniform heating conditions of the channel, as requested by high heat flux component designers.

The near-wall bubble crowding theory gives good results for medium-to-low subcooling or nearly saturated flow boiling, being the most suitable model to predict CHF with fluids other than water, especially refrigerants.

Currently, there is insufficient evidence to state definitely that either the near-wall bubble agglomeration or the liquid sublayer dryout theory correctly describe the true thermal-hydraulic behavior at the departure from nucleate boiling. The most logical conclusion may be that a unique mechanism does not exist and therefore we have to look for a specific mechanism that is typical of the given range of thermal-hydraulic conditions [87], reviewing models of boiling crisis under conditions of subcooled liquid boiling, recognizing that no universal model of critical heat flux of subcooled liquid in channels exists at present. Attempts at developing such a model will apparently be fairly artificial, because the diversity of the existing flow patterns objectively prevents one from developing a universal model with one or two dominating processes.

Fundamental aspects of subcooled flow boiling heat transfer must therefore be fully investigated and understood, as their knowledge is very important for correct and proper modeling of heat transfer and CHF. Finally, visual studies have to be performed to obtain direct information on the flow regime and heat transfer characteristics at high heat fluxes. We should consider this as a top priority in research planning. Visual studies are the key for a full understanding of basic mechanisms of subcooled flow boiling CHF at very high heat fluxes, and for the development of a mechanistic model based on them for the prediction of the CHF. Difficulties connected with high subcoolings (and therefore very small bubble diameters, about some tens of micrometers) and high liquid velocities need to be overcome, and this may represent an exciting challenge.

Although the situation from the point of view of themal hydraulics is rather clear, we would like to close this contribution with a general thought on fusion energy. In his opening remark at the 5th International Symposium of Fusion Nuclear Technologies, held in Rome in September 1999, Professor Carlo Rubbia, Nobel Prize winner and current president of ENEA, the Italian Board for the New Technologies, the Energy and the Environment, said: the expectation of producing "an abundant and inexhaustible energy for mankind with fusion, as condensed in the 'Gorbachov–Reagan' statement in Geneva is, in my view, still rather far away. There are physical and economic laws that politicians, even at such a level, cannot revert. Any new form of commercial energy, in order to be widely accepted, has to be economically rewarding. The task is not an easy one: after having succeeded in making it work, you have to make competitive with other kinds of energy sources. These question need a convincing answer and, in my view, they cannot be taken for granted."

In addition, environmental concerns associated with the generalized use of tritium, to be continuously extracted and mass-separated from the blanket and the activation due to the strong 14-MeV neutron flux, with the associated problems of waste and meltdown risks, are making public acceptance of this form of nuclear energy, as for many other methods based on fission, far from granted. We should be honest: the deuterium–tritium reaction is not the one occurring in the sun, which is without neutrons. But even if the practical use of this "nuclear" form of energy—at least in the present explicitation of the D-T reaction—may be questioned, fusion should be pursued vigorously as such because it is exploring a fundamental domain of basic science.

There are two main forms of high temperature matter in galaxies: (1) the low-density, high-temperature gases, mostly hydrogen, gravitationally confined in space and (2) the high-temperature and very high density compressed matter in the interior of stars. These domains correspond roughly, and respectively, to magnetically confined and inertially confined fusion. They must both be thoroughly studied to better understand the Universe. For me, fusion is and remains for the moment primarily an essential field of fundamental science worthwhile pursuing vigorously.

The future, although clouded by the lack of success of ITER, is to my view relatively bright with the continuation of the ongoing program—in particular, the construction of the Stellarator in Germany—and, in addition, the perspective of a new huge, internationalized machine capable of approaching or even attaining, at least, energy amplification.

Appendix

Nomenclature

a_t	tangential centrifugal acceleration, m/s^2
CHF	critical heat flux, MW/m^2
D	diameter, m
D_B	detachment bubble diameter, and vapor blanket thickness, μm
D_b	maximum width of the bubble, μm
ΔT	temperature difference, K
F	fraction of the heat flux-producing vapor that enters the core region, Eq. 5
f	friction factor, nondimensional
Fr	Froude number, defined in Eq. 2, nondimensional
f_{tt}	friction factor for swirl flow defined in Eq. 7, nondimensional
G	mass flux, kg/m^2s
g	gravitational acceleration, m/s^2
G'	mass flow rate into the bubbly layer, $kg/m^2 s$
h_b	maximum height of the bubble, μm
L	channel length, m
L_B	vapor blanket length, μm
n	coefficient in Eq. 1, nondimensional
p	pressure, MPa
q''	heat flux, MW/m^2
R	ideal gas constant on a unit mass basis

T	temperature, °C
t	tube wall thickness, m
u	velocity, m/s
U_B	vapor blanket velocity, m/s
U_L	liquid sublayer velocity, with respect to the respective movements of the vapor blanket and liquid sublayer
x	quality, nondimensional
y	distance from the wall, μm
y*	superheated layer, μm

Greek Symbols

α	angle, degree
δ	liquid sublayer thickness, μm
γ	twist tape ratio, the number of tube diameters per 180° twist in the tape, dimensionless
λ	latent heat of evaporation, J/kg
μ	dynamic viscosity, kg/s m
ϑ	parameter defined in Eq. 4, nondimensional
ρ	density, kg/m^3

Subscripts

1	pertains to the core region
2	pertains to the bubbly layer
B	pertains to the vapor blanket
cal	calculated value
CHF	pertains to critical heat flux conditions
ex	exit conditions
exp	experimental value
g	pertains to the gas phase
in	inlet conditions
L	pertains to the liquid sublayer
l	pertains to the liquid phase
m	average
max	maximum
s	smooth pipe
SAT	saturation
sub	pertains to subcooled conditions
tt	pertains to twisted tape inserts
w	pertains to the wall

References

1. Bergles AE, Collier JG, Delhaye JM, Hewitt GF, Mayinger F (1981) Two-phase flow and heat transfer in the power and process industries, Hemisphere, New York, pp 226–255
2. Collier JG (1981) Convective boiling and condensation, 2nd edn. McGraw-Hill, New York, pp 144–177

3. Hsu YY, Graham RW (1986) Transport processes in boiling and two-phase systems. American Nuclear Society, La Grange, pp 217–232
4. Boyd RD (1985) Subcooled flow boiling critical heat flux (CHF) and its application to fusion energy components. Part I: A review of fundamentals of CHF and related data base. Fusion Technology 7(1):7–30
5. Boyd RD (1985) Subcooled flow boiling critical heat flux (CHF) and its application to fusion energy components. Part II: A review of microconvective, experimental, and correlational aspects. Fusion Technology 7:31–52
6. Celata GP, Mariani A (1993) A data set of critical heat flux in water subcooled flow boiling. Presented at the 3rd specialists' workshop on the thermal-hydraulics of high heat flux components in fusion reactors. J. Schlosser, Cadarache
7. Bergles AE (1963) Subcooled burnout in tubes of small diameter. Paper 63-WA-182. American Society of Mechanical Engineers, New York
8. Nariai H, Inasaka F, Shimura T (1987) Critical heat flux of subcooled flow boiling in narrow tube. ASME-JSME Thermal Engineering Joint Conference, Honolulu, March 22–27
9. Vandervort CL, Bergles AE, Jensen MK (1992) The ultimate limits of forced convective subcooled boiling heat transfer. RPI Interim Report HTL-9 DE-FG02-89ER14019, Rensselaer Polytechnic Institute, Troy, New York
10. Celata GP, Cumo M, Mariani A (1993) Burnout in highly subcooled water flow boiling in small diameter tubes. International Journal of Heat and Mass Transfer 36(5):1269–1285
11. Boyd RD (1988) Subcooled water flow boiling experiments under uniform high flux conditions. Fusion Technology 13:131–142
12. Boyd RD (1989) Subcooled water flow boiling at 1.66 MPa under uniform high heat flux conditions. ASME winter annual Meeting, San Francisco, December 10–15 (HTD; vol 119, pp 9–15)
13. Boyd RD (1990) Subcooled water flow boiling transition and the L/D effect on CHF for a horizontal uniformly heated tube. Fusion Technology 18:317–324
14. Celata GP, Cumo M, Mariani A (1994) Enhancement of CHF water subcooled flow boiling in tubes using helically coiled wires. International Journal of Heat and Mass Transfer 37(1):53–67
15. Celata GP, Cumo M, Inasaka F, Mariani A, Nariai H (1993) Influence of channel diameter on subcooled flow boiling burnout at high heat fluxes. International Journal of Heat and Mass Transfer 36(13):3407–3410
16. Celata GP, Cumo M, Mariani A (1997) Geometrical effects on the subcooled Fw boiling critical heat flux. Revue Générale de Thermique 36:807–814
17. Celata GP, Cumo M, Mariani A (1996) The effect of the tube diameter on the critical heat flux in subcooled flow boiling. International Journal of Heat and Mass Transfer 39(8):1755–1757
18. Groeneveld DC, Leung LKH, Erbacher FJ, Kirillow PL, Bobkov VP, Zeggel W (1993) An improved table look-up method for predicting critical heat flux. Proceedings, NURETH-6, sixth international topical meeting on nuclear reactor thermal hydraulics 1:223–230
19. Doroshchuk VE, Levitan LL, Lantzman FP (1975) Investigation into burnout in uniformly heated tubes. Paper 75-WA/HT-22. ASME, New York
20. Doroshchuk VE, Levitan LL, Lantzman FP (1975) Recommendations for calculating burnout in a round tube with uniform heat release. Teploenergetika 22:66–70
21. Groeneveld DC, Cheng SC, Doan T (1986) AECL-UO critical heat flux look-up table. Heat Transfer Engineering 7:46–62
22. Smith RA (1986) Boiling inside tubes: critical heat flux for upward flow in uniformly heated tubes. ESDU data item no. 86032. Engineering Science Data Unit International, London
23. Nariai H, Inasaka F (1992) Critical heat flux and flow characteristics of subcooled flow boiling with water in narrow tubes. In: Jones OC, Michiyoshi I (eds) dynamics of two-phase flows. CRC Press, Boca Raton, pp 689–708
24. Merilo M (1977) Critical heat flux experiments in a vertical and horizontal tube with both freon-12 and water as coolant. Nuclear Engineering & Design 44(1):1–16

25. Cumo M, Fabrizi F, Palazzi G (1978) The influence of inclination on CHF in steam generators channels. CNEN Report RT/ING (78)11, CNEN, Rome
26. Celata GP, Cumo M, Mariani A, Simoncini M, Zummo G (1994) Rationalization of existing mechanistic models for the prediction of water subcooled flow boiling critical heat flux. International Journal of Heat and Mass Transfer 37(1):347–360
27. Nariai H, Inasaka F, Ishikawa A, Fujisaki W (1992) Critical heat flux of subcooled flow boiling in tube with internal twisted tape under non-uniform heating conditions. Proceedings, 2nd JSME-KSME Thermal Engineering Conference 3:285–288
28. Gaspari GP (1993) Comparison among data of electrically and E-beam heated tubes. Proceedings, 3rd international workshop on high heat flux components thermal hydraulics in fusion reactors. Schlosser, Cadarache
29. Celata GP (ed) (1992) Proceedings, of the 2nd specialists' workshop on high heat flux components thermal hydraulics, Rome, 9–11 September
30. Gambill WR, Lienhard JH (1989) An upper bound for the critical boiling heat flux. Journal of Heat Transfer 111:815–818
31. Gambill WR, Greene ND (1958) Boiling burnout with water in vortex flow. Chemical Engineering Progress 54(10):68–76
32. Gambill WR, Bundy RD, Wansbrough RW (1961) Heat transfer, burnout and pressure drop for water in swirl flow through tubes with internal twisted tapes. Chemical Engineering Symposium Series 57(32):127–137
33. Nariai H, Inasaka F, Fujisaki W, Ishiguro H (1991) Critical heat flux of subcooled flow boiling in tubes with internal twisted tapes. Proceedings, ANS winter meeting (THD), San Francisco, November 11–14, pp 38–46
34. Achilli A, Cattadori G, Gaspari G (1993) Subcooled burnout in uniformly and non-uniformly heated tubes. Paper C2, European two-phase flow group meeting, Stockholm, June
35. Cardella A, Celata GP, Dell'Orco G, Gaspari GP, Cattadori G, Mariani A (1992) Thermal hydraulic experiments for the NET divertor. Proceedings, 17th symposium on fusion technology, Rome, September 14–18
36. Sutherland WA (1967) Improved heat-transfer performance with boundary-layer turbulence promoters. International Journal of Heat and Mass Transfer 10:1589–1599
37. Kumar P, Judd RL (1970) Heat transfer with coiled wire turbulence promoters. Canadian Journal of Chemical Engineering 48:378–383
38. Sethumadhavan R, Raja Rao M (1983) Turbulent flow heat transfer and fluid friction in helical-wire-coil-inserted tubes. International Journal of Heat and Mass Transfer 26(12):1833–1845
39. Cattadori G, Gaspari GP, Celata GP, Cumo M, Mariani A, Zummo G (1993) Hypervapotron technique in subcooled flow boiling CHF. Experimental Thermal and Fluid Science 7:230–240
40. Fiori MP, Bergles AE (1970) Model of critical heat flux in subcooled flow boiling. 4th International heat transfer conference, Paris-Versailles, vol 4, B 6.3
41. Gunther FC (1951) Photographic study of surface-boiling heat transfer to water with forced convection. Transactions of the American Society of Mechanical Engineers 73:115–121
42. Kirby GJ, Stainforth R, Kinneir JH (1965) A visual study of forced convection boiling. Part 1. Results for a flat vertical heater. AEEW-R281. Atomic Energy Establishment, Winfrith
43. Kirby GJ, Stainforth R, Kinneir JH (1967) An investigation into a possible mechanism of subcooled burnout. AEEW-M725. Atomic Energy Establishment, Winfrith
44. Kirby GJ, Stainforth R, Kinneir JH (1967) A visual study of forced convection boiling. Part 2. Flow patterns and burnout for a round test. AEEW-R506. Atomic Energy Establishment, Winfrith
45. Celata GP, Cumo M, Mariani A, Zummo G (1995) Preliminary remarks on visualization of high heat flux burnout in subcooled water flow boiling, two-phase flow modelling and experimentation 1995, Celata GP, Shah RK, eds. Edizioni ETS 2:859–872

46. Celata GP, Cumo M, Gallo D, Mariani A, Zummo G (1998) Visual investigation of high heat flux burnout in subcooled flow boiling of water. Proceedings, 3rd international conference on multiphase flow, Lyon, 8–12 June
47. Katto Y (1994) Critical heat flux. International Journal of Multiphase Flow 20(1):53–90
48. Katto Y (1995) Critical heat flux mechanisms. Proceedings, engineering foundation convective flow boiling conference, keynote lecture V, Banff, May
49. Weisman J (1992) The current status of theoretically based approaches to the prediction of the critical heat flux in flow boiling. Nuclear Technology 99:1–21
50. Celata GP, Cumo M, Mariani A (1994) Assessment of correlations and models for the prediction of CHF in subcooled flow boiling. International Journal of Heat and Mass Transfer 37(2):237–255
51. Tong LS, Currin HB, Larsen PS, Smith OG (1965) Influence of axially non-uniform heat flux on DNB. American Institute of Chemical Engineers Symposium Series 64:35–40
52. Kutateladze SS, Leontiev AI (1966) Some applications of the asymptotic theory of the turbulent boundary layer. Proceedings, 3rd international heat transfer conference, Chicago, August 8–12, 1996, vol 3, pp 1–6. American Institute of Chemical Engineers, New York
53. Tong LS (1966) Boundary layer analysis of the flow boiling crisis. Proceedings, 3rd international heat transfer conference, vol. III, pp 1–6. Hemisphere, New York
54. Tong LS (1975) A phenomenological study of critical heat flux. Paper 75-1-HT-68. ASME, New York
55. Purcupile JC, Gouse SW Jr (1972) Reynolds flux model of critical heat flux in subcooled forced convection boiling. Paper 72-HT-4. ASME, New York
56. Hancox WT, Nicoll WB (1973) On the dependence of the flow-boiling heat transfer crisis on local near-wall conditions. 73-HT-38. ASME, New York
57. Thorgerson EJ, Knoebel DH, Gibbons JG (1974) A model to predict convective subcooled critical heat flux. Journal of Heat Transfer 96:79–82
58. van der Molen SB, Galjee FWBM (1978) The boiling mechanism during burnout phenomena in subcooled two-phase water flow. Proceedings, 6th international heat transfer conference, vol 1, pp 381–385. Hemisphere, New York
59. Hino R, Ueda T (1985) Studies on heat transfer and flow characteristics in subcooled flow boiling—Part 2. Flow characteristics. International Journal of Multiphase Flow 11:283–298
60. Mattson RJ, Hammit FG, Tong LS (1973) A photographic study of the subcooled flow boiling crisis in freon-113. Paper 73-HT-39. ASME, New York
61. Bergelson BR (1980) Burnout under conditions of subcooled boiling and forced convection. Thermal Engineering 27(1):48–50
62. Hebel W, Detavernier A, Decreton M (1981) A contribution to the hydrodynamics of boiling crisis in a forced flow of water. Nuclear Engineering and Design 64:433–445
63. Weisman J, Pei BS (1983) Prediction of critical heat flux in flow boiling at low qualities. International Journal of Heat and Mass Transfer 26:1463–1477
64. Weisman J, Ying SH (1983) Theoretically based CHF prediction at low qualities and intermediate flows. Transactions of the American Nuclear Society 45:832–833
65. Weisman J, Ileslamlou S (1988) A phenomenological model for prediction of critical heat flux under highly subcooled conditions. Fusion Technology 13:654–659 (Corrigendum in Fusion Technology 15:1463–1989)
66. Styrikovich MA, Newstrueva EI, Dvorina GM (1970) The effect of two-phase flow pattern on the nature of heat transfer crisis in boiling. Proceedings, 4th international heat transfer conference, vol 9, pp 360–362. Hemisphere, New York
67. Lee CH, Mudawar I (1988) A mechanistic critical heat flux model for subcooled flow boiling based on local bulk flow conditions. International Journal of Multiphase Flow 14:711–728
68. Katto Y (1990) A physical approach to critical heat flux of subcooled flow boiling in round tubes. International Journal of Heat and Mass Transfer 33:611–620
69. Katto Y (1990) Prediction of critical heat flux of subcooled flow boiling in round tubes. International Journal of Heat and Mass Transfer 33:1921–1928

70. Katto Y (1992) A prediction model of subcooled water flow boiling CHF for pressure in the range 0.1–20 MPa. International Journal of Heat and Mass Transfer 35:1115–1123
71. Celata GP, Cumo M, Katto Y, Mariani A (1998) A new mechanistic model for the prediction of the CHF in water subcooled flow boiling. Proceedings, 3rd international conference on multiphase flow, Lyon, 8–12 June
72. Thompson B, Macbeth R (1964) Boiling water heat transfer burnout in uniformly heated round tubes—a compilation of world data with accurate correlations. AEEW-R356. UK Atomic Energy Authority, London
73. Papell SS, Simoneau R, Brown DD (1966) Buoyancy effects on critical heat flux of forced convection boiling in vertical flow. TND-3672. National Aeronautics and Space Agency, Washington, DC
74. Coffield RD, Rohrer WM, Tong LS (1969) A subcooled DNB investigation of freon-113 and its similarity to subcooled water DNB data. Nuclear Engineering & Design 11:143
75. Fukayama Y, Hirata M (1982) Boiling heat transfer characteristics with high mass flux and disappearance of CHD following DNB. Proceedings, 7th International Heat Transfer Conference 4:273–278
76. Harmathy TZ (1960) Velocity of large drops and bubbles in media of infinite and restricted extent. American Institute of Chemical Engineers Journal 6:281–288
77. Ishii M, Zuber N (1979) Drag coefficient and relative velocity in bubbly, droplet or particulate flows. American Institute of Chemical Engineers Journal 25:843–854
78. Levy S (1967) Forced-convection subcooled boiling prediction of vapor volumetric fraction. International Journal of Heat and Mass Transfer 10:951–965
79. Cole R, Rohsenow WR (1968) Correlation of bubble departure diameters for boiling of saturated liquids. Chemical Engeenering Progress Symposium 92:211–213
80. Smogalev IP (1981) Calculation of critical heat fluxes with flow of subcooled water at low velocity. Thermal Engineering 28(4):208–211
81. Staub FW (1968) The void fraction in subcooled boiling—prediction of vapor volumetric fraction. Journal of Heat Transfer 90:151–157
82. Haramura Y, Katto Y (1983) A new hydrodynamic model of critical heat flux, applicable widely to both pool and forced convection boiling on submerged bodies in saturated liquids. International Journal of Heat and Mass Transfer 26:389–399
83. Martinelli RC (1947) Heat Transfer to molten metals. Transactions of the American Society of Mechanical Engineers 69:947–951
84. Katto Y (1997) Synthetic view of the CHF mechanism in subcooled flow boiling based on the recent mechanistic models. Proceedings, Convective Flow and Pool Boiling Conference, Irsee, May
85. Lopina RF, Bergles AE (1969) Heat transfer and pressure drop in tape-generated swirl flow of single-phase water. Journal of Heat Transfer 91:8–13
86. Mudawar I, Bowers MB (1995) Parametric study of ultra-high CHF in highly subcooled water flow inside small diameter tubes. Paper II.9, Proceedings, engineering foundation convective flow boiling conference, Banff, May
87. Zeigarnik YA (1996) Universal model of critical heat flux of subcooled liquid boiling in channels. High Temperature 34(1):48–52

The Importance of Thermophysical Properties in Optimum Design and Energy Saving

MARC J. ASSAEL

Summary. During the last two decades the sharp rise in the price of petrol and raw materials has made imperative the need for optimum design and energy saving. As a consequence of this, more and better values, as well as better predictive schemes for the thermophysical properties of fluids, were required. Thus research in the area of thermophysical properties increased, regardless of the generally low level of funding. In this chapter, a selection of some generalized methods more often employed in the prediction of equilibrium properties, vapor–liquid and liquid–liquid equilibrium, and transport properties are critically examined. From their discussion, the needs and the areas of measurements and prediction evolve. The importance of some of the properties in the theoretical understanding of the molecular theories are also discussed. Finally, through a discussion of international matters related to the thermophysical properties, such as custody and technology transfer or quality assurance, the need for properties for reference fluids are also shown. It is thus shown that the increased need for an optimum design will create a parallel need for accurate, internationally acceptable thermophysical properties that cover a wide range of conditions.

Key words. Thermophysical properties—Equilibrium properties—Transport properties

Introduction

The thermophysical properties are composed of two large groups of properties, equilibrium properties and transport properties. The equilibrium properties, also known as thermodynamic properties, refer to differences between two states of thermodynamic equilibrium and as such they constitute *P-V-T* properties, enthalpies or heat capacities, entropies, fugacities, etc. The transport properties refer to the rate of change of the state of a fluid, as a result of a change in the external condition. Among this group of properties are viscosity, thermal conductivity, diffusion, etc.

The mathematical formulation employed today for the calculation of the equilibrium properties historically starts with the evolution of the equations of state. Although it was Boyle in 1662 who first deduced that at a given temperature, the volume of a fixed mass of gas is inversely proportional to its pressure, its linear dependence in temperature at constant pressure was postulated by Gay-Lussac in 1802. Together with Dalton's law of

Faculty of Chemical Engineering, Aristotle University, 54006 Thessaloniki, Greece partial pressures, these observations resulted in the well-known equation of state of a perfect gas mixture, that is [1]:

$$P = \left(\sum_i n_i \right) RT \Big/ V \qquad (1)$$

where P is the pressure in the gas, n_i the number of molecules of species i, R is the universal gas constant, T is the absolute temperature, and V denotes the volume. Following the establishment of the perfect gas equation, attention was concentrated on the vapor–liquid phase transition. Special note must be made of the work of Andrews on carbon dioxide (1869) in which the volumetric and phase behavior of the pure fluid was established for the first time over appreciable ranges of temperature and density. This behavior is illustrated by the three-dimensional phase diagrams shown here, together with its projections in the *P-V* and *P-T* plane in Fig. 1. For comparison purposes, in the same figure the perfect gas behavior is also shown.

The experimental work of Andrews and others that followed helped in the evolution of the mathematical formulation of the modern equation of state and led van der Waals to propose, in his dissertation in 1873 "On the continuity of the gas and liquid states" [2], his famous equation of state:

$$P = \frac{RT}{V_{\mathrm{m}} - b} - \frac{a}{V_{\mathrm{m}}^2} \qquad (2)$$

The equation is cubic in the molar volume, V_{m}, and the two parameters a and b allow the necessary adjustments required to reproduce the critical temperature and pressure of a

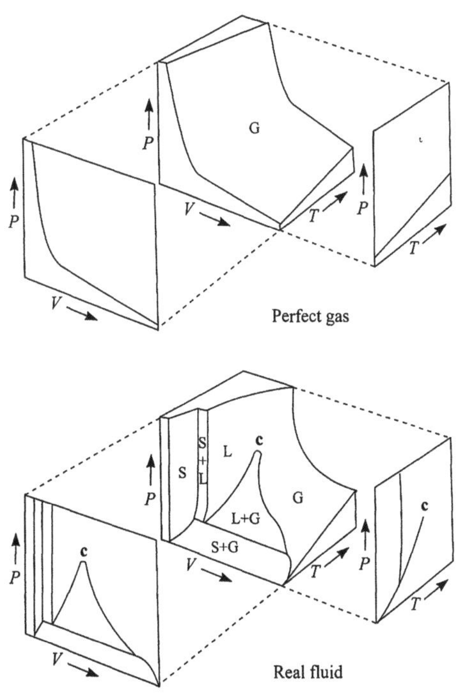

FIG. 1. Three-dimensional *P-V-T* diagram for a perfect gas and a real fluid. *G*, gas; *L*, liquid; *S*, solid; *P*, pressure; *T*, temperature; *V*, volume; *c*, critical point

pure fluid. van der Waal's equation of state, due to its semitheoretical background (proven later by a rigorous statistical-mechanical analysis by Ursell in 1912), was the first equation capable of predicting both gaseous- and liquid-phase behavior.

It is worthwhile pointing out that the majority of today's equations of state are simple mathematical modifications of this equation that retain its cubic dependence. Numerous modifications of the van der Waal's equation have been and are still being proposed. Some of the most notable ones, still employed today, are the equations of Redlich and Kwong (1949) [3], of Soave (1972) [4], and the widely employed equation of state of Peng and Robinson (1976) [5]. These are cubic equations whose pure fluid parameters are determined from the critical constants and the acentric factor, whereas for mixtures combining rules are used to express the parameters in terms of pure component values. Vapor properties and vapor–liquid equilibrium (VLE) conditions predicted by these models are usually in good agreement with experimental observations, although the prediction of liquid density is generally rather poor. The combination of simplicity of form with reasonably reliable predictions of VLE conditions has made cubic equations the most widely employed in design calculations and engineering problems. More accurate results can be obtained, but usually only by recourse to much more complicated equations of state, such as generalizations [6] of the Benedict–Webb–Rubin equation [7], with parameters fitted to wide-ranging and accurate experimental data.

The transport coefficients describe the rate at which a fluid relaxes toward its equilibrium state when it is perturbed by the application of a temperature, pressure, density, velocity, or composition gradient. The nonequilibrium statistical mechanics describing these processes is known as *kinetic theory* and aims to relate the macroscopic (observable) properties of a system to the microscopic properties and interactions of the individual molecules. Transport properties are expressed today as the sum of three contributions, a low-density contribution that is only temperature dependent, a critical enhancement term, and an excess contribution that describes the density dependence.

As a result of substantial advances in recent years, it is now possible to describe essentially exactly the low-density transport properties as well as the critical enhancement [8]. The excess contribution is less well understood theoretically, although various modifications of the Enskog theory can be used to establish effective correlations of experimental results. In particular, for dense gases and liquids remote from the critical region, methods based on the Enskog theory for hard spheres give an excellent representation of experimental data.

In this chapter, we concentrate on the application of the methods that are most widely employed by the majority of engineers or incorporated in most computer packages in performing optimum design or energy-saving calculations. Hence the methods that are discussed, for example, employ generalized equations of state rather than fluid-specific equations aiming to show what can be achieved by these methods.

The first part of this chapter concentrates on the ability to predict thermophysical properties. The discussion is illustrated by many examples of mass and energy balances, VLE and liquid–liquid equilibrium (LLE) calculations, as well as transport property calculations. In the second part of the chapter, the theoretical understanding of some properties and the insight that is gained by their study are discussed. Finally, the international dimension of thermophysical properties and factors such as custody transfer and quality assurance as well as the need for recommended values are presented. A parallel question that is examined is whether more thermophysical property measurements are required.

Optimum Process Design

In Fig. 2, a typical chemical process plant design is shown. The feeds are mixed in the mixer and heated to the optimum reaction temperature before entering the reactor. The reactor's products are cooled and then separated in the separator. This is a process diagram that applies more or less to most chemical processes. The design engineer is called upon to calculate, among other conditions, the optimum size of pipes, capacity of pumps, sizes of the exchangers and the reactor vessels, and the reaction and separation conditions. It should be absolutely clear that, even if one employs the best energy-optimized procedure, *the output of the design always depends upon the values of the thermophysical properties employed.* We can further demonstrate this statement with two examples.

In Fig. 3 we show a schematic diagram of a typical methanol catalytic reactor. The reaction takes place at 610 K and 20 MPa. The feed, hydrogen and carbon monoxide, enters at the bottom of the reactor vessel and is preheated to the required temperature by the hot product gases in the preheater. Following that, hydrogen and carbon monoxide react in two catalyst beds. To keep the reaction temperature low, as the reaction is

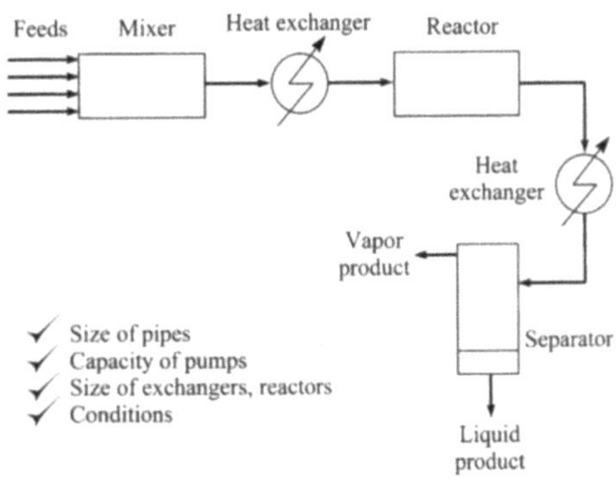

FIG. 2. A typical chemical process plant

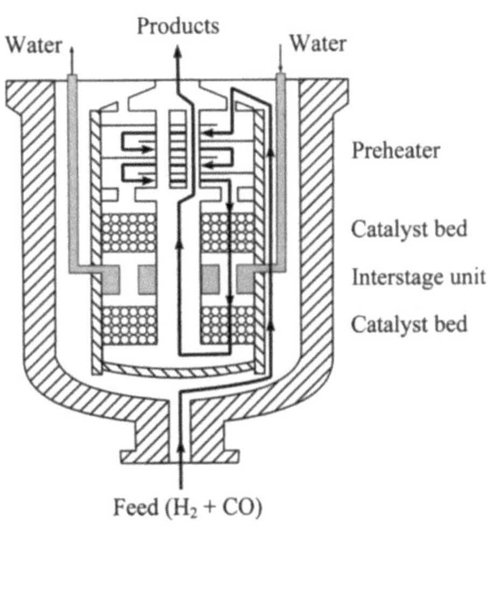

FIG. 3. Methanol catalytic reactor

exothermic, a water-cooled interstage unit is employed between the two catalyst beds. This type of interstage heat exchanger is very common as a means of keeping the temperature low when exothermic reactions take place. The area of the interstage heat exchanger is clearly a function of the viscosity and thermal conductivity of the gases (the area is approximately proportional to the ratio viscosity/thermal conductivity to the power 1/3 [9,10]). Hence, and irrespective of any optimized design procedure employed, if the gas viscosity was underestimated by 30% and its thermal conductivity was overestimated by an equal amount, then the area of the interstage heat exchanger will be underestimated by 20% and the reactor will fail. Obviously, the uncertainty of the viscosity and thermal conductivity of the feed and product gases at 610 K and 30 MPa can be quite large.

A similar case can easily be demonstrated by the Magnox nuclear power plant (Fig. 4). The typical Magnox reactor is a 14-m-diameter, 8-m-high enclosure shielded by thick concrete walls. The heat of reaction is removed by carbon dioxide gas that circulates at 670 K and 2 MPa and which in turn is employed to heat the steam that drives the steam turbines in the electricity generation plant. As in the previous example, for the accurate design of the plant it is imperative that the properties of carbon dioxide and steam are known to a high accuracy. In this case, however, and as is discussed later on, a very important role is also played by safety calculations, quality assurance, and validation of the plant. For the latter three factors, accurate values of the thermophysical properties do not suffice. Values must also be internationally accepted and validated. In the particular case of the Magnox reactor, the steam values employed are those proposed by IAPWS (International Association for the Properties of Water and Steam) while the carbon dioxide values are those proposed by IUPAC (International Union of Pure and Applied Chemistry).

The aforementioned two cases demonstrate the requirement for accurate values of the thermophysical properties. The question that directly arises from this requirement is how and how well properties are calculated. The following examples, which are based on the book by Assael et al. [1], attempt to answer this question by critically examining some characteristic cases. The cases chosen are very simple ones as their purpose is clearly to demonstrate capabilities and limitations in the best possible way. In the following exam-

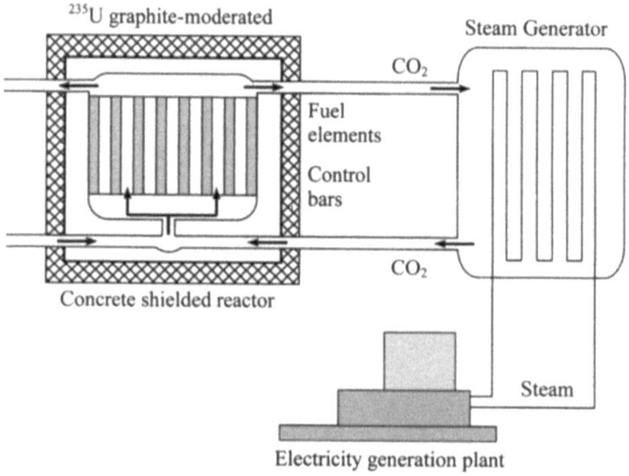

Fig. 4. Magnox nuclear power plant reactor

ples, the FORTRAN routines freely supplied by Assael et al. [1] (anonymous ftp with electronic mail address as password at *transp.eng.auth.gr*) were employed.

Equations of State and Corresponding States Predictions

For the calculation of equilibrium properties of pure fluids or mixtures, two approaches are extensively employed today.

1. The fastest and easiest way is to employ a generalized equation of state. In the case of pure fluids, equilibrium properties are calculated as a function of the critical parameters and the acentric factor, while in the case of mixtures, appropriate mixing rules of these parameters are usually incorporated. Additionally, for mixtures a binary interaction parameter, k_{ij}, needs to be specified. This is usually obtained by fitting VLE experiments. The most commonly employed generalized equations of state [1] are the equations of Soave [4], Peng and Robinson [5], and Benedict, Webb, and Rubin in the Han and Starling form [6]. This approach is highly recommended for nonpolar fluids because the equation constants were derived for such cases.
2. When a higher degree of accuracy is required, the corresponding states approach is preferred. One of the most widely employed corresponding states schemes is the three-parameter scheme proposed by Lee and Kesler [11] and its four-parameter modification for polar molecules proposed by Wu and Stiel [12]. This approach also requires critical parameters and acentric factors, whereas in the case of mixtures the Plöcker et al. [13] mixing rules for the critical parameters are usually employed. The corresponding states approach constitutes a more accurate scheme, employed also—but with care—with polar molecules, but it involves a higher degree of calculations.

The aforementioned two approaches are demonstrated in the following two examples [1].

Example 1. Calculation of the density and enthalpy of a mixture of 0.33 *n*-octane + 0.67 benzene at (1) 470 K and 1.4 MPa and (2) 590 K and 9.7 MPa.

For this example, the calculations have been performed with the Soave (RKS) equation, those of Peng and Robinson (PR) and of Benedict, Webb, and Rubin in the Han and Starling form (BWR), and the Lee and Kesler (LK) scheme. The results for the density, ρ, and the enthalpy, h, are shown in Table 1. In the same table, the enthalpy difference, Δh, between the two states is also given and compared with the experimental value. As the binary interaction parameter, k_{ij}, was not known, it was set equal to zero in all cases. For the calculations, the computer programs given in Assael et al. [1] were employed. It should be noted that enthalpy was arbitrarily set equal to zero at 0°C and 0.101325 MPa.

In Table 1, it can be seen that the density values predicted by the Benedict, Webb, and Rubin (BWR) equation of state and the Lee and Kesler (LK) corresponding states scheme, which are the most accurate ones, are very close. In the case of the enthalpy difference, the Lee and Kesler (LK) scheme produced the value nearer to the experimental one.

Example 2. Calculation of the density of R134a, at (1) 250 K and 2 MPa and (2) 450 K and 10 MPa, and its enthalpy difference between the two states.

Table 1. Density and enthalpy of 0.33 *n*-octane + 0.67 benzene

	470 K, 1.4 MPa		590 K, 9.7 MPa		Δh (kJ·kg^{-1})
	ρ (kg·m^{-3})	h (kJ·kg^{-1})	ρ (kg·m^{-3})	h (kJ·kg^{-1})	
RKS	511	21.8	381	363.3	342
PR	579	21.3	418	363.0	340
BWR	601	17.3	433	360.6	343
LK	597	15.2	462	347.3	332
Exp. [14]					315

ρ, density; h, enthalpy; Δh, enthalpy difference
RKS, Soave equation; PR, Peng and Robinson equation; BWR, Benedict, Webb, and Rubin equation; LK, Lee and Kesler scheme; Exp., experimental value

Table 2. Density and enthalpy of R134a

	ρ (kg·m^{-3}), 250 K, 2 MPa	ρ (kg·m^{-3}), 450 K, 10 MPa	Δh (kJ·kg^{-1})
LK	1322	464	319
WS	1371	473	318
Exp. [15]	1371	475	324

ρ, density; Δh, enthalpy difference
LK, Lee and Kesler scheme; WS, Wu and Stiel scheme; Exp., experimental value

Because R134a is a polar molecule, the corresponding states scheme of Lee and Kesler (LK) with the Wu and Stiel (WS) modification was employed. The results are shown in Table 2 together with the experimental value [15]. As expected, the predicted value employing the Wu and Stiel modification for polar fluids produces values for the density and enthalpy that are in very good agreement with the experimental ones.

VLE and LLE Predictions

To perform VLE calculations, two methods are usually employed.

1. The mole fractions in the vapor and liquid phases, y_i and x_i, are expressed as a function of the respective fugacity coefficients, $\Phi_i^{(L)}$, $\Phi_i^{(V)}$, as

$$K_i = \frac{y_i}{x_i} = \frac{\Phi_i^{(L)}}{\Phi_i^{(V)}} \tag{3}$$

In this case, an equation of state can be employed for the calculation. This approach is easy and fast, and is recommended for nonpolar fluids. As already discussed, for calculation through the equation of state, critical constants, acentric factors, and binary interaction parameters are required.

2. In this case, the mole fractions in the vapor and liquid phases, y_i and x_i, are expressed employing an activity coefficient, γ_i. The resulting equation [1] is

$$K_i = \frac{y_i}{x_i} = \frac{\gamma_i \Phi_i^\sigma P_i^\sigma}{\Phi_i^{(V)} P} \tag{4}$$

where Φ_i^σ, P_i^σ, and P are the fugacity coefficient at saturation pressure, the vapor pressure, and the total pressure of the system. In this case, the fugacity coeffients are usually obtained through an equation of state, the vapor pressure through an Antoine equation, and the activity coefficient through an activity coefficient model (Wilson, NRTL, UNIFAC, etc.). Because the parameters of the activity coefficient model are obtained by employing VLE data, it is obvious that their use will produce excellent results. This approach is a little more complicated than the previous one, but provided activity coefficient parameters do exist, the results are very good. It is also recommended for all kind of fluids.

We consequently employ the aforementioned two equations for the calculation of the boiling point of a mixture, for the construction of VLE diagrams, and for composition diagrams demonstrating the capabilities of these approaches and their limitations at the same time.

Example 3. Calculation of the boiling point of a mixture of 0.69 carbon dioxide + 0.31 ethane, at 3.025 MPa.

Equation 3 was employed together with the fact that the sum of mole fractions in the liquid phase must be equal to one. In this particular example, a generalized Peng and Robinson equation of state was employed. As already mentioned, for mixtures a binary interaction parameter, k_{ij}, is required. When this is not available, the binary interaction parameter is set equal to zero. To demonstrate the effect of this parameter, in this example the calculation has been performed with the binary interaction parameter equal to the calculated value of 0.124 and the value equal to zero.

When the binary interaction parameter is equal to 0.124, then the boiling point is calculated equal to 263.15 K, and the vapor mole fractions were found to equal 0.69 and 0.31 for carbon dioxide and ethane, respectively; that is, the azeotropic point was fully predicted. If, however, k_{ij} is set to equal zero, then the boiling point is calculated to equal 271.31 K and the vapor mole fractions to equal 0.73 and 0.27 for carbon dioxide and ethane, respectively.

The aforementioned results clearly indicate that in this case, if k_{ij} is not known and is thus set equal to zero, the boiling point obtained is about 8°C higher and the azeotropic point is not predicted.

This behavior is more obvious in the next example.

Example 4. Construction of the VLE diagram of a mixture of carbon dioxide + ethane, at 263.15 K.

In this example, one can adopt the same algorithm described in Example 3. In this case the pressure is calculated for the given temperature, for different liquid mole fractions. As in the previous example, the whole calculation was performed twice, first with $k_{ij} = 0.124$ and then with $k_{ij} = 0$ (Fig. 5). Is is interesting to see that when the value of the

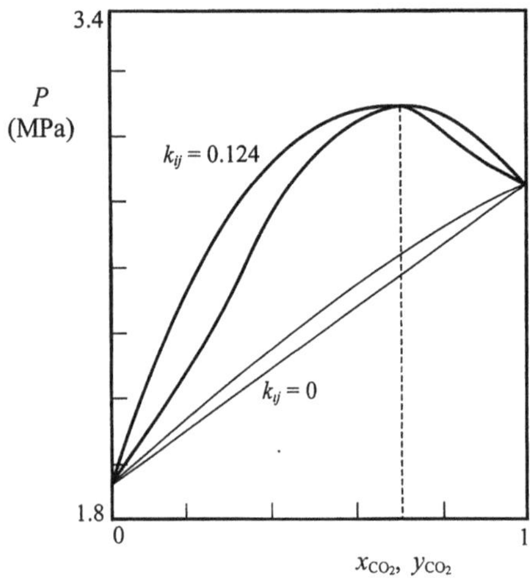

FIG. 5. Effect of the value of the binary interaction parameter on the vapor–liquid equilibrium (VLE) diagram of a mixture of carbon dioxide + ethane, at 263.15 K. P, pressure; x,y, mole fractions; k_{ij}, binary interaction parameter

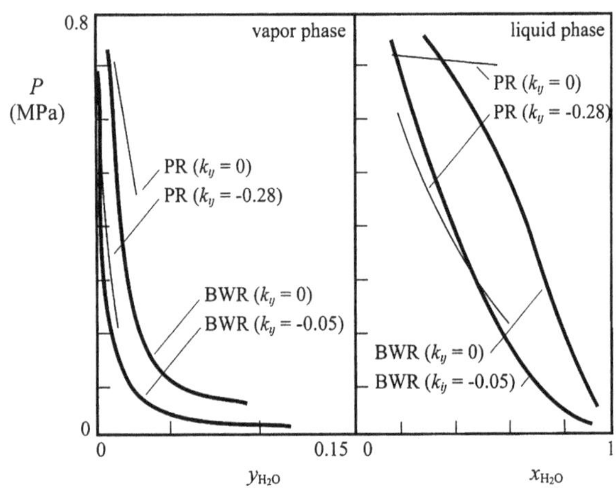

FIG. 6. Effect of the value of the binary interaction parameter on the VLE diagram of a mixture of water + ammonia, at 293.15 K. PR, Peng and Robinson equation; BWR, Bennedict, Webb, and Rubin equation; P, pressure; x,y, mole fractions; k_{ij}, binary interaction parameter

binary interaction parameter is not known, then the azeotropic behavior of this mixture cannot be obtained.

Example 5. Construction of the VLE diagram for the liquid and vapor phase of a mixture of water + ammonia, at 293.15 K.

In this case, a polar mixture is considered. The algorithm is identical with that described in Example 4. To further demonstrate the importance of the binary interaction parameter, however, the diagrams were constructed employing the Peng and Robinson equation of state and that of Benedict, Webb, and Rubin. In both cases, calculations were repeated with the calculated value for the binary interaction parameter and when this is set equal to zero.

The two diagrams are shown in Fig. 6. Although in the vapor phase the two approaches might look similar, in the liquid phase the differences are very large. It is clearly

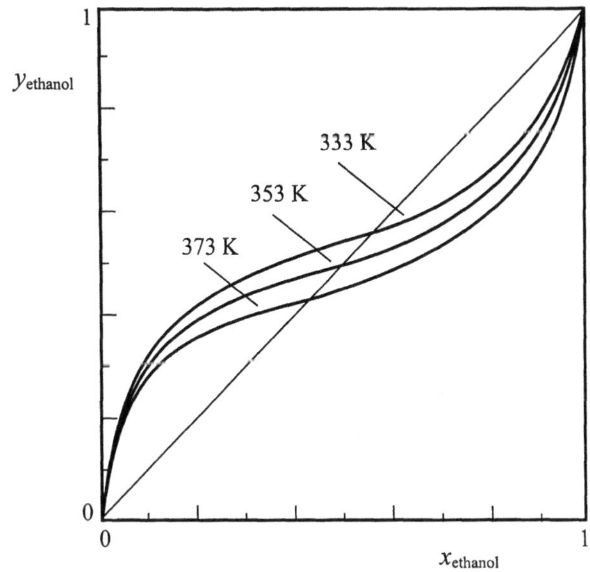

FIG. 7. VLE composition diagram of a mixture of ethanol + benzene, at 333, 353, and 373 K. x,y, mole fractions

evident that if the correct value of the binary interaction parameter is not employed, the behavior cannot be predicted. It should also be pointed out that in the case of polar mixtures, Eq. 4 should be preferred. This is demonstrated in the following two examples.

Example 6. Construction of the composition diagram of a mixture of ethanol + benzene, at 333, 353, and 373 K.

In this case, Eq. 4 was employed. The activity coefficient was obtained from the Wilson model, the vapor pressure from Antoine's equation, and the fugacity coefficient from the virial equation. All required parameters were obtained from Assael et al. [1], whereas the computer package employed for this calculation is described in the same book and supplied freely (anonymous ftp with electronic mail address as password at *transp.eng.auth.gr*).

Figure 7 shows the three isotherms, which are in excellent agreement with the real behavior of the mixture. The activity coefficient approach, although it requires longer computing time, it is the best approach today for VLE calculations.

Example 7. Construction of a LLE diagram of a mixture of 2-propanone + toluene + water, at 283.15 K and 0.1 MPa.

Here a ternary mixture is considered. The approach is similar to that employed in Example 6 with the activity coefficient obtained from the UNIQUAC model. All required parameters were obtained from the literature [1]. Figure 8 shows the resulting triangular diagram. Again, good values for the activity coefficient were required.

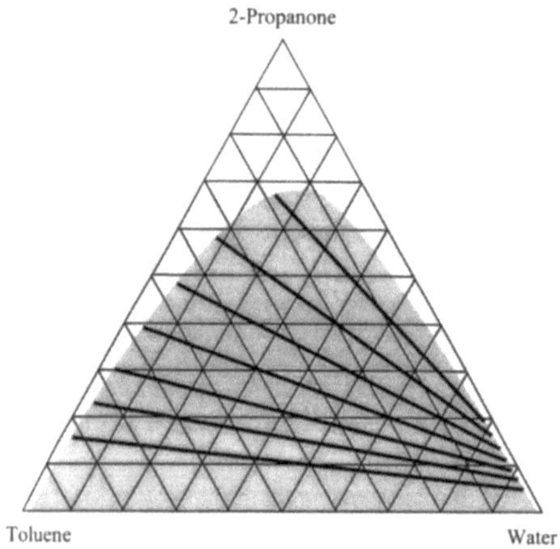

FIG. 8. Triangular VLE diagram of a mixture of 2-propanone + toluene + water

Transport Properties Prediction

As already mentioned, transport properties are expressed today [8] as the sum of three contributions: a low-density contribution that is only temperature dependent, a critical enhancement term, and an excess contribution that describes the density dependence.

The low-density contribution is well understood and is readily obtained [8]. The critical enhancement is now also understood and can be calculated in most cases [8]. The excess contribution, however, is more difficult to obtain. For dense gases and liquids away from the critical region, methods based on the Enskog theory for hard spheres give an excellent representation of experimental data [16,17]. A more generalized approach is adopted by the corresponding states schemes, which, although they lack a rigorous theoretical background, by employing so-called shape factors can very well describe the whole liquid- and vapor-phase behavior for pure fluids and their mixtures. A corresponding states approach has very successfully been applied to hydrocarbons [18] and, more recently, to refrigerants [19].

Finally we ought to mention the more recent promising scheme of Vesovic and Wakeham [20], which on a more theoretical basis predicts the viscosity and thermal conductivity of dense gases and liquid mixtures from the pure component values.

When conditions are more extreme, however, behavior cannot be predicted, and that in part is attributed to very large inconsistencies still in the measurement values. In Fig. 9 the thermal conductivity of KCl and NaCl is plotted as a function of temperature, as measured by different investigators. Although most investigators quote very low uncertainty in their measurement technique, the discrepancy in the values is very large.

Example 8. Calculation of the viscosity of an equimolar mixture of *n*-octane + *n*-dodecane, at 323.22 K and 0.1, 96.1, and 252.4 MPa.

In this example, two approaches were employed; the corresponding states approach as applied in the computer package SUPERTRAPP [18] and the hard sphere-based scheme as applied in the computer package TRANSP [21]. The results are shown in Table 3 together with the measured values [3]. The agreement is very good in this case. It should

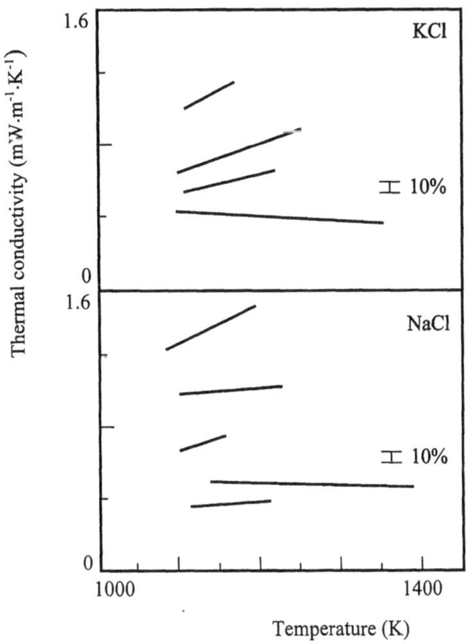

FIG. 9. Measurements of the thermal conductivity of potassium chloride (KCl) and sodium chloride (NaCl) as a function of temperature

TABLE 3. Viscosity of equimolar mixture of *n*-octane + *n*-dodecane

P (MPa)	η (μPa·s)	η (μPa·s)	Deviations (%)	η (μPa·s)	Deviations (%)
	Exp. [22]	SUPERTRAPP		TRANSP	
0.1	623	635	1.9	637	2.2
96.1	1482	1501	1.3	1482	0
252.4	4017	3280	−18.3	3914	−2.5

P, pressure; η, viscosity; Exp., Experimental value

be pointed out, however, that although SUPERTRAPP covers the whole liquid and vapor phases for a very large number of hydrocarbons and their mixtures, the application of TRANSP is limited to the liquid phase and to a small number of components and mixtures.

Calculation Discussion

In the previous three sections, examples have illustrated the capabilities of the generalized methods usually employed today in the calculation of the thermophysical properties. These capabilities are also summarized in Table 4. Although most generalized methods can be applied successfully to the prediction of properties or behavior of nonpolar mixtures, in the case of polar mixtures one should be careful. It should also be pointed out, that all methods presented rely on having values either for the binary interaction parameter, k_{ij}, or for the activity coefficient, γ_i, or for shape factors. All the aforementioned parameters are obtained by fitting experimental measurements. But, can we

TABLE 4. Prediction power of different schemes applied for the calculation of thermophysical properties of mixtures

	Method	For	Against	Nonpolar	Polar
ρ, Δh	EoS	Fast–easy	k_{ij}	Yes	No
	C-S	Slow	k_{ij}	Yes	No
	C-S + Pol	Slower	k_{ij}	Yes	Yes
VLE	EoS	Fast–easy	k_{ij}	Yes	No
VLE + LLE	ACT	Slow	γ_i, k_{ij}	Yes	Yes
λ, η	C-S	Large range	Shape factors	Yes	No[a]

ρ, density; Δh, enthalpy difference; VLE, vapor–liquid equilibrium; LLE, liquid–liquid equilibrium; λ, thermal conductivity; η, viscosity
EoS, equation of state methods; C-S, corresponding states methods; C-S + Pol, corresponding states corrected for polar fluids methods; ACT, activity-coefficient methods
k_{ij}, binary interaction parameter; γ_i, activity coefficient
[a] Corresponding states methods have been applied to refrigerant mixtures but with many difficulties

measure everything, and how accurate should the measurements be? The answer to these two questions is now discussed.

Let us assume that we need to measure only 10 properties at just 10 temperatures and 10 pressures, for 15 pure fluids and all their mixtures, at 5 compositions in the liquid and gas phases. Then, the total number of measurements required is 3.3×10^8 ($10 \times 10 \times 10 \times 32766 \times 5 \times 2$). If one further assumes that 3 measurements per day can be performed, then it is obvious that for the foregoing program of measurements 300 000 man-years will be required. And this program considered only 15 fluids!

The second question posed was related to the accuracy of the measurement itself. In relation to the answer to this question, one ought to remember that it was a small discrepancy in the measurement of the weight of nitrogen that led to the discovery of argon by Lord Rayleigh (Nobel Prize, 1904). Lord Rayleigh wrote in 1895 [23]:

"One's instinct at first is to try to get rid of a discrepancy, but I believe that experience shows such an endeavor to be a mistake. What one ought to do is to magnify a small discrepancy with a view to finding out the explanation; and, as it appeared in the present case that the root of the discrepancy lay in the fact that part of the nitrogen prepared by the ammonia method was nitrogen out of ammonia, although the greater part remained of common origin in both cases, the application of the principle suggested a trial of the weight of nitrogen obtained wholly from ammonia."

In that case the difference in the weight obtained was less than 0.5% (atmospheric nitrogen 2.3102, chemical nitrogen 2.2990).

Thus, we can conclude that to obtain all necessary parameters for the aforementioned schemes, *we need a few accurate measurements of selected properties, of selected fluids that will help us predict the rest.*

Theoretical Understanding

The selection of thermophysical properties to study in depth is also very important as in some cases it can give us a very good theoretical understanding. Let us consider the case of the low-density transport properties. In a low-density gas, the diffusion of a group of molecules in the gas characterizes the mass, momentum, and heat transport and con-

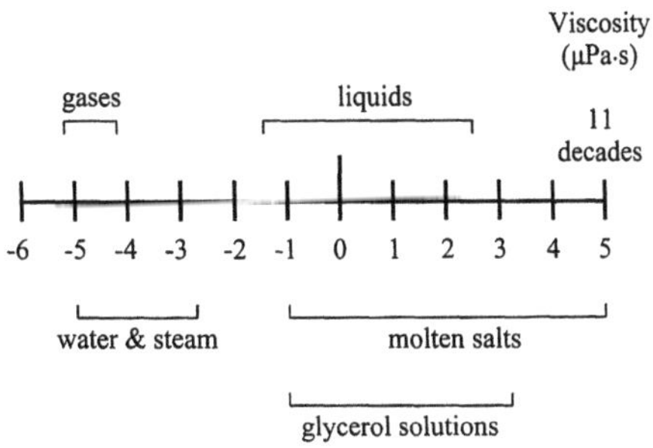

FIG. 10. Viscosity ranges encountered in various working fluids

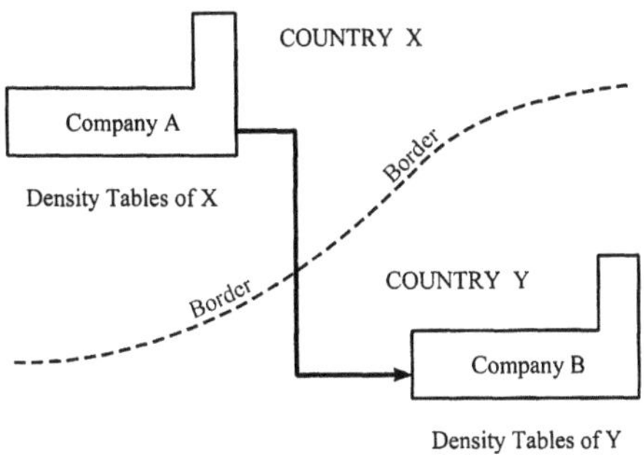

FIG. 11. Custody transfer of a product between company A (country X) and company B (country Y)

sequently the diffusion, the viscosity, and the thermal conductivity coefficients, respectively. In this case the kinetic theory is well defined and although its mathematical formulation is complicated, in 1950 a team of 30 clerical assistants with mathematical calculators succeeded in calculating the transport properties of monatomic gases from the Lennard–Jones potential. Hence, from the intermolecular potential, through the kinetic theory formulations, the transport properties were calculated. In the 1980s moreover, and in the case of low-density monatomic gases, the theoretical progress allowed the reverse procedure [24]. Thus, from measurements of one property at one temperature and pressure, the reverse calculation was performed and the intermolecular potential was obtained. Having obtained the potential, then the other transport properties at many temperatures and pressures could be calculated. It is unfortunate that this kind of approach is still restricted to monatomic gases at low density only.

Furthermore, because transport properties depend on the intermolecular potential and in the case of monatomic gases this is a function of the length and energy scaling parameters, then by scaling these parameters a corresponding states procedure was developed [24] whereby the transport properties of monatomic gases can be obtained. This indeed is an excellent example in which few accurate measurements, combined with a good understanding of theory, helped to develop a procedure by which many properties at different conditions can be predicted.

One further point is related to the values of the selected property to study. In the case of transport properties, the study of the viscosity is very important as the value of this property changes by 11 orders of magnitude in various working fluids (Fig. 10).

Thus, in addition to the conclusion of the previous section, we can add that *the choice of properties to study must take into consideration properties able to give a good understanding of the processes involved.*

International Dimension

One factor that has not really been discussed so far, except in the case of the Magnox nuclear reactor, is the international dimension of the thermophysical properties. This can easily be illustrated by considering Fig. 11.

We can assume that company A in country X sells a fluid to company B in country Y. In both cases, the quantity delivered by company A and the quantity received by company B is measured by volume but calculated and paid by mass. Because the same volume quantity crosses the border:

a. If both countries employ the same density tables, then the mass calculated in both countries is the same and hence payments requested will be equal to payments to be paid.
b. If, however, different density tables are employed, different masses will be calculated and clearly payments requested and paid will not agree and a payment dispute will result.

It is thus evident that, in the case of custody transfer, the accuracy of the density correlation itself is not of primary importance but whether this is internationally accepted.

The same can be said in the case of technology transfer. A process or plant developed in one country and sold to another must meet detail specifications of design methodology and design data employed. This was discussed in the beginning of this chapter in the case of the Magnox nuclear power reactor. The international dimension of the thermophysical properties of steam was long recognized, as in electric power plants the properties of steam are an essential part of the design and the claimed and real efficiency of the plant. Traditionally, each country had her own Steam Tables. It took 40 years for the International Association for the Properties of Water and Steam (IAPWS) to develop internationally acceptable steam tables and formulations.

A final point that ought also to be presented is quality assurance, that is, the requirement to satisfy regulatory frameworks concerned with safety and environmental issues. Traditionally, only national regulatory bodies had to be satisfied. Now however, international regulatory bodies have to be satisfied while national regulatory bodies must in many cases accept the regulations of other nations. Quality assurance of a plant or a process can often require a demonstrable pedigree for each number in the calculation of some aspect of the plant, for example, calculation of heat transfer during meltdown of a nuclear reactor.

The foregoing discussion clearly demonstrates the need for internationally acceptable thermophysical data, that is, data that are accepted by the majority of the world as a basis for trade, regulation, or standardization. Two implications are apparent from this statement. The first refers to supranational bodies that will propose such standards. Such bodies are the International Union of Pure and Applied Chemistry (Subcommittee on Transport Properties), the International Association for the Properties of Water and Steam, and others. The second implication is the establishment of a procedure for approval of such propositions. It is my belief that much work will be accomplished in the next decade along these lines.

Future

Efficient design, energy savings, and stricter environmental restrictions will produce a need for thermophysical property measurements that are more accurate, internationally accepted, and cover a wide range of conditions. The fluids to be studied at the selected conditions will be dictated by coordination of international bodies.

At the same time, work on predictive schemes will continue. The work on the generalized equations of states needs improvements and so does the description of the liquid-phase vapor–liquid equilibrium (VLE). In the case of the transport properties, we still do not have a kinetic theory for the liquid phase, while theories for mixtures need a lot of improvement.

As the need for an optimum design increases, this lack of accurate, internationally acceptable thermophysical properties that cover wide ranges of conditions will become more apparent. It is hoped that industry will have the foresight to see it and finance such research in time. As Professor Joseph Kestin wrote sometime ago [25], the road to prosperity is measurements → evaluation → formulation → validation → data bank → optimized design → prosperity.

Acknowledgments. The author would like to acknowledge the financial support and help of the Tokyo Electric Power Company and Keio University in Japan.

References

1. Assael MJ, Trusler JPM, Tsolakis TF (1996) Thermophysical properties of fluids. An introduction to their prediction. Imperial College Press, London
2. van der Waals JD (1873) Over de Continuitet van den Gas en Vloeistoftoestand. Doctoral dissertation, Leiden
3. Redlich O, Kwong JNS (1949) On the thermodynamics of solutions. Chemical Review 44:233–244
4. Soave G (1972) Equilibrium constants from a modified Redlich-Kwong equation of state. Chemical Engineering Science 27:1197–1203
5. Peng DY, Robinson DB (1976) A new two-constant equation of state. Industrial and Engineering Chemistry Fundamentals 15:59–64
6. Starling KE, Han MS (1972) Thermo data refined for LPG. Part 14: Mixtures. Hydrocarbon Processing 51:129–132
7. Benedict M, Webb GB, Rubin LC (1940) An empirical equation for thermodynamic properties of light hydrocarbons and their mixtures. I. Methane, ethane, propane and butane. Journal of Chemical Physics 8:334–345
8. Millat J, Dymond JH, Nieto de Castro CA (Eds) (1996) Transport properties of fluids. Their correlation, prediction and estimation. Cambridge University Press, Cambridge
9. Kern DQ (1950) Process heat transfer. McGraw-Hill, Tokyo
10. Assael MJ, Nieto de Castro CA, Wakeham WA (1978) The estimation of physical properties of fluids. Part II. The economic advantages of accurate transport property data. Proceedings of CHEMPOR '78:16.1–16.9
11. Lee BI, Kesler MG (1975) Generalized thermodynamic correlation based on three-parameter corresponding states. Journal of the American Institute of Chemical Engineers 21:510–527
12. Wu GZA, Stiel LI (1985) A generalized equation of state for the thermodynamic properties of polar fluids. Journal of the American Institute of Chemical Engineers 31:1632–1644

13. Plöcker U, Knapp H, Prausnitz JM (1978) Calculation of high pressure vapor-liquid equilibrium from a corresponding-states correlation with emphasis on asymmetric mixtures. Industrial and Engineering Chemistry Process Design and Development 17:324–332
14. Lenoir JM, Hayworth KE, Hipkin HG (1971) Enthalpies of benzene and mixtures of benzene with *n*-octane. Journal of Chemical and Engineering Data 16:280–284
15. Sato H, Higashi Y, Okada M, Takaishi Y, Kafawa N, Fukushima M (1994) JAR Thermodynamic tables, vol 1: HFCs and HCFCs. Japanese Association of Refrigeration, Tokyo
16. Assael MJ, Dymond JH, Papadaki M, Patterson PM (1992) Correlation and prediction of dense fluid transport coefficients. I. *n*-Alkanes. International Journal of Thermophysics 13:269–281
17. Assael MJ, Dymond JH, Patterson PM (1992) Correlation and prediction of dense fluid transport coefficients. V. Aromatic hydrocarbons. International Journal of Thermophysics 13:895–905
18. Huber M (1998) SUPERTRAPP V.2.0: a computer package for the calculation of the transport properties of nonpolar fluids and their mixtures. Available from National Institute of Standards and Technology, Gaithersburg, MD
19. Gallagher J, McLinden M, Morrison G, Huber M (1999) REFPROP V.5.0: a computer package for the calculation of the thermodynamic properties of refrigerants and refrigerant mixtures. Available from National Institute of Standards and Technology, Gaithersburg, MD
20. Vesovic V, Wakeham WA (1989) Prediction of the viscosity of fluid mixtures over wide ranges of temperature and pressure. Chemical Engineering Science 44:2181–2189
21. Assael MJ, Dymond JH (1999) TRANSP V.2.0: a computer package for the calculation of high-pressure liquid phase transport properties. Available from the authors
22. Dymond JH, Robertson J, Isdale JD (1981) Transport properties of non-electrolyte liquid mixtures. III. Viscosity coefficients for *n*-octane, *n*-dodecane, and equimolar mixtures of *n*-octane + *n*-dodecane and *n*-hexane + *n*-dodecane from 25–100°C up to 500 MPa. International Journal of Thermophysics 2:133–145
23. Rayleigh MA (1970) Argon. In: Bragg L, Porter G (eds) The Royal Institution Library of Science, vol 4. Applied Science, London
24. Maitland GC, Rigby M, Smith EB, Wakeham WA (1981) Intermolecular forces. Their origin and determination. Clarendon Press, Oxford
25. Kestin J (1981) Thermophysical properties in science and industry. In: Sengers JV (ed) Proceedings of the Eight Symposium on Thermophysical Properties. American Society of Mechanical Engineers, New York

Part III: Gas Hydrates: Their Impact on Energy and Environmental Prospects

Energy, Environment, and Naturally Occurring Methane Gas Hydrate: Connections

PETER ENGLEZOS

Summary. Methane gas hydrate is a crystal formed by the enclosure of methane molecules into a hydrogen-bonded network of water molecules under appropriate temperature and pressure conditions. Sufficient methane of biogenic and/or thermogenic origin exists in the earth's crust and over the years has combined with water and formed hydrate. This naturally occurring gas hydrate is regarded as a future energy resource. At the present time, hydrate energy recovery technology is not available. Engineering limitations and economic barriers have prevented the exploitation of this resource. At the same time, naturally occurring methane hydrate is potentially threatening to the global environment if it decomposes as a result of the current climate warming. Calculations suggest that this is possible to occur in the distant future as it happened 55 million years and 13 500 years ago. This chapter discusses the connections between energy, environment, and methane gas hydrate. The significance to our society is also discussed.

Key words. Methane—Gas hydrates—Energy—Environment—Global warming

Introduction

Methane, the main constituent of natural gas, is known to physically combine with water under appropriate temperature and pressure to form gas hydrate. Methane hydrate crystals can easily be formed in the laboratory by mixing water and methane at temperatures and pressures shown in Fig. 1. Figure 1 also shows equilibrium hydrate formation pressures for the formation of hydrate from gaseous carbon dioxide and liquid water. Suitable conditions, that is, sufficient amounts of biogenic or thermogenic methane gas and water at the temperatures and pressures shown in Fig. 1, exist in the earth's crust for methane hydrate formation. Russian researchers were the first to confirm that gas hydrates, containing mostly methane, were formed naturally in the earth and exist in vast quantities within and below the permafrost zone and in subsea sediments in many locations around the globe [1–4]. It should be noted that samples recovered from the earth during Leg 164 of the Ocean Drilling Program have revealed that the gas from the decomposition of the hydrate is mostly methane (99.9% volume) plus some C-2 to C-5 hydro-

Department of Chemical and Biological Engineering, The University of British Columbia, 2216 Main Mall, Vancouver, B.C. V6T 1Z4, Canada

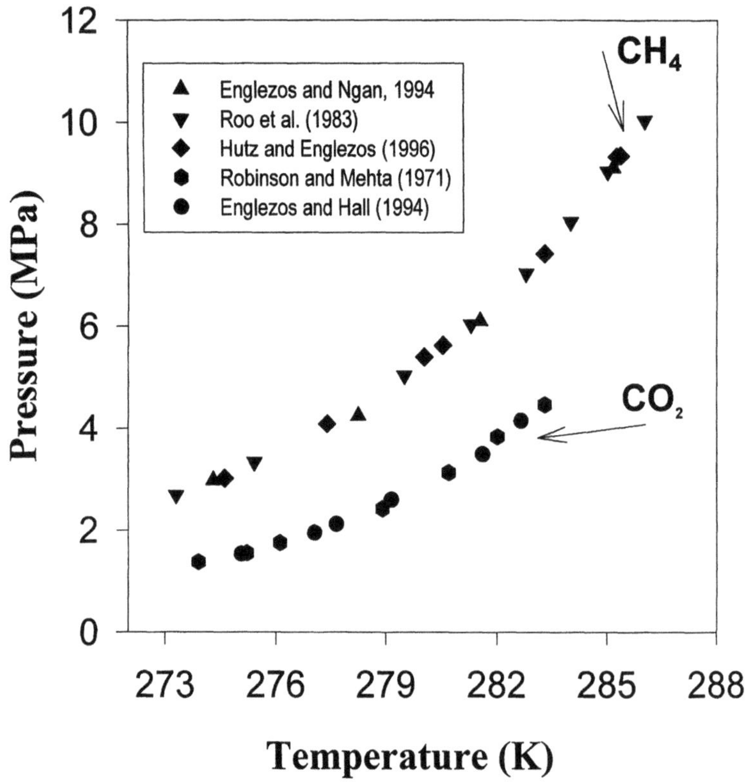

FIG. 1. Experimentally determined methane hydrate formation conditions from methane and liquid water (data from [13])

carbons [5,6]. In addition, samples from the Mallik 2L-38 gas hydrate research well showed 98% to 100% methane. The balance in some of the samples was propane and carbon dioxide [7].

Gas hydrates are nonstoichiometric crystals formed by the entrapment of molecules such as methane, ethane, and carbon dioxide within cavities formed by hydrogen-bonded water molecules. There is no chemical interaction between the water and the enclosed molecules. The first hydrate that was formed in the laboratory was that of chlorine by Sir Humphry Davy in 1810. Faraday confirmed Davy's experiments in 1823. Information on hydrates is widely available in several books [8–13], review papers [2,3,14–22], three hydrate conference proceedings [23–25], and proceedings from other meetings [26,27]. Table 1 shows the most important milestones in the history of gas hydrates, and Tables 2 and 3 summarize the various technological and environmental aspects of gas hydrates.

In this chapter, we discuss the connections between methane hydrate, energy, and the global environment. This relationship is both direct and indirect. Naturally occurring hydrates in the earth contain mostly methane and therefore constitute a potential energy resource. On the other hand, the indirect relationship arises because hydrates may plug

TABLE 1. Major milestones in the history of gas hydrates

Year	Milestone
1810–1823	Sir Humphry Davy discovered that a solid is formed when an aqueous solution of Cl_2 was cooled below 9°C. Faraday confirmed this in 1823 and suggested $Cl_2 \cdot 10\ H_2O$.
1823–1934	Identification of substances that may form gas hydrate.
1934	Hammerschmidt in the United States announced that plugging of natural gas pipelines was not due to ice but was caused by hydrate formation. This sparked intense research efforts that continue to the present to develop methods to prevent hydrate formation.
1949–1951	Elucidation of structures (structure I and structure II) by X-ray diffraction in Germany and United States.
1959	Statistical thermodynamics model at Shell by van der Waals and Platteeuw.
1965	Makogon in U.S.S.R. establishes the fact that gas hydrates, mainly of methane, exist naturally in the earth's crust.
1987–1994	New hydrate structure (structure H) at National Research Council of Canada (NRC). In 1994 it was found that structure H hydrates were found to occur naturally in the Gulf of Mexico.
1993	1st International Hydrates Conference in New Paltz, NY.
1996	2nd International Hydrates Conference in Toulouse, France.
1999	3rd International Hydrates Conference in Salt Lake City, UT.
1998	An 1150-m-deep gas hydrate research well was completed in the Northwest Territories of Canada (Mallik 2L-38) by Japanese and North American researchers.
1999	New hydrate structure was identified at NRC.

TABLE 2. Technological and environmental aspects of gas hydrates

Topic	Comment
Prevention of hydrate formation in pipelines and other hydrocarbon transport and processing facilities	R&D work in progress for continuous improved inhibition methods
Drilling through regions in the earth's crust containing naturally occurring gas hydrate may create safety problems	Field experience
Hydrate formation in the drilling mud may occur and will be detrimental to the drilling operation	Prevention of hydrate formation via thermodynamic inhibition
Hydrate formation from the light fraction of oil may render an oil well blowout in the ocean an environmental disaster by spreading the oil over a wide area	Physical and mathematical model
Storage and/or transport of natural gas by converting it into hydrate is considered more economical and safer than compressed or liquefied natural gas	R&D work in progress
Storage of natural gas by creating an impermeable gas reservoir through gas hydrate formation	Concept illustrated by computer simulation

pipelines due to their solid nature and create problems during drilling, production, transportation, and processing of oil and gas. Furthermore, hydrate formation can be a means to store and transport natural gas more economically. In this work, the direct linkage is explored. Furthermore, hydrate occurrence is linked to the state of the global environment. Methane is a strong greenhouse gas, and possible decomposition of naturally occurring methane hydrates from current global warming will strengthen the current

TABLE 3. Other technological and environmental aspects of gas hydrates

Topic	Comment
Development of separation technology:	
Aqueous organic solution concentration	Lab stage
Recovery of water from seawater	Lab and pilot plant
Recovery of water from industrial effuents	Lab stage
Fractionation of gases and liquids	Lab stage
Applications in bioengineering and biotechnology:	
Modification of enzyme activity	Lab stage
Recovery of proteins encapsulated in reversed micellar solutions	Lab stage
Various other applications:	
Thermal energy storage ("cool storage") systems for electric power load leveling	Work mainly in Japan
Hydrates are a *geological hazard*: unstable marine hydrates may have triggered catastrophic slumps	Offshore Norway (e.g.)
Capture of the CO_2 produced in thermal power plants and disposal in the ocean might alleviate the increasing atmospheric concentration of carbon dioxide	CO_2 hydrate may interfere with the process or be part of it intentionally
Gas hydrates within the foundation sediments of deepwater offshore structures may present a hazard to the foundation of pipelines and other production facilities	
Gas hydrate-microcrystal theory of anesthesia	Discussed by Pauling
Gas hydrate occurrence in the solar system; Bermuda triangle secrets	By Makogon By Davidson
Energy resource and global environmental implications of methane gas hydrate	Discussed in this work

global warming. There is now evidence suggesting that methane hydrate has played a key role in the evolution of the global environment at least twice in the distant past.

Methane Hydrate

Methane forms a cubic structure referred to as structure I. Ethane and carbon dioxide also form structure I hydrate, but propane, a larger molecule, forms another cubic structure known as structure II. Methane together with a larger molecule like neohexane can form structure H hydrate crystal [28]. It is interesting to note that the hydrate formation pressures are lower than the corresponding pure methane hydrate formation pressures at the same temperature. Methane hydrate consists of two types of cavities. One is a polyhedron with 12 pentagonal phases (pentagonal dodecahedron; 5^{12}), and the other is a larger cavity that is a polyhedron with 12 pentagonal faces and 2 hexagonal ones (tetrakaidecahedron; $5^{12}6^2$). The 5^{12} and $5^{12}6^2$ cavities are 7.9 and 8.7 Å, respectively, in diameter. The methane molecule has a diameter of 5.5 Å and therefore is able to fit in both cavities. If all the water cavities are assumed to be filled with methane molecules, then 1 molecule of methane is combined with 5.75 water molecules. However, the ratio of water to gas molecules (hydration number) is larger because not all cavities are filled. The hydration number is a function of the formation conditions, and several values ranging from 5.77 to 7.4 have been reported [13]. Hydrate formation is an exothermic process. The enthalpy change is approximately 64 kJ/mole of hydrate formed or 8.8 kJ/mole of water, assuming the following hydrate formula: $CH_4 \cdot 7.2H_2O$.

Methane Gas Hydrate in the Earth

Hydrates in the earth are located within the pores of sediment. They may exist within a zone known as the theoretical methane hydrate stability zone (TMHSZ). This zone is determined if the geothermal gradient, that is, temperature profile T (x,t), is known in a particular location in the earth. One then superimposes on this diagram the equilibrium pressure–temperature hydrate formation conditions after converting those conditions to a temperature–depth diagram. One such diagram is shown in Fig. 2. It should be kept in mind that the T–x profile changes from location to location whereas the T–x curve corresponding to methane hydrate equilibrium is affected by the presence of salts in the porewaters and the porous medium. These effects can be calculated if we know the salt type and concentration in the porewater [29] and the pore size [30,31]. The presence of liquid hydrocarbons may also affect the equilibrium curve.

Methane hydrate may form if sufficient amounts of gas and water exist at a point whose T–x coordinates are to the right of the curve UML and to the left of the lines UB and BL (see Fig. 2). Thus, the points U and L are the upper and lower point of the TMHSZ, respectively. It is noted that the line AB changes slope at point B because it has been assumed that a permafrost layer exists that extends from the earth's surface down to a depth corresponding to point B. Further down, the properties are those of sediment, and the temperature drops at a smaller rate with depth. One may construct a similar diagram for hydrates below the ocean floor.

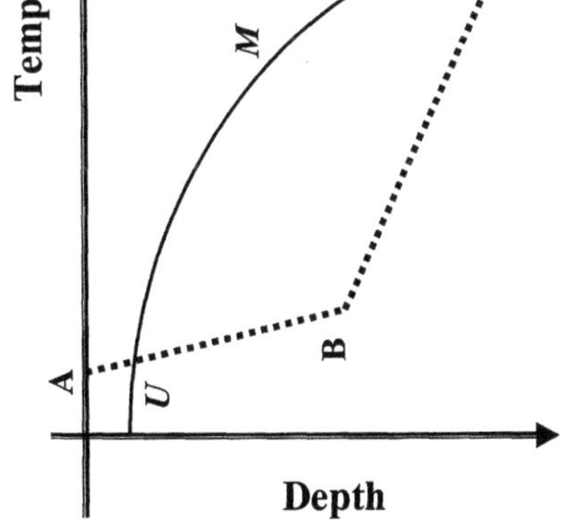

FIG. 2. Schematic illustrating the theoretical methane hydrate stability zone (TMHSZ) concept for hydrates within and below permafrost. *Line ABC* is the local geothermal gradient; *the curve UML* corresponds to the equilibrium methane hydrate formation conditions

Methane Hydrate Energy Stored in the Earth

Taking into account the fact that 1 m^3 of methane hydrate will release approximately 160 m^3 of methane gas at standard temperature and pressure (STP) conditions, one recognizes immediately the potential of methane hydrate in the earth to become a source of natural gas. It should also be kept in mind that approximately 0.8 m^3 of water are also released when 1 m^3 of methane hydrate dissociates.

Estimates of the amount of in situ gas hydrates vary considerably [2]. A consensus value of $21 \times 10^{15} m^3$ of methane has emerged, but any value between 2×10^{15} and $20 \times 10^{15} m^3$ is considered acceptable [3]. In our opinion, any estimate should be viewed with extreme caution because of the porous nature of the hydrate. Hydrates have the tendency to occlude significant amounts of water, and thus the amount of water converted to hydrate is significantly overestimated [19]. It appears, however, that the estimated amount of methane in solid hydrate form ("frozen gas") exceeds the amount of methane that exists as a conventional natural gas reserve. It is therefore considered to be a huge energy resource ("frozen energy"). This reserve has sparked an intense effort by earth scientists to generate a detailed map of hydrate locations and methane content in several places around the world and especially in countries lacking conventional oil and gas resources. The state of the art on the subject of hydrate resource assessment was discussed at the last international hydrate conference [25].

The detection of the earth's locations where gas hydrates deposits are found and the subsequent estimation of the quantity of gas included are important pieces of information for assessing the energy potential of the hydrate reserves. Thus, earth scientists spend considerable effort on these techniques. Bottom simulated reflectors (BSR) on seismic reflection profiles have provided a means for identifying hydrate reservoirs. However, this has proven to be a not so robust indicator [32].

While earth scientists pursue the detailed mapping of hydrate reservoirs, engineers are confronted with the task of examining whether the recovery this "frozen energy" is possible. The issue is whether methane hydrates are a recoverable reserve or an expected resource. This question is addressed next.

Methane Hydrate Energy Recovery

In theory, one has to mine the hydrate and subsequently decompose the hydrate and obtain almost pure methane gas. Based on experience from oil and gas reservoirs and knowledge of the methane hydrate partial phase diagram, the following three conceptual or laboratory-tested models have emerged [13].

1. Injection of inhibitors (methanol and/or electrolytes): Injection of inhibitors alters the methane hydrate phase diagram; that is, the hydrate equilibrium curve shifts to higher pressures. Therefore, this renders the pressure–temperature conditions in the reservoir favorable for the hydrates to decompose. The inhibitors act as freezing point depressants. The problem of circulating these inhibitors in impermeable formations arises.

2. Thermal stimulation: Thermal stimulation techniques include steam or hot gas injection, injection of hydrothermal waters, and in situ combustion of some of the methane gas. The objective is to raise the temperature above the equilibrium formation temperature at the given reservoir pressure. Radiofrequency or electromagnetic (RF/EM)

heating has also been proposed. The temperature rise should be sufficient for hydrate to decompose. Large heat inputs are required.

3. Decompression: Depressurization alters the reservoir conditions to a point at which gas hydrate decomposition is favored thermodynamically.

As seen, all methods alter the phase diagram and shift the reservoir conditions inside the hydrate instability region. All these methods are extremely costly. Research on these techniques has been intensive but has not yet produced any concrete technological path or basis for a production technology. However, significant knowledge has been gained. For example, it is believed that a hydrate reservoir must have high in situ *permeability* or be located above a free gas zone for potential energy recovery schemes to work. The existence of geothermal aquifers nearby is also helpful. It was also found that reservoir porosity and the thermal properties of the hydrates and the reservoir are determining factors for the production of gas from hydrate dissociation. Hydrate decomposition is an endothermic process and hence rapid decomposition could cause a temperature drop such as would allow water from the decomposing hydrate to form ice. In this case, decomposition will practically stop. It should also be kept in mind that hydrate dissociation is expected to generate a considerable amount of water.

Recently, Kamath gave a perspective on gas production from hydrates at an international symposium on methane hydrate as an energy resource [27]. Assuming that there is sufficient gas either as a hydrate and/or trapped as free gas in a conventional reservoir beneath the seal of a hydrate-bearing sediment, several technical issues must be resolved before production could begin. Those challenges were investigated at Shell in the Netherlands and the findings presented at the Japan National Oil Corporation (JNOC) symposium in Chiba, Japan [33]. The analysis indicated the following:

1. Knowledge of the spatial distribution of hydrate within a reservoir is not possible without a calibrated geological/geophysical/petrophysical rock model
2. Well engineering issues that must be resolved include the ability to predict and manage borehole stability in shallow wells drilled in hydrate-containing sediments
3. Production issues to be resolved include water production and disposal, and energy supply for heat and gas permeability
4. It was concluded that there are no major technical obstacles if suitable hydrate accumulations are found but that economics could very well be the major barrier.

Kvenvolden listed a number of factors that paint a negative outlook for the exploitation of methane hydrate energy in the near future [3]. Among the most important factors are (i) low permeability of the sediment; (ii) lack of interest from the gas industry; (iii) decreasing of the gas hydrate enrichment factor with increasing depth; and (iv) lack of comprehensive field experience.

Field Experience

Based on information from Makogon's extensive work, Sloan reviewed the history surrounding gas recovery from the Messoyakha gas hydrate field [13]. This field in Siberia is believed to be the only example to date of methane recovery from the decomposition of in situ hydrates. The production was based on hydrate inhibitor injection. It is estimated that about 36% of the gas recovered from this field came from hydrate decomposition. It should be noted, however, that the data from this reservoir have been viewed with caution.

An 1150-m-deep gas hydrate research well known as JAPEX/JNOC/GSC Mallik 2L-38 was completed in the Mackenzie Delta, Northwest Territories of Canada as a collaborative effort by Canada, Japan, and the United States. This scientific effort is well documented in a volume produced by the Geological Survey of Canada [34]. The results obtained from the Mallik drilling program indicated the occurrence of concentrated hydrate accumulations. Extrapolations indicated that 0.17 trillion m^3 (standard) of gas may be present. In addition, the reservoir properties were characterized as favorable for production. Thus, these results are rather optimistic as opposed to those from the Ocean Drilling Program legs 146 and 164 off the coast of Vancouver Island and North Carolina, respectively. Finally, it should be noted that the results from the Mallik well would serve to generate valuable experience for further exploratory drilling in Japan and elsewhere.

Hydrate Reservoir Simulation

Reservoir simulation is expected to aid the successful design of gas recovery schemes from hydrate reservoirs. A comprehensive reservoir simulation model will facilitate the assessment of the bounds of the energy efficiency of a gas recovery process. The intrinsic rate of dissociation together with heat and mass transfer phenomena relevant to a particular system as well as the sediment and hydrate reservoir properties are essential elements of a reservoir model. It is noted that such a simulator is not available yet, not because efficient numerical methods or computational power are lacking but only because of the complexity of the relevant physical and chemical phenomena. For example, knowledge of the rate of hydrate decomposition is required.

Two fundamental questions must be answered when time is a factor to be considered in hydrate decomposition:

1. How much time is required to begin dissociating hydrate crystals (induction time)?
2. What will be the rate of dissociation of hydrate crystals?

Both questions are challenges to experimentalists because hydrate formation and dissociation experiments are characterized by limited observability, that is, only few state variables are observed. Gas hydrate kinetics is a subject of intense ongoing research efforts that have been extensively reviewed [13,18,19].

History of Technology

It is clear that the exploitation of hydrate reserves has inherent technological difficulties and economic barriers. However, complex technological problems have been solved in the past, enabling the satisfaction of human energy needs. Furthermore, the need to find solutions provided the opportunity for the expression of human ingenuity. For example, the lack of wood to satisfy energy needs in ancient Greece and Rome led to the use of solar energy and innovations in architecture [35]. More recently, the need to find an alternative to wood led to two notable technological achievements in England. First was the development of the reverbatory furnace that allowed a cleaner burning of coal. The second achievement was the Newcomen pumping device, powered by steam, that enabled pumping water from deep mines, thus allowing deep coal mining to satisfy the increasing need for coal as a fuel [36].

Thus, the question whether hydrate energy technology will be developed is really a question about the existence of a driving force to develop such technology. This is a "mega project" that is perhaps invasive to the marine environment. Furthermore, it is difficult for our society to make a strong commitment for the required huge capital investment. Finally, there are several alternative solutions to energy needs. Haseltine discussed sustainable technologies that are expected to reduce energy costs and hence can be easily adopted by society [37]. In our opinion, these concerns create a climate of a very low probability for the development of hydrate energy technology in the foreseeable future.

Connections with Global Climate

The methane hydrate stored in the earth's crust is linked with changes in the state of the global climate that have occurred in the past and with changes anticipated in the future.

A "Runaway Greenhouse Effect"

It is well known that the atmosphere's near transparency to visible radiation and strong trapping of infrared energy is responsible for a warming phenomenon that is commonly referred to as the greenhouse effect. It is the trace atmospheric gases that exist in the atmosphere that enable it to trap the longer-wavelength infrared energy. It is noted that the main constituents, nitrogen and oxygen, are transparent to the energy emitted by the sun and to that reradiated from the earth's surface (infrared). It is also well established that the concentrations of CO_2 and CH_4 in the atmosphere have increased from 280 to 350 ppmv and from 0.8 to 1.72 ppmv, respectively, since the Industrial Revolution [38]. This continuous increase in the concentration of the trace atmospheric gases has strengthened the atmosphere's ability to trap infrared energy and has caused the current concern that the climate will enter an unstable warming trend. Although the physical processes involved in the greenhouse effect are well validated, the anticipated global temperature rises cannot be easily estimated [38]. Thus, annual temperature rises of 0.006°C (low-impact global warming scenario), 0.03°C (moderate warming scenario), and 0.08°C (catastrophic scenario) have been proposed [39].

A rise in the global atmospheric temperature resulting from the increase in the concentration of trace atmospheric gases just described raises an important issue related to the methane hydrate in the earth. Is this temperature rise sufficient to cause the decomposition of the hydrates? If such decomposition occurs, it will result in a substantial increase in the atmospheric concentration of methane. Hence, global warming will be enhanced and the possibility of a "runaway" greenhouse effect (RGE) (Fig. 3) becomes realistic. It should be noted that methane's contribution to global warming is about 15% but is important because its global warming potential is 21 times that of carbon dioxide.

To assess the probability of the RGE, one needs to determine these factors:

1. The time required for the onset of hydrate decomposition
2. The rate of methane release from the dissociation of hydrate
3. The impact of this released methane on the global climate

These are complicated multistep phenomena. For example, one has to account for the formation of ice covering the hydrate layer during hydrate decomposition and for the transformation reactions that methane might undergo after its release from the hydrate.

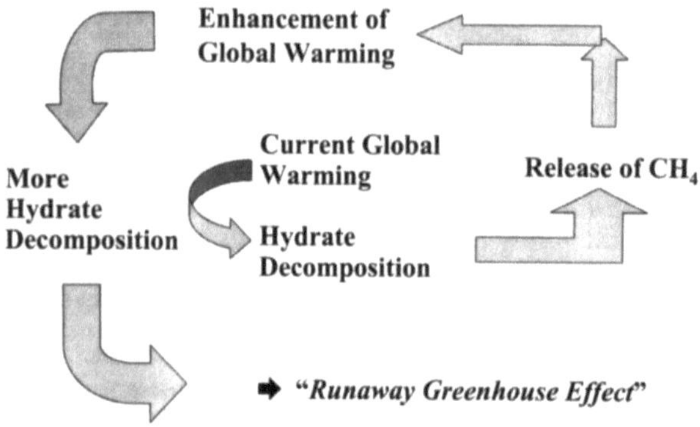

FIG. 3. Evolution of a runaway greenhouse effect caused by methane hydrate decomposition in the earth's crust

The impact of this scenario has been evaluated in recent years by assuming instantaneous hydrate decomposition and heat transfer in a single medium without permafrost phase change [40–42]. Hatzikiriakos and Englezos presented a heat transfer in composite media model that took into account permafrost phase change and a time-dependent boundary condition [43]. This model is the most comprehensive one on the RGE phenomenon. The model computes the critical time for the onset of methane hydrate decomposition. The boundary condition arises from the fact that the temperature on the surface of the earth changes as the result of current global warming.

The results from the simulations were found to depend on the type of global warming scenario, the thermophysical properties of the earth, and the required driving force for methane hydrate dissociation. Under the catastrophic scenario, the temperature at the top of a typical methane hydrate zone will begin to rise within the next 100 years. However, under a global temperature rise of 0.006°C/year, these phenomena will be substantially delayed. The permafrost phase change was found to have a shielding effect, that is, it increases the time required for the onset of hydrate dissociation. Suboceanic hydrates were found to remain stable within the next 1000 years; this reduces the risk of a RGE because the amount of suboceanic hydrates is estimated to be much greater than that below permafrost. However, hydrates within permafrost could begin decomposing within the next 100 years, and in fact there is evidence that such decomposition has already started in the Arctic [3].

Based on the foregoing analysis, it is evident that hydrate decomposition is associated with long time constants and hence it is safe to say that the possibility of RGE is very small. Furthermore, although hydrate dissociation is expected to impact global climate, its effect will be preceded by other direct implications from the current global warming. Therefore, it appears, from an anthropocentric point of view, that hydrate dissociation caused by global warming is not significant.

Connections with Past Climate Changes

Methane hydrate dissociation has had an effect on global climate in the past. It is believed that decomposition of submarine and continental hydrates could have triggered a RGE and thus contributed to the rapid warming at the end of the last glaciation 13 500 years ago [44]. The destabilization of methane hydrate in the earth's crust is attributed to a pressure reduction from the melting of ice crystals (end of glaciation). Lower sea levels during a period of global cooling may have also caused decomposition of the submarine methane hydrate.

Recently, methane hydrate decomposition in the earth has been linked to a variety of phenomena that occurred 55.5 million years ago. There is evidence that at that time the temperature in the deep ocean increased by about 6° Celsius [45–47]. This phenomenon, known as the latest Paleocene thermal maximum (LPTM), had a significant impact on life and its evolution [48]. This rapid warming is attributed to several factors because it is not clear what is the cause [46]. Among those factors cited are changes in tectonism, thermohaline circulation, and atmospheric CO_2. It was suggested that the initial methane hydrate dissociation might have been caused by submarine seismicity, volcanism, or simple gravitational slumping that induced catastrophic slope failure on continental margins containing methane hydrate reservoirs [49].

Relative carbon isotopic compositional changes provided the link with the phenomena in that distant time in the earth's history. In particular, a large influx of ^{12}C occurred that resulted in a 0.25% drop in the ^{13}C values. This conclusion is reached from the characterization of carbonate and organic matter deposited during the latest Paleocene thermal maximum. The relative compositional imbalance is known as a carbon isotope excursion and lasted approximately 2×10^5 years [4,46].

Based on mass balance calculations it was hypothesized that the source of the lighter carbon was methane hydrate that decomposed from sinking warm waters [50]. In particular, it was suggested that the isotopic imbalance was created by the release and oxidation of 1.4 to 2.8×10^{18} g methane from oceanic hydrates. Subsequently, Dickens and coworkers presented numerical simulation results from a carbon cycle model [46]. Those results were able to match the carbon isotopic record. The carbon isotope excursion was caused by an addition of approximately 1.12×10^{18} g biogenic methane over 10^4 years. Furthermore, they demonstrated that those results are consistent with observations concerning deep-sea carbonate dissolution and global temperature.

Dickens based the hydrate dissociation hypothesis on the following three assumptions:

1. An estimated amount of 11×10^{18} g C is trapped as oceanic hydrate
2. Methane hydrate formation and decomposition processes that occur today are of similar nature to those that occurred during the Paleocene period
3. Carbon transfer can occur between oceanic hydrates and the ocean–atmosphere inorganic carbon reservoir during deep-sea warming

Katz et al. (1999) suggested that the Blake Nose east of the Florida coast was identified as the site where most of the methane release took place [51].

Epilogue

Evidence and results from computer simulations of carbon cycle models suggest that dissociation of naturally occurring suboceanic methane gas hydrate and release of the methane in the atmosphere has altered the global climate 55 million years ago as well as 13 500 years ago. The possibility of enhancing the current global warming from a similar destabilization of the earth's methane hydrate was evaluated, suggesting that while decomposition of hydrates below permafrost may start within this century, suboceanic hydrates will not be affected within the next 1000 years. Thus, although the impact of hydrate dissociation in the past presents a template of future events, the impact on humans will not be felt immediately because long time constants are associated with the relevant hydrate transformation processes. In the meantime, the impact of current global warming from increased concentration of trace atmospheric gases is expected to precede that from methane hydrate.

On the other hand, it is estimated that 2 to $20 \times 10^{15}\,m^3$ methane exist in the earth's crust in the form of solid hydrate. The recovery of this methane to satisfy human energy needs presents an enormous challenge to engineers. Presently, economic and safe technology to accomplish this task is not available because of the nature of the hydrate reservoirs and the fact that hydrates are dispersed and occur more frequently below the ocean floor. Furthermore, it should be noted that energy needs might be more easily satisfied with sustainable technologies that do not require the level of capital and risk associated with hydrate energy development.

Acknowledgments. The financial assistance provided by Tokyo Electric Power Company (TEPCO), Natural Sciences and Engineering Research Council of Canada (NSERC), and Keio University is greatly appreciated.

References

1. Makogon YF, Trebin FA, Trofimuk AA, Tsarev VP, Chersky NV (1972) Detection of a pool of natural gas in a solid (hydrate gas) state. Doklady of the Academy of Sciences of the USSR—Earth Science Sections 196:197–200
2. Kvenvolden KA (1993) Gas hydrates—geological perspective and global change. Reviews in Geophysics 31(2):173–187
3. Kvenvolden KA (1999) Potential effects of gas hydrate on human welfare. Procedings of the National Academy of Sciences of the United States of America 96(7):3420–3426
4. Suess E, Bohrmann G, Greinert J, Lausch E (1999) Flammable ice. Scientific American 11:76–83
5. Dickens GR, Paull CK, Wallace P, ODP Leg 164 Scientific Party (1997) Direct measurement of in situ methane quantities in a large gas-hydrate reservoir. Nature (London) 385:426–428
6. Holbrook WS, Hoskins H, Wood WT, Stephen RA, Lizarralde D (1996) Methane hydrate and free gas on the blake ridge from vertical seismic profiling. Science 273(5283): 1840–1843
7. Tulk CA, Ratcliffe CI, Ripmeester JA (1999) Chemical and physical analysis of natural gas hydrate from the JAPEX/JNOC/GSC Mallik 2L-38 gas hydrate research well. In: Dallimore SR, Uchida T, Collett TS (eds) Geological Survey of Canada Bulletin 544:251–262
8. Davidson DW (1973) Gas hydrates. In: Frank F (ed) Water: a comprehensive treatise, vol 2. Plenum Press, New York, pp 115–234
9. Makogon YF (1981) Hydrates of natural gas. Penn Well Publishing, Tulsa
10. Berecz E, Balla-Achs M (1983) Studies in inorganic chemistry, vol 4: Gas hydrates. Elsevier, Amsterdam
11. Jeffrey GA (1984) Hydrate inclusion compounds. In: Inclusion compounds. Academic Press, London, pp 135–190
12. Ginsburg GD, Soloviev VA (1998) Submarine gas hydrates (in Russian). VNII Okeangeologia, St. Petersburg
13. Sloan ED (1999) Clathrate hydrates of natural gases. Dekker, New York
14. van der Waals JH, Platteeuw JC (1959) Clathrate solutions. Advances in Chemical Physics 2:1–57
15. Byk SS, Fomina VJ (1968) Gas hydrates. Russian Chemical Reviews 37(6):469–491

16. Parrish WR, Prausnitz JM (1972) Dissociation pressures of gas hydrates formed by gas mixtures. Industrial Engineering Chemistry Process Design Development 11:26–34
17. Holder GD, Zetts SP, Pradham N (1988) Phase behavior in systems containing clathrate hydrates. Reviews in Chemical Engineering 5(1):1–69
18. Englezos P (1993) Clathrate hydrates. Industrial Engineering Chemistry 32(7):1251–1274
19. Englezos P (1996) Nucleation and growth of gas hydrate crystals in relation to kinetic inhibition. Revue de l'Institut Français du Pétrole 51(6):789–795
20. Sloan ED (1998) Gas hydrates: review of physical/chemical properties. Energy & Fuels 12:191–196
21. Ohmura R, Mori YH (1999) Physical chemistry of clathrate hydrates—an overview and a state-of-the-art review of fundamental studies contributing to energy and environmental technology. Thermal Science & Engineering 7(2):35–58
22. Ripmeester JA, Ratcliffe CI, Klug DD, Tse JS (1994) Molecular perspectives on structure and dynamics in clathrate hydrates. In: Sloan ED, Happel J, Hnatow MA (eds) Annals of the New York Academy of Sciences 715:161–176
23. Monfort JP (1996) Proceedings of 2nd international conference of gas hydrates, June 2–6, 1996, Toulouse, France
24. Sloan ED, Happel J, Hnatow MA (1994) Annals of the New York Academy of Sciences 715
25. Holder GD, Bishnoi PR (1999) Proceedings of 3rd international conference on gas hydrates, book of abstracts, July 18–22, 1999, Salt Lake City, UT
26. JNOC (Japan National Oil Corporation) (1998) Proceedings of the international symposium on methane hydrates—Resources in the near future, October 20–22, 1998, Chiba City, Japan
27. Warzinski RP, Holder GD (1998) Gas clathrate hydrates. Energy & Fuels 12(2):189–190
28. Ripmeester JA, Tse JS, Ratcliffe CI, Powell BM (1987) A new clathrate hydrate structure. Nature (London) 325:135–136
29. Englezos P, Bishnoi PR (1988) Prediction of gas hydrate formation conditions in aqueous electrolyte solutions. American Institute of Chemical Engineers Journal 34(10):1718–1721
30. Uchida T, Ebinuma T, Ishizaki T (1999) Dissociation condition measurements of methane hydrate in confined small pores of porous glass. Journal of Physical Chemistry B 103(18):3659–3662
31. Clarke MA, Pooladi-darvish M, Bishnoi PR (1999) A method to predict equilibrium conditions of gas hydrate formation in porous media. Industrial Engineering Chemistry Research 38(6):2485–2490
32. Paull CK (1999) Sensing marine gas reservoir. In: Holder GD, Bishnoi PR (eds) Proceedings of 3rd international conference on gas hydrates, book of abstracts, July 18–22, 1999, Salt Lake City, UT
33. BeMent WO, Bil K, Drenth AJJ, Klomp UC, Roodhart LP, Sambell RH, Swinkels WJA, White D, Zoons CW (1998) Are there "show stoppers" to commercial gas hydrate production. In: Proceedings of the Japan National Oil Corporation (JNOC) international symposium on methane hydrates-resources in the near future, October 20–22, 1998, Chiba City, Japan
34. Dallimore SR, Uchida T, Collett TS (1999) Scientific results from JAPEX/JNOC/GSC Mallik 2L-38 gas hydrate research well, Mackenzie Delta, Northwest Territories, Canada. Geological Survey of Canada Bulletin 544
35. Botkin DB, Keller EA (1998) Environmental science, 2nd edn. Wiley, New York, pp 308–309
36. Cassedy ES, Grossman PZ (1990) Introduction to energy. Cambridge University Press, Cambridge, pp 1–105
37. Haseltine DM (2000) Exploring our energy future. Chemical Engineering Progress 1:75–83
38. Taylor FW (1991) The greenhouse effect and climate change. Reports on Progress in Physics 54:881–918
39. Schneider SH (1990) The global warming debate: science or politics. Environmental Science & Technology 24:432–435

40. Bell PR (1982) Methane hydrate and the carbon dioxide question. In: Clark WC (ed) Carbon dioxide review. Oxford University Press, New York, pp 401–406
41. Revelle RR (1983) Methane hydrates in continental slope sediments and increasing atmospheric carbon dioxide. In: Changing climates. National Academy Press, Washington, DC, pp 252–261
42. MacDonald GJ (1990) Role of methane clathrates in past and future climates. Climatic Change 16:247–281
43. Hatzikiriakos SG, Englezos P (1993) The relationship between global warming and methane gas hydrates in the earth. Chemical Engineering Science 48(23):3963–3969
44. Nisbet EG (1992) Sources of atmospheric CH_4 in early postglacial time. Journal of Geophysical Research 97(D12):12859–12867
45. Zachos JC, Lohmann KC, Walker JCG, Wise SW (1993) Abrupt climate change and transient climates during the paleocene: a marine perspective. Journal of Geology 101:191–213
46. Dickens GR, Castillo MM, Walker JCG (1997) A blast of gas in the latest Paleocene: simulating first-order effects of massive dissociation of oceanic methane hydrate. Geology 25(3):259–262
47. Simpson S (2000) Methane fever. Scientific American 2:24–27
48. Kerr R (1997) Did a blast of sea-floor gas hydrates usher in a new age? Nature (London) 275:1267
49. Bains S, Corfield RM, Norris RD (1999) Mechanisms of climate warming at the end of the paleocene. Science 285:724–727
50. Dickens GR, O'Neil JR, Rea DK, Owen RM (1995) Dissociation of oceanic methane hydrate as a cause for the carbon isotope excursion at the end of the paleocene. Paleoceanography 10:965–971
51. Katz ME, Pak DK, Dickens GR, Miller KG (1999) The source and fate of massive carbon input during the latest paleocene thermal maximum. Science 286:1531–1533

The Phase Changes in $CH_4 + C_2H_6$ Hydrates, and Their Impact on Oil and Gas Production

Sivakumar Subramanian[1], Adam Ballard[1], Ramesh A. Kini[1], Steven F. Dec[2], and E. Dendy Sloan, Jr.*[1]

Summary. Although methane and ethane form sI hydrates by themselves, the binary gas mixtures of these components form sII hydrates at certain compositions. In our earlier work, it was reported that the hydrate structure in methane + ethane hydrates changes from sI to sII over a methane vapor composition range (y_{CH4}) of 0.72 to 0.75. Recently, Raman spectroscopic measurements on the $CH_4 + C_2H_6$ system suggested that the hydrate structure reverts to sI from sII over a y_{CH4} range of 0.993 to 0.994. ^{13}C-NMR measurements on this system suggested that hydrate composition changes, by about 50%, as the hydrate structure changes from sI to sII at the lower transition point. A new hydrate prediction model suggested that the hydrate structure at pressures higher than the incipient hydrate formation conditions might be different than that at the incipient pressures. The major industrial applications of these structural transitions in the $CH_4 + C_2H_6$ hydrates are discussed.

Key words. Hydrate—Methane—Ethane—Raman—NMR

Introduction and Background

Natural gas hydrates are crystalline inclusion compounds with a well-defined structure that consists of a cage network of water molecules with hydrocarbon guests trapped inside the cages. The three known natural gas hydrate structures—sI, sII, and sH—have unique combinations of various cavity types. The size and shape of the guest molecules determine the hydrate structure formed in the system [1].

Typical natural gases encountered in oil and gas production contain methane in excess of 85 mole% [2], with ethane and higher paraffins comprising the balance. Because methane and ethane are two major components of natural gases, their hydrates are of practical significance. The hydrate phase equilibria of these pure components and their binary mixtures have been measured over a number of years [3–5]. Kinetic studies of hydrate formation from pure methane and ethane as well as their binary mixtures can

[1] Center for Hydrate Research, Colorado School of Mines, Golden, CO 80401-1887, USA
[2] Department of Chemistry and Geochemistry, Colorado School of Mines, Golden, CO, USA
* Corresponding author

also be found in the literature [6–8]. However, measurement of the hydrate structures was neglected in these studies. For example, it may be difficult to combine the kinetics of pure methane and pure ethane sI hydrates if the gas mixtures form sII, a different hydrate with different kinetics.

Because methane and ethane are relatively small molecules, they each form sI hydrates, with methane residing in both type cages (5^{12} and $5^{12}6^{2}$) and ethane in the large $5^{12}6^{2}$ cage [9]. However, it was shown in our previous work that the binary gas mixtures of methane and ethane form sII hydrates at certain compositions with a structural transition from sI to sII over the range of 72.2 to 75 mole% methane at 274.2 K [10]. This composition range was referred to as the lower transition point. Because pure methane, situated at one of the extremes of the composition range, forms sI hydrate, the structure of methane + ethane hydrates must revert from sII to sI at a methane composition above 75 mole%. This composition was called the upper transition point.

Raman spectroscopic measurements indicated that, at 274.2 K, the upper transition point in methane + ethane hydrates occurs at a y_{CH4} of 0.992–0.994 [11]. Also, ^{13}C-NMR spectroscopic measurements of the guest composition of the hydrate phase on a water-free basis at various vapor compositions near the lower transition point showed that the hydrate composition changes by about 50% with the structure change (unpublished results).

The measured lower structural transition point was used to model the entire hydrate phase diagram of the methane + ethane + water system. Most of the hydrate models developed to date assume that the binary mixtures of methane and ethane form sI hydrates over the entire composition range [11]. In this work, a different approach was adopted to develop a model to predict hydrate phase equilibria under excess water conditions at both transition points and intermediate conditions.

Modeling

The model used to calculate phase equilibria for methane + ethane + water in this work is a multiple phase Gibbs energy minimization method, modified from the proposed methodology of Bishnoi's laboratory [12]. The current model accounts for the three hydrate structures: sI, sII, and sH, liquid water (L_w), liquid hydrocarbon (L_{hc}), and vapor (V). Details of the model are given elsewhere [14].

All fluid phases (V, L_w, L_{hc}) are described by the Soave–Redlich–Kwong equation of state while the solid hydrate phases are described by the theory of van der Waals and Platteeuw [13]. The Kihara spherical core potential is used to calculate the Langmuir constants. The Kihara parameters, σ and ε, were fit to incipient hydrate equilibria data of pure components as well as the lower structural transition point obtained from Raman spectroscopy. The spectroscopic data were used to optimize the Kihara parameters of methane and ethane by forcing the model to predict the experimentally determined lower transition point at 274.2 K. The new set of Kihara parameters for CH_4 and C_2H_6 along with the reference properties, $\Delta\mu$, Δh, and Δv, are shown in Tables 1 and 2. The prefix Δ refers to the difference in chemical potential (μ), molar enthalpy (h), and molar volume (v) of water in the empty hydrate lattice and pure water. Details of this procedure are given by Ballard and Sloan [14].

Previous hydrate prediction methods, such as the method developed by Parrish and Prausnitz [15], are capable of calculating only the incipient hydrate structure. The model

TABLE 1. Parameters used for hydrate modeling

Parameter	Units	Structure I		Structure II	
$\Delta\mu_w^0$	J/mole	1263.60		883.82	
Δh_w^0 (ice)	J/mole	1389.08		1025.08	
Δh_w^0 (water)	J/mole	−4623.27		−4987.27	
Δv_w^0 (ice)	Cm^3/mol	3.0		3.4	
Δv_w (water)	Cm^3/mol	4.6		5.0	
ΔC_{pw}^0	J/mole K		−38.12		
Cavity type	—	5^{12}	$5^{12}6^2$	5^{12}	$5^{12}6^4$
	nm	0.395	0.43	0.391	0.473

$\Delta\mu_w^0$, reference chemical potential difference between empty hydrate lattice and ice at 273.15 K and zero absolute pressure; Δh_w^0 (ice), enthalpy difference between empty hydrate lattice and ice; Δh_w^0 (water), enthalpy difference between empty hydrate lattice and water; ΔV_w^0 (ice), volume difference between empty hydrate lattice and ice; Δv_w^0 (water), volume difference between empty hydrate lattice and water; ΔC_{pw}^0, heat capacity difference between empty hydrate lattice and water

TABLE 2. Optimized Kihara parameters

Component	a (Å)	σ (Å)	ε/κ (K)
Methane	0.3834	3.1695	154.1815
Ethane	0.5651	3.3404	180.0164

a, Kihara spherical core radius, Å; σ, distance between molecular core surfaces at zero potential energy, Å; ε, depth of intermolecular potential well, J/mol; κ, Boltzmann's constant, J/K

used in this work, however, can generate the complete hydrate phase diagram with the relative amounts of all phases.

Results and Discussion

Figure 1 shows the complete pressure versus composition (P–x) phase space for methane + ethane + water at 277.6 K, generated using this model. Hydrate equilibrium data at L_w-H-V equilibrium conditions by Deaton and Frost [3] are also plotted to compare the model predictions with experimental data. The model predicts these incipient hydrate formation data well, including the sI–sII transition points; thus, we have some confidence to predict the hydrate phase behavior at pressures above the incipient conditions.

At 277.6 K, these predictions show that the incipient hydrate structure concentration and that the upper transition point from sII to sI is predicted to occur at approximately 99.3 mole% methane. The upper transition point predictions guided the experimental work in preparing gas mixtures with high methane mole fractions, as later confirmed.

As shown in Fig. 1, in the composition range of 0% to 74% methane, sI hydrates are predicted to form at 277.6 K and pressures greater than 0.8 MPa. However, for initial com-

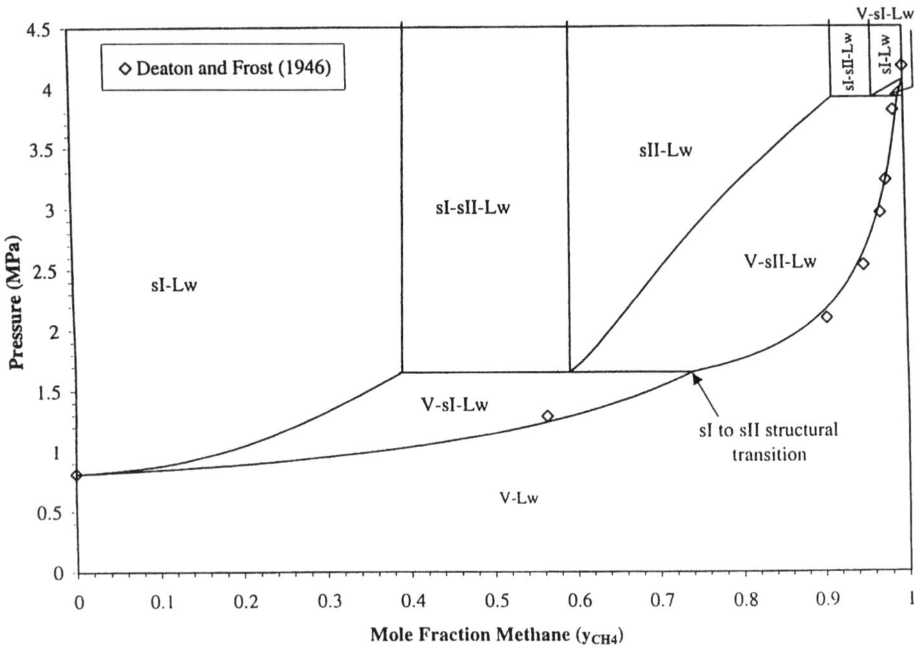

FIG. 1. The complete pressure versus composition (P–x) phase diagram for methane + ethane + water system at 277.6 K. Note that the incipient hydrate equilibrium data by Deaton and Frost [3] (*diamonds*), also plotted in the diagram, compare well with the model predictions. The lower and upper transition points were predicted to be at $y_{CH4} = 0.74$ and 0.993, respectively

positions of 40% to 60% CH_4, as the pressure is increased to about 1.7 MPa and above, sII is predicted to coexist with sI hydrates. A similar region is predicted in the composition range of 90 to 95 mole% methane and pressures above 4 MPa. In the composition ranges of 40 to 74 and 90 to 99.3 mole% methane, the hydrate structure at higher pressures is predicted to be different than the incipient hydrate structure.

Figure 2 shows a plot of water-free hydrate composition (z_{CH4}) against vapor composition (y_{CH4}) for eight different $CH_4 + C_2H_6$ hydrate samples used in the NMR study. It can be seen that there is a sudden increase in the hydrate composition, of about 50%, as vapor composition increases from $y_{CH4} = 0.18$ to 0.83. This change is caused by a change in the hydrate structure. This observation is similar to the one reported in our earlier paper [10] based on the Raman results at 274.2 K. The vapor compositions (y_{CH4}) at 0.74 and 0.99 were estimated using Raman spectra as described earlier, because ^{13}C-NMR peaks for the gas phase were too weak to derive any quantitative information. For similar reasons, the vapor composition at y_{CH4} of 0.18 is assumed to be the same as feed gas composition.

It should also be noted that both sI and sII coexist in samples with compositions close to the lower and upper transition points. Here we present two reasons as to why this should happen. According to the Gibbs' phase rule, in a three-component (CH_4, C_2H_6, and

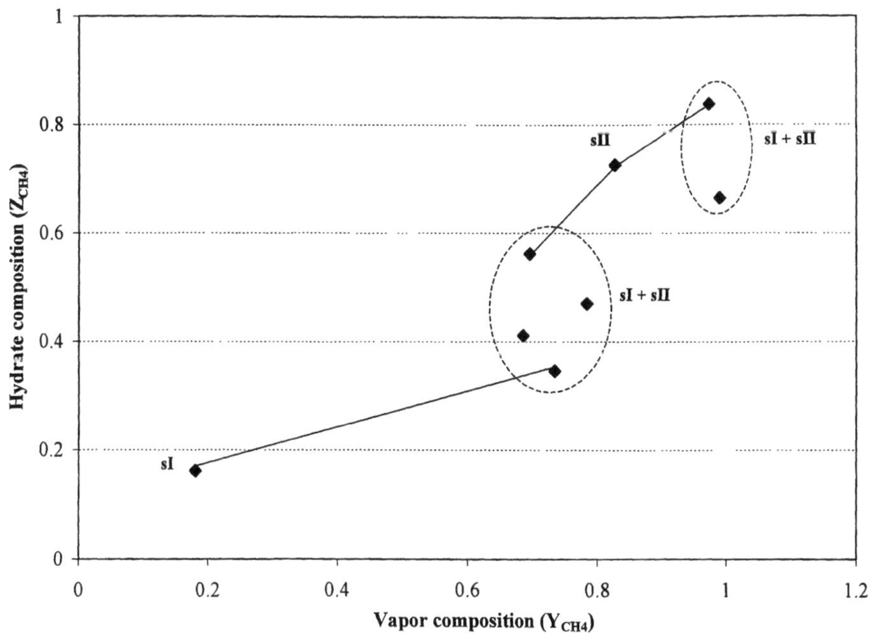

FIG. 2. Hydrate compositions (z_{CH4}) obtained from ^{13}C-NMR spectra for eight different samples are plotted versus equilibrium vapor compositions (yCH4). Note the sudden change in hydrate composition over the yCH4 range of 0.69 to 0.78 as the result of the change in hydrate structure from sI to sII

water) and four-phase (gas, sI, sII, and ice) system, there is only one degree of freedom, which means that these four phases can coexist at only one pressure at a given temperature. Because the sample pressures are not the same, it can be concluded that the samples with both structures are metastable in nature. On the other hand, if we assume that all ice is converted to hydrates, the additional degree of freedom can be used to vary the sample pressures. The hydrate samples are now stable with both sI and sII hydrates. Nevertheless, experimental evidence to support either of these arguments is awaited.

The entire incipient hydrate equilibrium literature data set for the methane + ethane + water system is plotted in Fig. 3 in a natural logarithm of pressure versus temperature diagram. The sII stability envelope, which is the locus of the structural transition points, is shown within two intersecting dark lines in the diagram (dashed region). It can be seen that more than 50% of the methane + ethane hydrate literature data are within the sII stability envelope, even though these mixtures were previously assumed to form sI hydrates at the incipient conditions.

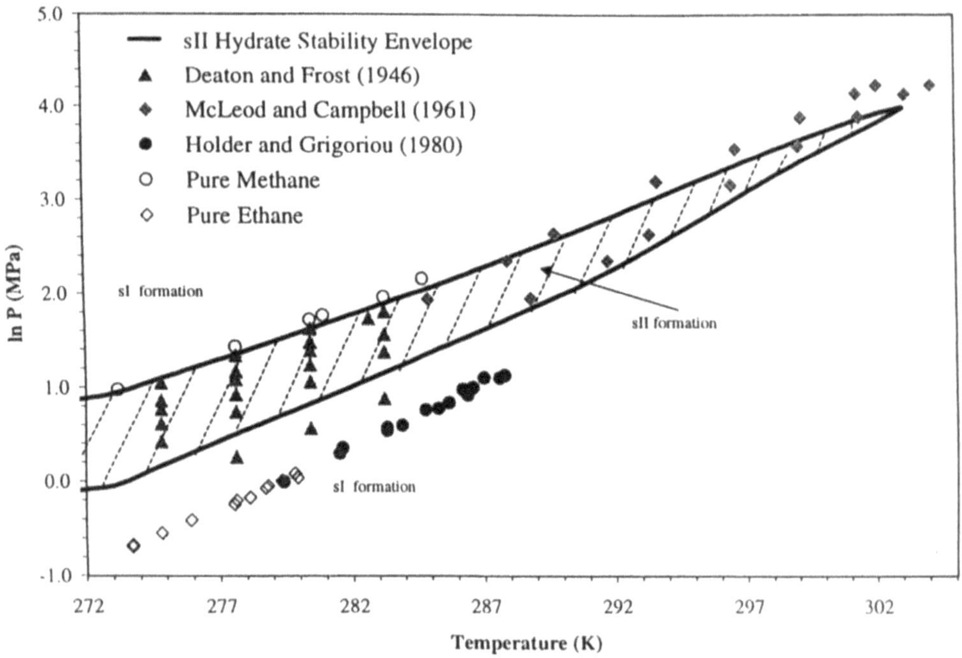

FIG. 3. The pressure versus temperature (P-T) diagram for methane + ethane + water system. Note that the hydrate structure changes from sI to sII for a small change in pressure; the *arrow* shows the structure II data envelope

Industrial Applications

Most natural gases in pipelines are predominantly methane, up to about 85%, and ethane, the next most prominent component [2]. Due to their high concentration of methane, a sI former, natural gases with CH_4 + C_2H_6 have been almost always assumed to form sI hydrates. However, this work suggests that if methane gas contains as little as 0.7% ethane, the system no longer forms sI hydrates, but rather forms stable sII hydrates. Although small amounts of propane normally cause sII hydrate formation [2], this work suggests that, even without propane or heavier components, sII hydrates are likely to predominate in pipeline applications.

In the oil industry, major challenges occur in the area of gas exploration and recovery, principally requiring mass and energy balance calculations. Because the hydrate composition is a strong function of hydrate structure, these calculations essentially rely on the determination of hydrate structure.

For example, in a multiphase system containing hydrates, oil, gas, and water, the amount of gas released on hydrate dissociation can be calculated only by knowing the hydrate structure present in the system. Similarly, the hydrate structure is crucial also in energy balance calculations. The amount of heat required for hydrate dissociation is directly related to the structure of the hydrate lattice [1]. Thus, an incorrect assumption of the hydrate structure might lead to serious errors during plug remediation in pipelines.

As discussed in the previous section, a small increase in pressure above incipient conditions can cause a change in the hydrate structure. This suggests that if the operating pressure in the pipelines exceeds the incipient hydrate equilibrium pressure by a small amount, as is often the case, the structure of the hydrate phase may not be the same as the one formed at the incipient conditions. Inaccurate prediction of hydrate stability regions might thus mislead pipeline operations. These predictions therefore indicate the necessity to predict the correct hydrate structure not only at the incipient hydrate conditions but also at conditions well into the hydrate region.

In the case of kinetic inhibition of hydrates, the additives that are developed to inhibit hydrate growth are designed for each hydrate structure. One of the common industrial kinetic inhibitors, polyvinylcaprolactam, has been shown to better inhibit sII than sI hydrates [16]. The critical amount of the additive required to prevent hydrate growth was shown to depend on the hydrate structure. Thus, the knowledge of the hydrate structure in the system is crucial in developing effective kinetic inhibitors.

The pressure versus methane composition (P–x) diagram in Fig. 4 shows the stable sI and sII equilibrium lines and the metastable sI and sII equilibrium lines for the methane + ethane + water system at 277.6 K. The metastable lines are predicted assuming one of

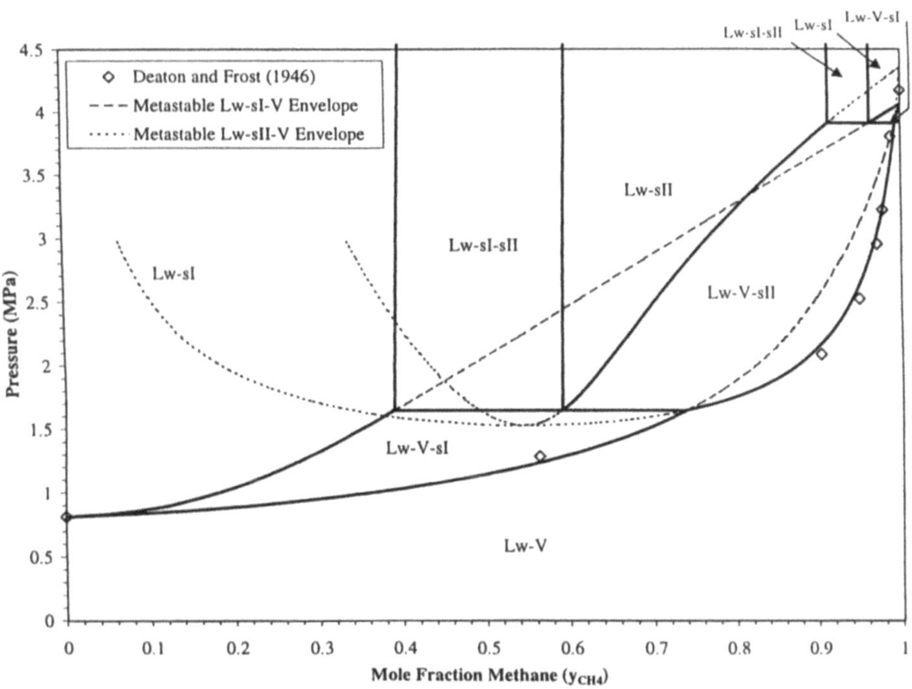

FIG. 4. P–x diagram for methane + ethane + water at 277.6 K, with metastable L_w-sI–V and L_w-sII–V envelopes. Note that at values above $y_{CH4} = 0.74$, the L_w-H–V equilibrium pressure for sII hydrate is less that that for metastable sI hydrate. *Solid lines*, stable sI and sII equilibrium lines; *dashed lines*, metastable sI and sII equilibrium lines

the unstable structures, sI or sII, to be present in the system in place of the stable structure. These predictions have some important implications to the stability of natural hydrates.

In nature, hydrates occur in both oceanic and permafrost environments. In the oceans, the hydrates are mostly methane up to 99% or more. For instance, the gas found close to the hydrates offshore Brazil was mostly methane and only 0.1% to 1% ethane and the higher paraffins [17]. In such cases, the hydrate structure is usually assumed to be sI. Our results, however, suggest that sII hydrates are stable at such gas compositions. From Fig. 4 it can be seen that sII hydrates can be stabilized at lower pressures than metastable sI hydrates, and therefore the hydrate stability region might actually extend beyond the earlier estimation based on sI hydrate formation. This result implies that the hydrate formation potential is in fact greater than expected.

Conclusions

Our previous work provided evidence for the occurrence of sII in methane + ethane hydrate system and measured the lower composition bound at which the gas mixtures of methane and ethane, two sI formers, form sII. Recent Raman results indicated that, at 274.2 K, the upper transition point in methane + ethane hydrates occurs at a y_{CH4} of 0.992–0.994. This result suggests that a trace amount of ethane can be sufficient to drive the hydrate structure from sI to sII. The upper transition point has some serious implications for hydrates in pipelines.

Recent ^{13}C-NMR measurements of water-free guest composition of hydrate phase at various vapor compositions suggested that the hydrate composition changes significantly, at least by about 50%, with the change in hydrate structure at the lower transition point.

In our modeling efforts, a complete phase diagram for methane + ethane + water system has been generated. The lower transition point obtained from Raman experiments was used to obtain a new set of Kihara parameters for methane and ethane. Predictions showed that the hydrate structure can change from sI to sII and vice versa for very small changes in the system pressure. The possibility of the coexistence of both sI and sII at pressures above the incipient conditions was also predicted. The modeling results indicate that the hydrate structure formed at the incipient conditions may be different than that formed at conditions away from the incipient conditions.

This study also suggests that Raman and NMR spectroscopy are very powerful, nondestructive tools to directly measure the microscopic properties of the hydrate phase and the vapor phase in equilibrium with the hydrate.

Acknowledgments. The National Science Foundation provided funding for the Raman and NMR part of this work through research grant CTS-9634899. A.B. is supported through a consortium of oil companies: ARCO, Chevron, Department of Energy, Mobil, Phillips, Shell, and Texaco.

References

1. Sloan ED (1998) Clathrate hydrates of natural gases, 2nd edn. Dekker, New York, P 705 and software

2. Katz DL, Cornell D, Vary JA, Kobayachi R, Elenbaas JR, Poettmann FH, Weinaug CF (1959) Handbook of natural gas engineering. McGraw-Hill, New York
3. Deaton WM, Frost EM Jr (1946) Gas Hydrates and Their Relation to the Operation of Natural Gas Pipe Lines. Monograph 8. U.S. Bureau of Mines, Bartlesville, UK
4. Holder GD, Grigoriou GC (1980) Hydrate Dissociation Pressure of Methane + Ethane + Water. Existence of a Locus of Minimum Pressures. Journal of Chemical Thermodynamics 12:1093–1104
5. McLeod HO, Campbell JM (1961) Natural Gas Hydrates at Pressures to 10,000 psia. Journal of Petroleum Technology 222:590–594
6. Englezos P, Kalogerakis N, Dholubhai PD, Bishnoi PR (1987) Kinetics of Formation of Methane and Ethane Gas Hydrates. Chemical Engineering Science 42(11):2647–2658
7. Englezos P, Kalogerakis N, Dholubhai PD, Bishnoi PR (1987) Kinetics of Gas Hydrate Formation from Mixtures of Methane and Ethane. Chemical Engineering Science 42(11):2659–2666
8. Skovborg P, Rasmussen P (1994) A Mass Transport Limited Model for the Growth of Methane and Ethane Gas Hydrates. Chemical Engineering Science 49(8):1131–1143
9. von Stackelberg M, Jahns W (1954) Feste gashydrate. VI: Die gitteraufweitungsarbeit. Zeitschrift fur Elektrochemie 58:162–164
10. Subramanian S, Kini RA, Dec SF, Sloan ED (2000) Evidence of Structure II Hydrate Formation from Mehane + Ethane Mixtures. Chemical Engineering Science 55(11):1981–1999
11. Holder GD, Hand JH (1982) Multiple-Phase Equilibria in Hydrates from Methane, Ethane, Propane, and Water Mixtures. American Institute of Chemical Engineers Journal 28:440–447
12. Gupta AK (1988) Steady state simulation of chemical processes. Ph.D. thesis, University of Calgary, Calgary, Canada
13. van der Waals JH, Platteeuw JC (1959) Clathrate Solutions. Advances in Chemical Physics 2:1–57
14. Ballard A, Sloan ED (2000) 3rd international hydrate conference proceedings, Salt Lake City, UT, July 19–24, 1999. Annals of the New York Academy of Sciences 912:702–712
15. Parrish WR, Prausnitz JM (1972) Dissociation Pressures of Gas Hydrates Formed by Gas Mixtures. Industrial Engineering Chemical Process Design and Development 11(1):26–35
16. Larsen R (1997) Clathrate hydrate single crystals: growth and inhibition. D. Ing. thesis, Norwegian University of Science and Technology, Trondheim, Norway
17. Fontana RL, Mussumeci A (1994) International conference on natural gas hydrates, June 20–24, 1993, New Paltz, NY, Annals of the New York Academy of Sciences 715: 106–113

Part IV: Multiphase Mechanics for Energy and Environmental Technology

Size Reduction of Solids: An Energy-Intensive and Highly Inefficient Process

Haim Kalman* and Evgeny Grant

Summary. The size reduction of solid particles is a major process used in industries—chemical, mining, pharmaceutical, metal, food, and coal power plants, among others. However, little success has been obtained in designing and optimal operation systems for this process without first performing full-scale experimentation. This lack is mainly due to the complex phenomena involved in operation, as well as inadequate characterization of particle strength (grindability). In this chapter, some of the available strength measurement procedures and standards for coal particles are examined. We show that the most common index for coal grindability (the Hardgrove Index, HGI) fails to predict the behavior of some of the coals (mainly the "younger" ones) in power plant grinders. Therefore, we introduce and examine some other strength measurement procedures such as compression of single particles (crush strength) and rotating drum tests.

Key words. Size reduction—Grinding—HGI—Coal—Crush strength—Rotating drum

Introduction

According to Rhodes, "The size reduction of solids is an energy intensive and highly inefficient process: 5% of all electricity generated in the world is used in size reduction. Based on the energy required for the creation of new surface, the industrial-scale process is generally less than 1% efficient." [1] The main reason for this depressing situation is that the phenomena involved in size reduction processes are very complex. Moreover, despite the many attempts to determine particle strength, which would enable the engineer to design a size reduction process, no reliable or accurate way is as yet available [2]. Without a reliable way of measuring particle strength, it is impossible to design a size reduction system without performing full-size experiments that will optimize its operation.

Size reduction of particles occurs whenever the applied loads are higher than the particle strength. Four types of loads could be applied to a particle (see Fig. 1):

1. Compression can be applied to a particle by pressing two plates. When a single particle is inserted between the plates, the compression force is applied directly. However,

Department of Mechanical Engineering, Ben-Gurion University of the Negev, P.O. Box 653, 84105 Beer Sheva, Israel
* Corresponding author

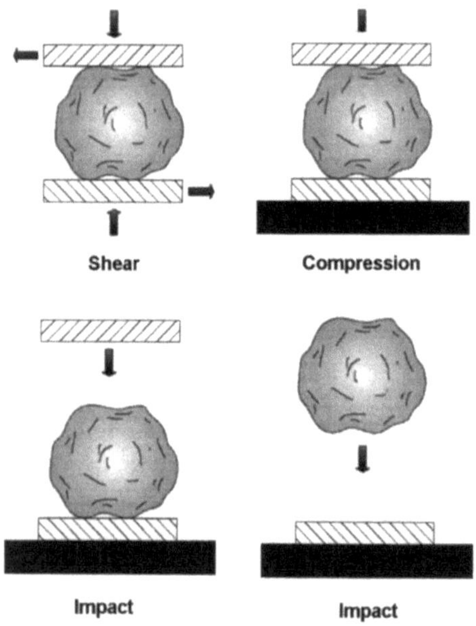

FIG. 1. Size reduction mechanisms

when a particulate bed is compressed, the compression force to any particle is transferred by its surrounding particles.
2. Sometimes shear loads can accompany the compression load to enhance the breakage of the particles.
3. Impact loads can be applied by a particle that is accelerated to a high velocity before collision with a rigid static target.
4. Impact loads can also be applied by moving rigid walls that are colliding with stationary particles.

In terms of practical systems, size reduction of particles is divided into comminution and attrition systems. Attrition is defined by the British Materials Handling Board [3] as "the unwanted breakdown of particles," while comminution is defined as "the intentional size reduction of particles by any means." Although these two terms are defined differently, the same size reduction mechanisms might be used for both. For example, particle size can be reduced significantly as a result of impacts to rigid walls and between the particles in pneumatic conveying [2,4–7], in a jet mill [8], or in a rotating drum [2,9]. In the case of a rotating drum, the particles are lifted at each revolution by the shelf and fall down due to gravity so that they collide with the drum wall (Fig. 2). In pneumatic conveying, particles collide against the pipe walls mainly at the bends (Fig. 2). In a jet mill, the particles are accelerated to high velocities by air streams and collide mainly with other particles but also with the mill walls (Fig. 2). Although these three systems provide impact loads, the range of the loads and the particle velocity is significantly different. In rotating drums, the particle velocity range is between 2 and 3 m/s, whereas in pneumatic conveying the velocity might be 20 to 30 m/s and in jet mills the velocity can reach more than 200 to 250 m/s. Whether the collision is with the wall or with other particles depends on the particles concentration in the three systems.

Any type of load applied to a particle can cause either abrasion or fragmentation. In abrasion, the material removed from the particle is very much smaller than the original

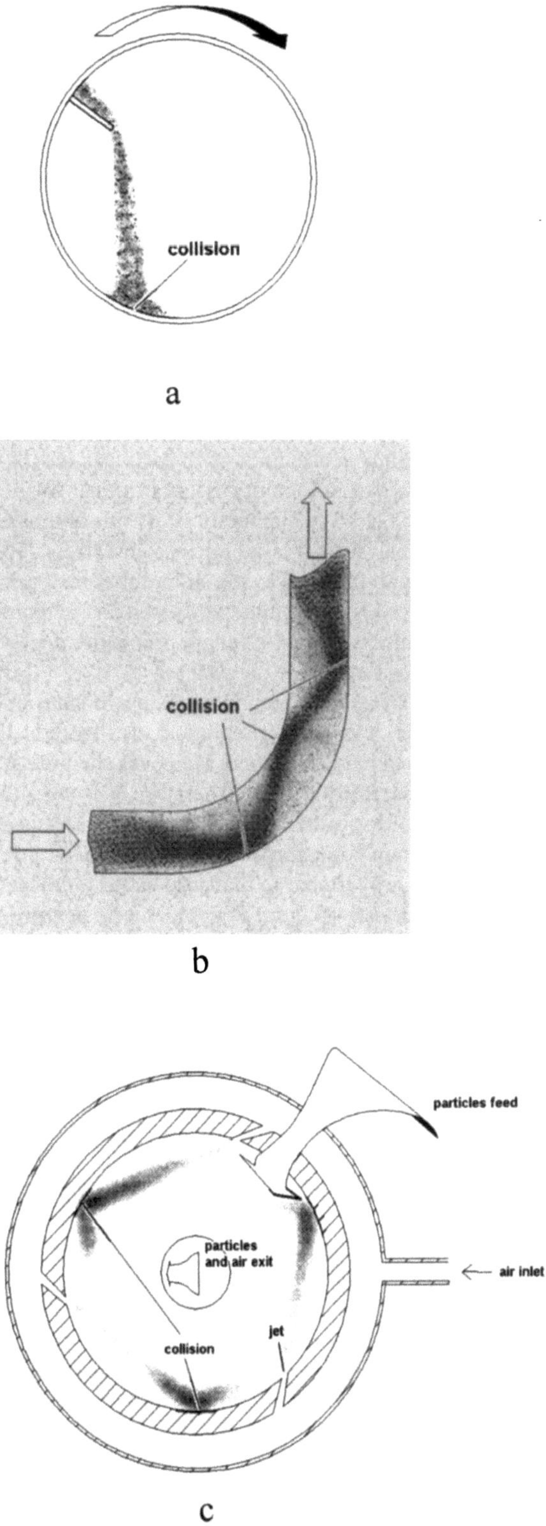

FIG. 2a–c. Impact of particles to walls in (**a**) a rotating drum; (**b**) pneumatic conveying; (**c**) a jet mill

particle. In fragmentation, the particle splits into two or more smaller parts. Because in comminution systems the aim is to reduce the particle size to a predefined level, a classifier must always be employed. The classifier, which can be a builtin system or a separate entity, should check the particle size and move it to the next process if the size is as required. If the particle is too large, it must be returned to the comminution system. For the latter case, the material experiences extra loads and energy that necessarily cause a significant reduction in the operation efficiency and the net mass flow rate.

Size reduction (either by comminution or attrition) is actually a failure of the particles to withstand applied loads (in this sense comminution and attrition are distinguished only by the user application and demonstrate the same physical phenomena), and in this sense can be compared to failure design procedures of metal parts. Most of the failure design procedures of metal parts are based on the requirement that the maximum stress at any point be less or equal to the permitted stress for the part to remain undamaged. The maximum stress side of the design equation is a result of a stress distribution analysis that takes into account stress concentration and other factors. This side of the equation refers to the system that the part serves and takes into account the strength properties of the material as well as a safety factor. The main advantage of this design procedure is that the strength properties depend only on the material and are not affected by the system that the part serves.

In the field of particle technology, the two sides of such a desirable design equation are ambiguous. It is not a simple task to theoretically analyze the maximum applied stresses within a particulate bed. Moreover, the pure strength of particles also has not as yet been satisfactorily defined. Therefore, it is a common practice to conduct standard experiments with similar size reduction mechanisms and to define various indexes that are no more than comparative. In any case, the size reduction design procedure should be defined in a way similar to that established for metal parts [2]. However, more research is needed before any attempt to apply this design procedure.

Performing simulation experiments with the system in question or a similar one is a common practice for size reduction of particles. Many different types of tests (for assessing the attrition tendency of particulate materials) have been described by the British Material Handling Board [3] and in studies such as those of Bemrose and Bridgwater [10] and Kalman and Goder [2], which include both single-particle and multiparticle tests. A variety of test equipment and procedures for each type of test have been reported [10], such as single-particle crushing tests, single-particle impact tests, fluidized-bed tests, shear tests, drum tests, grindability tests, vibration tests, drop shatter tests, confined compression tests, paddle wheel tests, enhanced sieving tests, viscometer tests, decapitation (thermal shock) tests, and fluid transport tests (by slurry or pneumatic). However, methods that ensure and quantify the occurrence of size reduction are likely to be inaccurate or restricted in the range of applicability. To comprehend the abundance of so-called standard size reduction tests, one can refer to a review of particle attrition published by the British Materials Handling Board [3] in which 16 drum tests (differing by geometry, size, and test procedure) for measuring friability are reported. The situation is not considerably different for other testing procedures, most of which fit into Clift's [11] category of "ritualized" tests: "Ritualized tests are easy to spot: the dimensions of the equipment and the procedure are set in stone, with no evident rationale." As a consequence, most known size reduction tests provide various indexes that are, at most, comparative in nature, but certainly cannot be relied upon to provide well-defined engineering design procedures.

In this chapter, various simulation and test procedures for particle strength are analyzed and compared. These procedures are applied to various coals taken from various mines and are used in a coal power plant. Some of the test results are compared to full-size measurements of power plant pulverizers. Although the results are not sufficiently complete that a reliable grindability index can be defined, they indicate a clear tendency and clarify various problems and questions concerning commonly used indexes.

Coal Grinders

Close grinding systems are commonly used for size reduction of coal in power plants [12]. The systems contain grinders, mills, and classifiers. The outlet coal from the size reduction system enters the classifier and there is divided into two streams. The fine particles are conveyed to the combustion system and the coarser particles return to the size reduction system. Hard particles might have to return to the size reduction system several times. The energy consumption of the grinder is in proportion to the rate of the coal passing through the grinder. The coal has to be milled into a specific size as required by the combustion process. Generally speaking, the specific surface area (surface area per weight) of any particulate material increases as the particle size decreases.

Size reduction systems usually used for coal grinding can be divided into four categories: ball tube, impact and attrition, ball race, and ring roll [12–14]. A schematic diagram of a ring roll mill is shown in Fig. 3. Some rolling elements, such as balls (for a

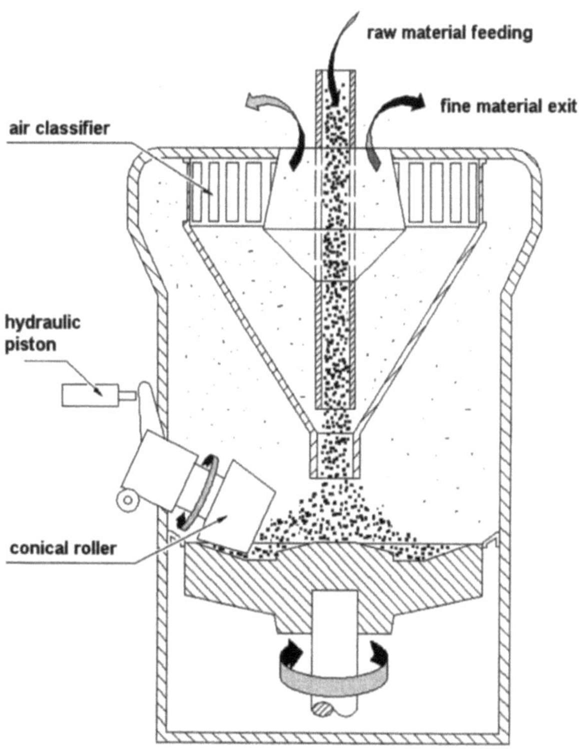

FIG. 3. Schematic diagram of a ring roll mill

ball race mill) or cylindrical or conical (for ring roll as shown in Fig. 3) are pressed to the lower ring (by the hydraulic piston in Fig. 3). The centrifugal forces acting on the coal that arrives from the feeder force the coal against the grinding roll, and the hot air stream passing through the mill carries the fine particles through the classifier. Coarser particles return to the grinding chamber at the bottom of the system.

Grindability Index and Grinder Performance

As previously described, an index should be defined by an experimental test rig that applies similar loads that occur at the full-scale apparatus, after which an index that describes the results is defined. The same procedure of experimentation should be repeated many times. The results are compared to real system results, and the index is then fixed for a specific case and for a narrow range of applications. One of the most widely used indexes (since 1951) for describing the grindability of coal is the Hardgrove Index (HGI).

Hardgrove Index for Coal

The procedure for HGI testing has been defined in detail [15]. First, 1 kg of particles smaller than 4.75 mm is sampled by a riffler from a gross coal sample of 80 kg. The sample is dried at room temperature for 12 h and then crushed by a laboratory crusher at 200 rpm to 0 to 0.6 mm. After this, a sample of 120 g of coal particles under 0.6 mm is separated. Only 50 g of the sample is inserted into the grindability measure machine (Fig. 4) operating at 20 rpm with a 29-kg load for 60 revolutions. Finally, the pulverized coal is sieved to measure the weight of the amount of coal of particles less than 75 μm. This weight determines the HGI.

The above-described HGI procedure should be able to simulate coal grinders that are used in coal power plants. The HGI machine operates principally in the same way as coal power plants grinders; only the rolling element, the size scale, and the loads are different. Full-size grinders should be able to grind different coals of various sizes and moisture contents whereas the HGI grinder tests only a fraction of the coal that was dried. This fact alone emphasizes a major difference between the HGI procedure and the

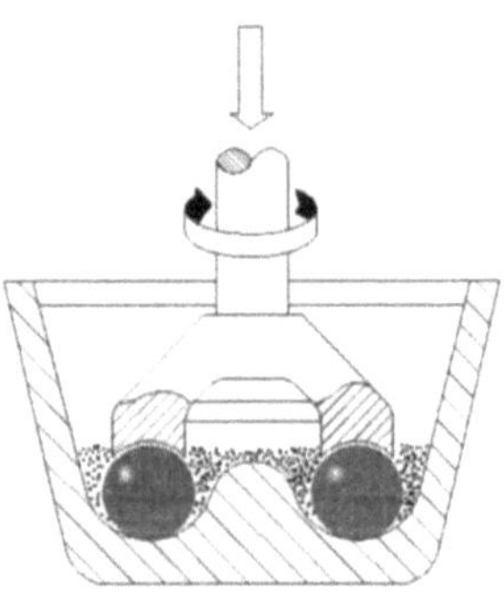

FIG. 4. Schematic diagram of the Hardgrove Index (HGI) machine

grinders that it should simulate. In addition, full-size grinders can operate with various compression loads and revolution speeds while the load and revolution speed is fixed by the HGI standard. Such handicaps and others as well have cast doubt on the ability of the HGI to simulate the process of the real grinding of coal in power stations [13,16–19].

Another major difference not mentioned previously in the literature relates to the classifying system. Usually, power station grinders have pneumatic classifiers. Hot air flows through the grinder and carries the small particles to the burner. Larger particles fall down and are milled again. Even if all the other parameters are the same in the HGI simulation, the HGI is determined by sieving. These two ways of defining the particle size are the same only for spherical particles. However, for nonspherical particles the sieve size relates to the smallest cross-section diameter of the particle and the swiping size relates to the flow diameter, which is the drag force equivalent diameter. Even for spherical particles, lower-density particles will be swept by the air stream more easily than particles with higher density at the same size. Therefore, particle shape and density are parameters that could significantly affect real grinding systems but are not represented in the HGI simulation.

Performance of Power Station Grinders

We studied the performance of full-size grinders operating at a coal power plant. Four grinders of the same type (as shown in Fig. 3) grinding four different types of coals were analyzed. The air flow rate versus the coal flow rate is shown in Fig. 5. The coal flow rate varied between 41 to 58 ton/h while the air flow rate, varied between 98 and 107 ton/h. The measured values present a clear linear dependency. The operating conditions of each grinder are equivalent to the maximum possible coal flow rate, which very much depends on the coal quality and grindability. If the coal grindability is low then the rate, of recycled particles for repeated grinding increases and the feed coal rate decreases. The air flow rate required for classification depends on the coal flow rate and coal properties. Figure 5 clearly indicates that the grinders are well controlled in terms of coal and air flow rates.

The pressure drop over the grinders linearly depends on the coal flow rate, but with higher scattering (Fig. 6). Obviously, the pressure drop depends on the air flow rate, which depends on the coal flow rate. The coal flow rate additionally affects the pressure drop by increasing the concentration of the coal in the grinder, which increases the resistance to the air flow.

A sample of the pulverized coal after grinding was taken for size distribution measurements. The size distribution was measured in two ways: by sieving (Fig. 7) and by fluidization (Fig. 8). The fluidization measurements were conducted to simulate the classifier of the power station grinders. The sample of the coal was inserted into a fluidized-bed column 32 mm in diameter and air was inserted from below. The fluidized-bed column was 1 m high, and the air flowed into a filter after passing through the fluidized bed to separate the particles from the air stream. Whenever the air velocity was increased, the fluidized bed was maintained so as to operate under constant conditions until all those particles that were sufficiently small were entrained into the air stream. Then the weight of the swept particles was measured, after which the air flow velocity was increased to entrain larger particles. Figures 7 and 8 also demonstrate the size distribution of coal particles after grinding in the HGI tester.

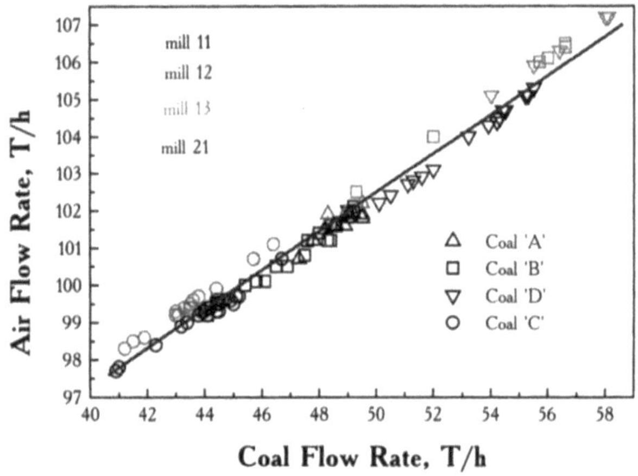

FIG. 5. Air versus coal mass flow rates

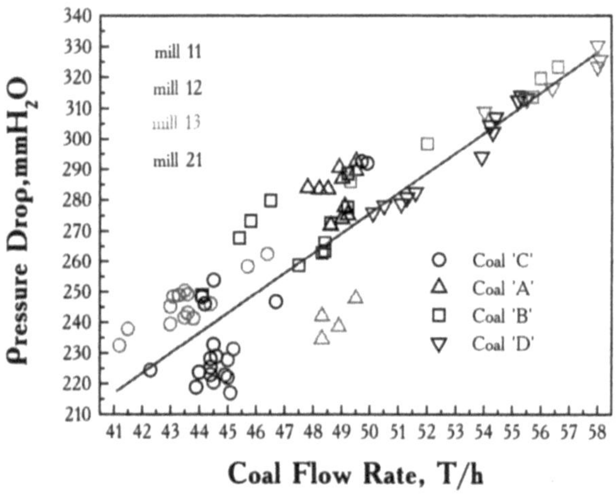

FIG. 6. Pressure drop over the mill versus the coal mass flow rate

In both Fig. 7 and Fig. 8, the size distribution of the samples after size reduction in full-size grinders is very narrow, which clearly indicates that the air classifier is performing well. The size distribution measured by sieving is slightly wider. It is difficult to make any conclusions based on these few measurements, however the wider distribution observed while sieving could be related to different shape and density of the particles. The size distribution of the HGI mill is much wider. Note that the HGI is defined by the weight of the particles under 75 μm. This turns out to be only a small part of the sample, reflecting the insensitivity of this procedure. Air velocity of 0.5 m/s is sufficient to sweep away most of the coal after pulverizing. The same velocity might yield different portions

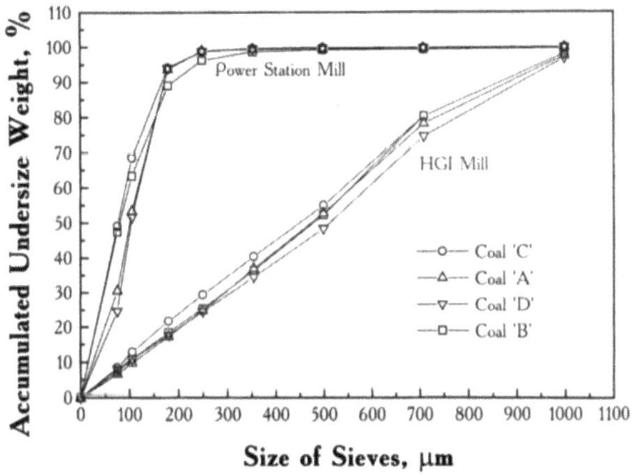

FIG. 7. Size distribution by sieving coal powder after grinding in the power station and after size reduction in the HGI machine

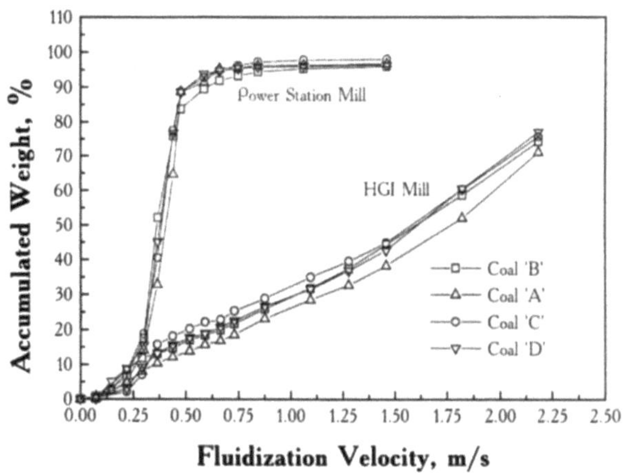

FIG. 8. Size distribution by pneumatic classification of coal powder after grinding in the power station and after size reduction in the HGI machine

of the coal after HGI milling of high sensitivity. However, this possibility has to be further explored by testing many more types of coals.

Measuring coal grindability should be aimed at conforming to the grinding operation conditions. The operation conditions for this study took into account the average energy consumption of the four mills investigated per coal mass flow rate. Figure 9 presents the energy consumption of our four mills versus HGI. Although the HGI of three of four coals reflecting the energy consumption very well, one of the coals is different. Although coals "A" and "B" have the same HGI, the last requires much higher energy for grinding. Therefore, it is clear that the HGI does not describe the real grindability of all types of coals.

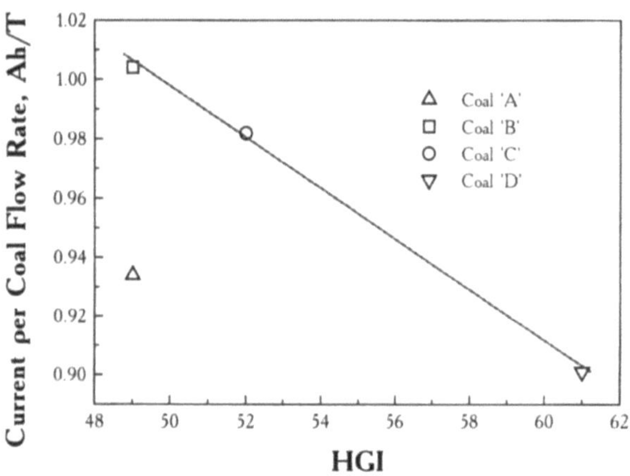

FIG. 9. Energy consumption of power plant grinders versus HGI

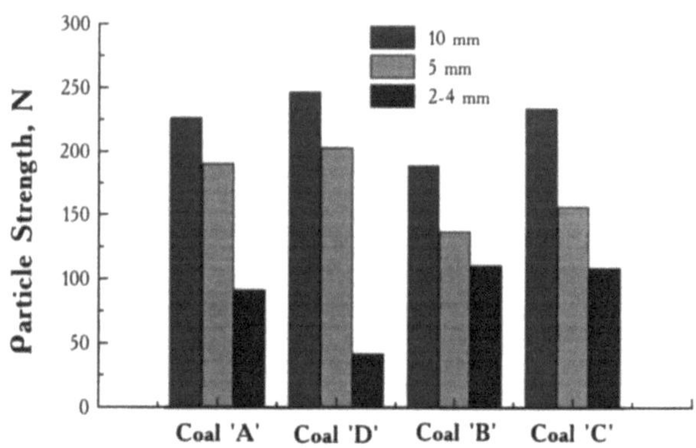

FIG. 10. Crush force of three size fractions of coals: "A", "D", "B", and "C"

Other Strength Measurements

Crush Strength

Figure 10 presents results on the crushing strength of three size fractions for each of the four coals. The tests were performed in a crush strength analyzer apparatus in which a particle is compressed between two platens. Each value in the figure represents an average of more than 20 measurements. Because of the orientation of the particles between the platens, a significant difference was found among the 20 measurements. It is clear from Fig. 10 that larger particles are able to withstand larger forces before crushing. However, this does not mean that their strength is higher, as their contact area also has to be larger. Therefore, the stress might even be lower. In fact, the larger the particles, the lower their strength. In any case, it is interesting to note that the relative crushing force is different for various particle sizes. For example, the crush force of 10-mm particles of coal "D" is higher than the crush force of the same size particles of coal "B",

FIG. 11. Energy requirement by power plant grinders versus crush force of 2 to 4-mm particles

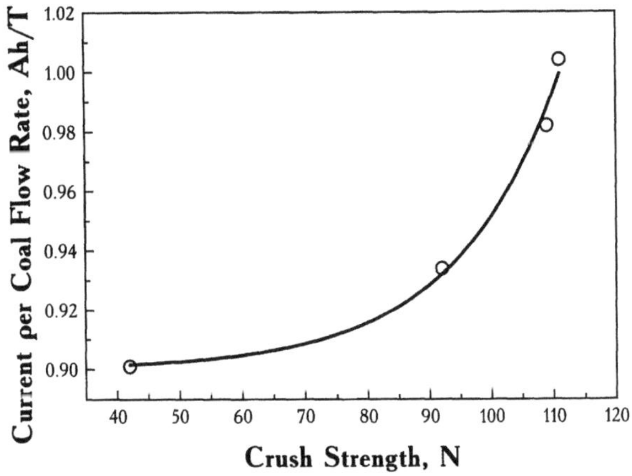

and vice versa for particle size of 2 to 4 mm. This discrepancy can only be accounted for by the different concentrations of the various minerals in different particle sizes. This result again emphasizes the limits of the HGI simulation in which only a fraction of the particles is tested.

A comparison of the crush forces of the four coals to the energy consumption of the power plant grinders showed a good correlation (see Fig. 11). As the crush force increased the energy required for grinding the particles to a defined size also increased. Although a good agreement is found for this very simple strength measurement, the practical implications should be further strengthened by testing other coal types.

Drum Tests

The British Materials Handling Board [3] has collected 16 different standards of drum tests to measure particle friability. They differ by the machine's dimensions, number and shape of shelves, number of revolutions, speed of rotation, material tested, and sample weight. The devices and procedures are defined very well to derive comparative indexes. Due to gravity the particles fall and collide with the drum wall at each rotation. It is assumed that the presence of the shelves leads to attrition, which occurs as a result of the impact, and that the attrition from shear forces is negligible. The diameters of the drums are too small to cause considerable breakage to most of the materials in one collision. Nevertheless, the number of revolutions (impacts) is sufficient to cause damage due to fatigue. Therefore, if the rotation speed is low enough (to avoid adherence of particles to the wall due to centrifugal forces) and the sample is small enough in relation to the length of the drum (to neglect collisions between particles), one can use any drum diameter, drum material (rigidity), rotation speed, and testing duration. All affect the impact load and the number of cycles, which are the basic parameters for defining fatigue strength [2].

In the present case, we used a steel drum 260 mm in diameter and 320 mm in length (see Fig. 2). The tested sample of the material was inserted into the drum and then rotated

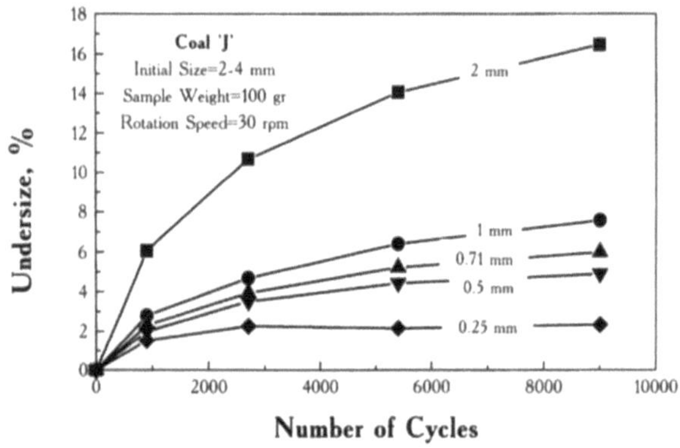

FIG. 12. Size distribution of coal "J" versus the number of cycles

at a predefined speed for a certain time. At the end of this period of time, the size distribution of the sample was measured and the sample was inserted back into the drum for another period of rotation. Because the drum had one shelf, it is assumed that at each rotation the particles experienced one impact load. It is also assumed that the particles collided with the drum walls but not with other particles, as the sample size was very small.

Figure 12 shows a typical size distribution in a rotating drum test. The sample weight of coal "J" was 100 g. It rotated in the drum at 30 rpm for up to 5 h. The test time is presented by the number of cycles until 9000 is reached. Because at some time intervals the size distribution was measured by sieve tests, the accumulated weight under each sieve size versus the number of cycles is presented. It is clear from the figure that the weight of particles under the bottom initial size (2 mm) is increasing with the number of cycles. However, the increase is sharper at the beginning and becomes moderate until about 17% of the coal was size reduced.

A comparison between the size reduction of four coals at a drum speed of 60 rpm and eight coals at drum speed of 30 rpm is shown in Figs. 13 and 14, respectively. For some coals the size reduction process becomes stabilized only after a long period of time, which might be related to two phenomena. First, the change of the size distribution changes the material and its behavior to yield lower impacts. Second, the bulk is not homogeneous in terms of particle strength. Therefore, the weaker particles break first and the stronger particles are only degradated slightly if at all. This was confirmed by measuring the crush force of the coal particles that remained at the initial size after the drum test. The crush force was significantly higher after the test than before it. A comparison between the results of coals "B" and "C" tested at 30 and 60 rpm reveals one of the system's effects. Coals "B" and "C" reach about 10.5% and 6.5% of damage at 60 rpm and 17% and 13% of damage at 30 rpm, respectively. Furthermore, damage intensity is almost twofold if the rotation speed of the drum is reduced, which is most likely related to the centrifugal forces that change the particle free fall path, resulting in their collision velocity.

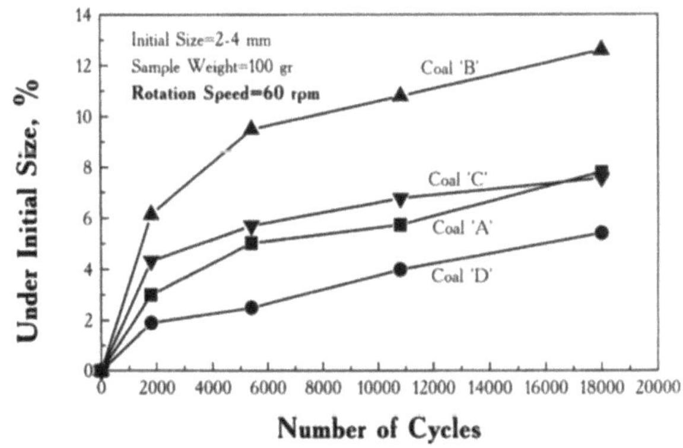

FIG. 13. Size reduction rate of four coals at 60 rpm

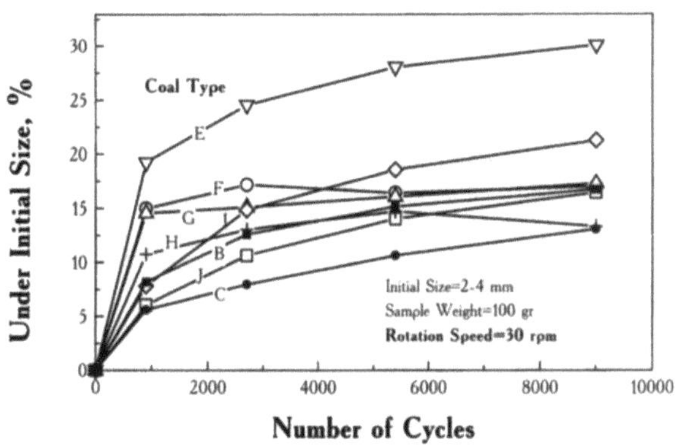

FIG. 14. Size reduction rate of eight coals at 30 rpm

Conclusions

This chapter presents a preliminary study and a new evaluation of the ability of the common grindability index (HGI) to predict the grinding process of coals in full-scale power plant grinders. We demonstrated that the HGI fails to predict the behavior of, at least, the "younger" coals. Therefore, we introduced several different strength measurement procedures. Totals of ten coals were tested in rotating drums and the particle crush force was measured. Although crush strength measurements showed good agreement with the performance of four coals in the full-scale grinders, the results do not provide a well-defined solution. However, our results point to a direction of further study that could help to establish a new and reliable grindability index. Future work should consider the mineralogical content of differently sized coal particles, and any new standard should test all the size spectrum of the coals. The test procedure should also take into account the classifier type of the full-scale grinders to be designed or optimized.

References

1. Rhodes M (1998) Introduction to particle technology. Wiley, New York
2. Kalman H, Goder D (1998) Design criteria for particle attrition. Advanced Powder Technology 9:153–167

3. British Materials Handling Board (1987) Particle attrition: state-of-the-art review. Trans Tech Publications, London
4. Kalman H, Goder D (1996) Pressure drop, wear and attrition in various bends for pneumatic conveying pipe-lines. Proceedings of the Fifth World Congress of Chemical Engineering VI:411–416
5. Aked C, Goder D, Kalman H, Zvieli A (1997) Attrition of very fine powders during pneumatic conveying. Powder Handling and Processing (The International Journal of Storing, Handling & Processing Powder) 9:345–348
6. Kalman H (1999) Attrition control by pneumatic conveying. Powder Technology 104:214–220
7. Kalman H (2000) Attrition of powders and granules at various bends during pneumatic conveying. Powder Technology (in press)
8. Eskin D, Voropayev S, Kalman H (2000) Research of vortex milling processes. In: Proceedings of the third Israeli conference for conveying and handling of particulate solids joined with the tenth international freight pipeline society symposium Deadsea, Israel, 6:12–17
9. Kalman H, Goder D, Targan S (1998) Preliminary investigation on the effect of production parameters on the strength of big tablets. Particle & Particle Systems Characterization 15:150–155
10. Bemrose CR, Bridgwater J (1987) A review of attrition and attrition test methods. Powder Technology 49:97–126
11. Clift R (1994) Powder technology and particle science. Proceedings of the 1st international particle technology forum. II, AICHE, Denver, USA
12. Unsworth JF, Barratt DJ, Roberts PT (1991) Coal quality and combustion performance. Elsevier, Amsterdam
13. Leonard JW, Humphreys KK, Lawrence WF, McMillan BG, Muter RB (1979) Coal preparation. The American Institute of Mining, Metallurgical and Petroleum Engineers, New York
14. Singer JG (1981) Combustion fossil power systems. Combustion Engineering, Pulverizers and Pulverized Coal System 12.1–12.35
15. Standard D 409–93a (1994) Standard test method for grindability of coal by the Hardgrove-machine method. ASTM standards 05.05. American Society for Testing and Materials, Philadelphia
16. Oder RR, Gray RJ (1992) The effect of coal characteristics on fine grinding in a pit mill. In: Kawatara SK (ed) Comminution—theory and practice, pp 661–675
17. Bailey JG, O'Brien G, Esterle J, Sexty G, Beath H, Chalmers G (1996) Insights into the grindability of coal microlithotypes. In: Boyd RL, MacKenzie GA (eds) Proceedings of the 30th Newcastle symposium on advances in the study of the Sydney Basin, pp 135–142
18. Bailey JG, Esterle J, O'brien G, Beath H, Chalmers G (1997) Reflections on the HGI in the light of microtextural analysis of milled coal. In: Waring P (ed) Proceedings of the 7th Australian Coal Science Conference, Australian Institute of Energy, Gippsland Group, Morwell, Vic., Australia, pp 277–284
19. Barton WA, Condie DJ, Lynch LJ (1994) Coal grindability: relationships with coal composition and structure. In: Proceedings of the 7th Australian Coal Science Conference, Australian Institute of Energy, Wahroonga, NSW, Australia, pp 55–64

Experiments on Turbulent Particle-Laden Flows

ELLEN K. LONGMIRE*, DANIEL A. KHALITOV, and STAMATIOS POTHOS

Summary. Turbulent gas flows containing dilute loadings of solid particles were investigated. These flows occur in many energy and environmental applications including coal combustion, sorbent injection, and material and contaminant transport. A method is presented for simultaneous determination of planar gas and particle velocity fields. The method is applied in streamwise-spanwise planes of a fully developed turbulent channel flow with Reynolds number of 9000 to determine single-point particle and fluid velocity statistics. Such statistics are employed to test numerical models for particle-laden flows. Additional experiments on flow control are presented in which a piezoelectric actuator is applied to generate normal wall motion at the exit of a turbulent channel. Single-point velocity measurements document the time-averaged flow behavior for several control conditions and demonstrate possibilities for particle control that is either coupled or decoupled with the turbulence. The control method can be applied to alter particle distribution and velocity near and downstream of solid surfaces.

Key words. PIV—Particle-laden flow—Piezoelectric actuator—Turbulent jet—Turbulent channel flow

Introduction

In this chapter, we consider turbulent gas flows containing solid particles. Our efforts focus on dilute flows where the particles are small enough to feel motion in the gas continuum, yet large enough to cross fluid streamlines. Particle-laden flows in this regime are important in many energy-related applications including coal and droplet combustion and transport of materials or contaminants. Some current issues concern how particle distribution and velocity affect transport efficiency, pollutant formation, erosion, agglomeration, and surface deposition. Understanding these physical issues is important for several reasons. First, improved numerical models can be developed for engineering use. Such models are needed to predict device behavior before construction. Second, improved understanding of the flow physics will lead to better methods for controlling flow behavior either in steady-state operating conditions or instantaneously.

Department of Aerospace Engineering and Mechanics, University of Minnesota, 107 Akerman Hall, 110 Union St. SE. Minneapolis, MN 55455, USA
* Corresponding author

In many cases, it is desirable to achieve some optimal distribution of particles in a flow stream to maximize process efficiency. For example, in a sorbent injection stream, the distribution of the particles is very important in determining the local reaction rate and the overall reaction efficiency. To optimize the local conditions, it might be desirable to focus or defocus the injected stream in real time. In ducts or pipes where particles are transported, one might wish to destabilize or break up concentrations of particles. For example, it has been documented that particles traveling around bends become concentrated into "ropes" that are relatively stable downstream of the bend [1]. Trailing particles are attracted to low-pressure or reduced drag regions in the wakes of downstream particles. When the cross section of a "rope" contains many particles, it is difficult to destabilize and may require an external perturbation.

In addition to optimizing particle distribution across a flow, it is often desirable to avoid or minimize particle accumulation on solid surfaces. In particular, particle buildup can cause deterioration of performance on chamber or duct walls. On heat transfer surfaces, stagnant particle layers act as insulators, decreasing the heat transfer rate. At sensor or measurement system locations, particle buildup may obscure optical access or decrease measurement accuracy. In both cases, it would be advantageous to clear particles from the surface (or to prevent them from reaching the surface in the first place).

In the work discussed here, we investigate and attempt to control particle-laden flows that contain a wide range of eddy scales. In the control studies, our goal is to modify the flow near walls, a region where strong gradients are present and where, in applications, heat transfer and particle accumulation may be important. Previous experiments and numerical simulations in turbulent channel flows have shown that particles often accumulate in the near-wall region, sometimes causing particle–particle interactions in otherwise dilute flows [1,2]. Thus, control strategies aimed at altering gas or particle flow in this region could yield significant effects.

Experiments on particle-laden channel flows are described. In one case, we apply particle image velocimetry and particle tracking to obtain statistical correlations important for testing and closure of numerical models. In another, we employ a moving wall section attached to a piezoelectric actuator to perturb the fluid and particles exiting the channel. The actuator frequency and amplitude can be tuned independently to vary the resulting behavior.

Experimental Facilities and Methods

Particle-laden velocity fields are measured in streamwise-spanwise planes of the vertical channel shown in Fig. 1. The Plexiglas channel has a constant cross section of 153 × 14.7 mm. The flow, driven by a frequency-controlled centrifugal blower, passes through a series of grids and honeycombs to achieve a uniform mean velocity profile. The flow is subsequently tripped by two plates mounted on the side walls. A flow control system was implemented to limit any low-frequency fluctuations in the bulk flow rate to less than 1%. A small amount of air seeded with smoke (less than 1% of the total mass flow rate) is added to the flow through a thin slit. The smoke, which consists of glycerin droplets with diameter approximately 1 μm, is used to provide tracers for particle image velocimetry (PIV) measurements of gas velocity. Because of their small size, the tracer droplets are expected to track the gas flow faithfully. Particles (solid glass beads) are fed into the flow upstream of the channel entrance by a pressurized screw-feeder. The grids

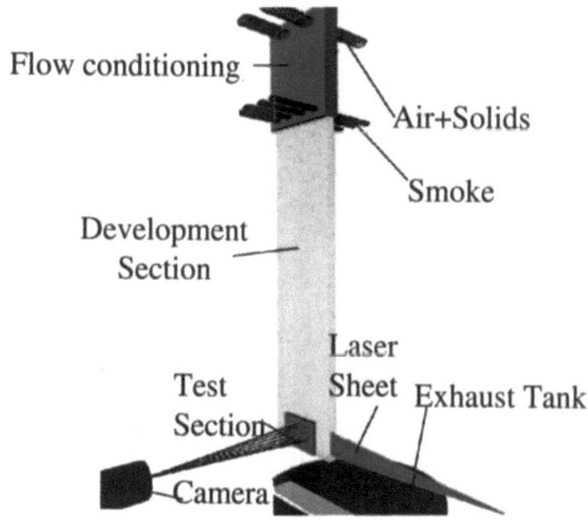

FIG. 1. Turbulent channel flow facility and particle image velocimetry (PIV) setup

in the flow conditioning section serve also to produce an even distribution of particles over the channel area.

The length of the development region of the flow facility is 50 h, where h represents the channel width of 14.7 mm. This section is made long to achieve "fully developed" turbulent conditions upstream of the test section. The fully developed condition is used frequently in numerical flow simulations. This condition was tested by measuring velocity in single-phase flow with hot wire and laser Doppler anemometers. Hot wire measurements in the center plane at the downstream end of the test section yielded a turbulence level of 4%. The mean and RMS velocities were uniform to within 1% over 60% of the channel span. Independent LDA measurements over the streamwise range of the test section yielded mean velocity variations in the midplane of less than 1%. These qualifications show that the mean air flow in the PIV measurement region is locally homogeneous, fully developed, and steady. Hence, the flow is suitable for the computation of one- and two-point turbulence statistics based on a large number of instantaneous velocity fields. The facility and single-phase velocity statistics are described in more detail in Anderson et al. [3]. Additional qualification measurements were performed on particle-laden flow using image processing and particle counting routines on large sets of images of flow planes in the test section. The distributions of particles and particle velocities were found to obey symmetry and the fully developed condition.

To generate pairs of images for planar field velocity (PIV) measurements, two double-pulsed Nd:YAG lasers were used to illuminate the flow. The laser sheet thickness in the measurement region was approximately 0.2 mm, and the laser energy output was 50 mJ per pulse. The time delay was 24 μs with approximately 200-ns jitter. Pairs of images were captured with a dual-channel digital camera (Kodak Megaplus ES 1; San Diego, CA, USA) that was synchronized with the laser pulses. The camera resolution was 1024 $\times$ 1024 pixels, and the magnification was 9.67 μm per pixel, resulting in a physical frame size of about 10 mm on each side.

A second facility, that relies on the same blower and particle feeder as described, is used for control studies (see Fig. 2). This facility has a development length of 70 h, where

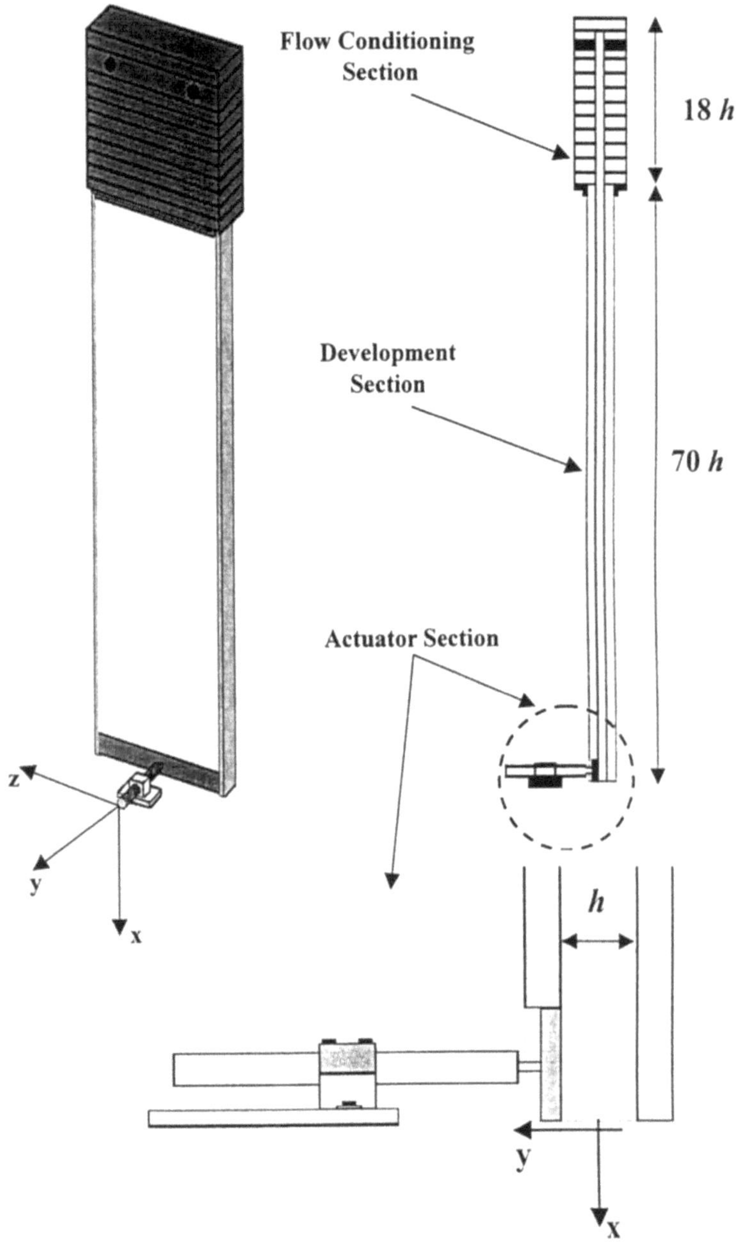

FIG. 2. Turbulent channel facility with moving wall section

h represents the channel width of 15.75 mm. The cross-sectional area is constant with an aspect ratio of 11.45. The development section consists of two aluminum side walls and two glass end walls. The glass walls are pressed against the aluminum ones with spacers. The lengths of the aluminum walls differ by 40 mm, which corresponds to the height of the moving wall section.

The moving wall consists of a rectangular aluminum plate (mass = 96 g) attached to a piezoelectric actuator. A detail of the actuator section can be seen in Fig. 2. In the present configuration, the activated wall is located at the downstream end of the channel, oriented to generate wall-normal motions. The actuator is a PZT "stack" (model P-245.77; Polytec-PI, Costa Mesa, CA, USA) with a maximum displacement of 120 μm. A linear amplifier (model P-270.10; Polytec-PI) is used to drive the piezoelectric translator. The maximum output voltage is 1000 V, and the maximum output current is 500 mA. The maximum current required for sinusoidal operation of the translator can be calculated from the equation $i_P = \pi \cdot f \cdot C \cdot V_p$, where f is the frequency, V_p is the peak voltage of the input signal, and C is the capacitance of the translator. For the P-245.77 translator used, $C = 0.5\,\mu$F. Because the required input current is proportional to the product of frequency and peak input voltage, it scales directly with the peak velocity of the translator. To avoid uncertainties in translator location, it is necessary to operate the translator using a closed-loop feedback circuit. This circuit limits the maximum velocities achievable with the translator as described below. A laser vibrometer system was employed to monitor both the velocity and displacement of the moving plate under various control conditions.

Results

Two-Phase Velocity Measurements

A sample section of an original picture, taken with the Kodak camera (Fig. 3), contains images of both smoke tracers (smaller and darker) and solid glass beads (larger and brighter). It also contains relatively large and dim images of out-of-focus glass beads located near the edges of the laser sheet. To obtain independent velocity measurements for the gas and the solid particles, each image must first be separated into two files, one containing only tracers and one containing only solids. The separation procedure requires several software routines. First, both solid particles and tracers are identified as "objects" and separated from the image background by using local peak detection and region growing algorithms [4]. The result of the first separation stage (Fig. 4) shows that most of the glass particles and tracers are identified correctly, while the out-of-focus glass particles are broken up.

In the next step, the objects are sorted according to their area in pixels ("size") and average pixel intensity ("brightness"). A contour plot of size and brightness distribution is shown in Fig. 5. Plots like this one are used to choose separation parameters for solids and tracer particles. Based on these parameters, all relevant objects are rewritten into solid-only and tracer-only files (Figs. 6, 7). All eight of the bright glass particles are rewritten into the glass file. The smoke particles are also rewritten faithfully to the smoke file. In addition, some of the "fragments" from the out-of-focus particles are rewritten to the smoke file. We assume that these fragments do not cause any bias in determination of gas velocity vectors as is explained below.

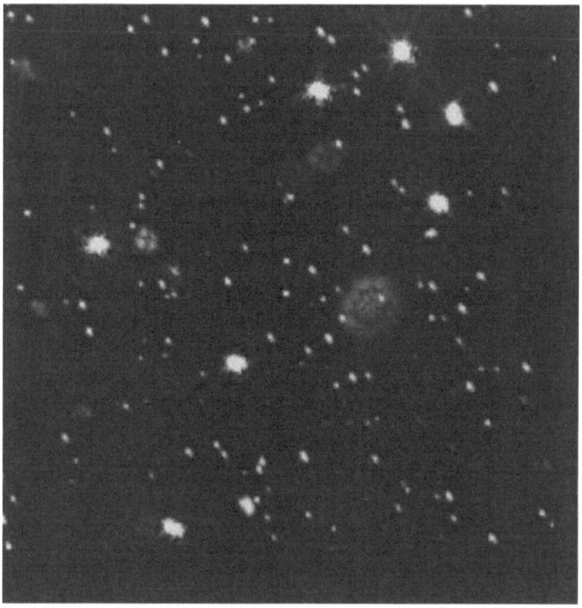

FIG. 3. Section of unprocessed image of particle-laden flow including smoke tracers

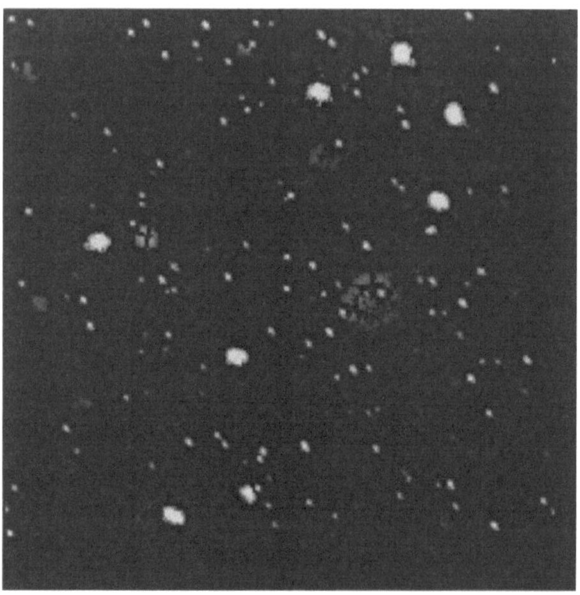

FIG. 4. Section of processed image showing identified objects

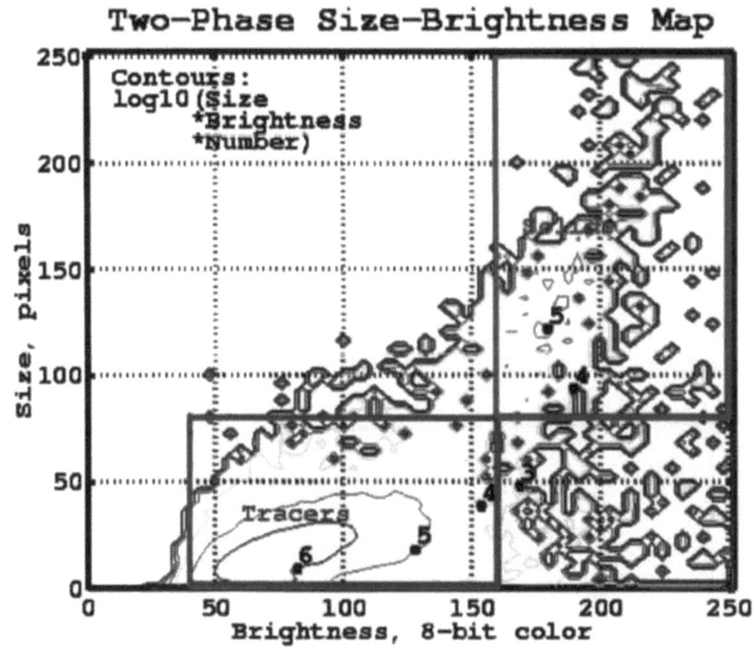

FIG. 5. Contour plot of object probability distribution

All velocity vectors are found by comparing pairs of images taken with a very small time delay. Solid velocity vectors are computed by a custom particle tracking routine that uses 8-bit gray scales and a center of gravity fit to locate particle centers. Particle "pairs" are identified based on a range of possible displacements between the first and second images, and each pair yields a velocity vector based on the measured center displacement divided by the time delay between images. Particle vectors thus are located at each particle position. Gas velocities are computed by a commercial cross-correlation routine (Visiflow; Oxford Lasers, Littleton, MA, USA). The gas velocities are computed on a rectangular grid where the size of each interrogation region is 128 × 128 pixels. Neighboring interrogation regions are overlapped by 75% on the grid. Each region typically contains about 20 smoke tracers, and the cross-correlation yields the dominant displacement in the given region.

A sample plot of gas velocity vectors is shown in Fig. 8. In the plot, the global mean velocity (10 m/s) has been subtracted to accentuate the spatial variations caused by eddies in the flow. In fields like this one, 98% of all interrogation areas yielded valid vectors. To compute joint gas particle quantities, such as slip velocities and mixed single-point correlations, the fluid velocity at each particle location was found using a bilinear interpolation scheme. Measurements of slip velocity are important to understand the amount of momentum being exchanged between the particles and the fluid. A typical slip velocity field is shown in Fig. 9.

The accuracy of the object identification routine and separation limits was tested in the following way. Streamwise and spanwise velocity statistics were computed in flow laden with smoke only, glass particles only, and smoke and particles, and the resulting

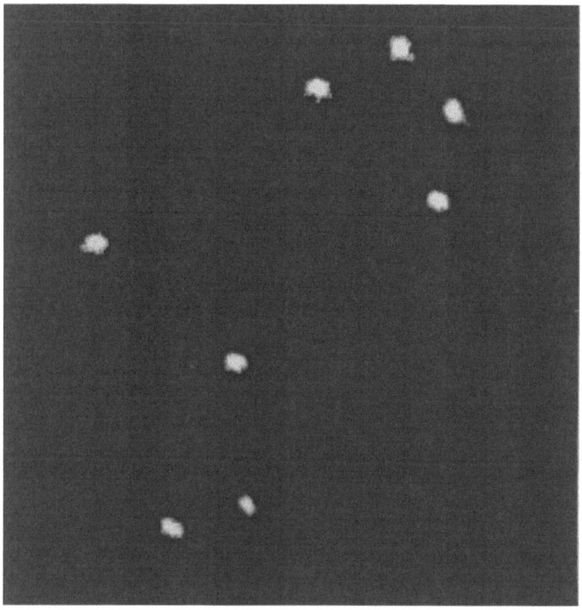

FIG. 6. Section of glass particle file found from processed image

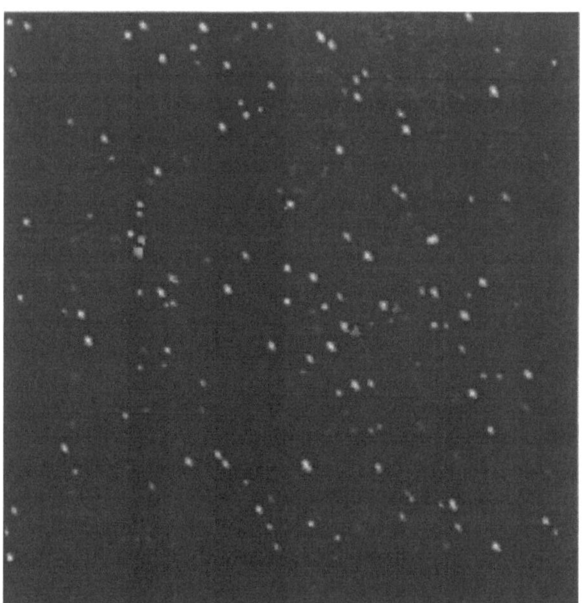

FIG. 7. Section of smoke tracer file found from processed image

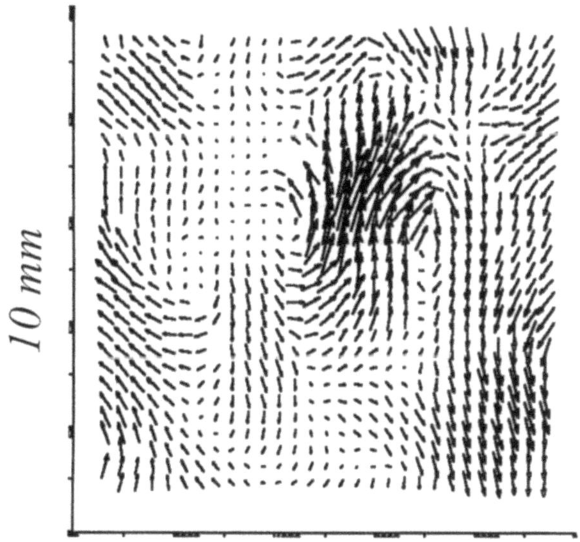

FIG. 8. Instantaneous gas velocity vectors at channel center plane with mean velocity subtracted. Flow moves from *top* to *bottom*

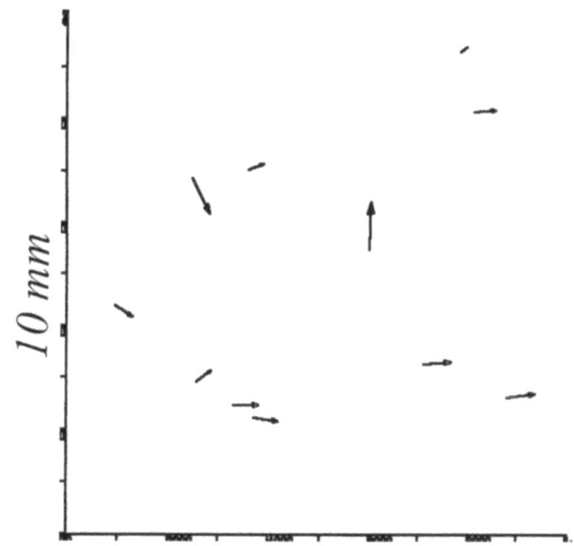

FIG. 9. Glass particle slip velocity vectors at channel center plane. Flow moves from *top* to *bottom*

velocity distributions were plotted (see Fig. 10). From the plots, it is clear that the smoke-only distribution matches the distribution found from images containing both particles and smoke. Also, the glass-only distribution matches the distribution found from images containing glass particles and smoke. Based on these results, we are confident that the separation software does not introduce systematic biases into ensemble-averaged velocity statistics.

Two preliminary sets of data were chosen as test cases for the separation and interrogation processing routines. In both cases, the measurements were taken in the

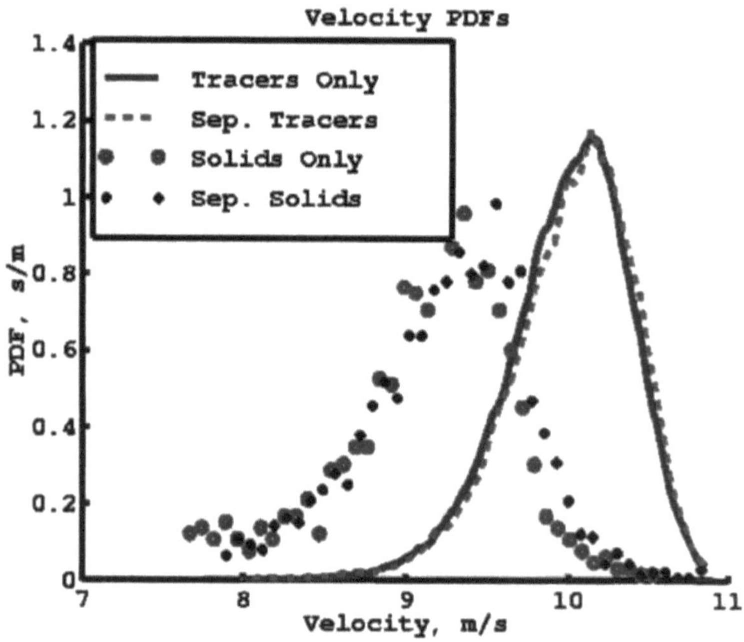

FIG. 10. Particle and gas velocity distribution functions found by processing sets of images with solids (glass) only, tracers only, separated solids (glass), and separated tracers

streamwise-spanwise plane at the midplane of the channel. The Reynolds number based on the channel width and the mean flow velocity of 10 m/s was 9000. Two sets of solid glass beads were employed. In both cases, the beads were sized to a narrow distribution (~15 μm) about the mean diameter. The mean diameters were 30 μm and 60 μm for the two cases, and the mass loading ratios of particles to air were 1.8% and 4.8%, respectively. For these experimental conditions, the particle motion does not have a significant effect on gas turbulence, as is discussed further. The Stokes numbers relating partial inertial time constants to the integral turbulent time scale were estimated as 0.35 and 1.4 for the two particle sets.

A total of 320 pairs of images were acquired of each flow case to compute single-point statistics. The average number of solid particle velocity vectors computed per image was 7 and 3.5 for the first and second case, respectively. The uncertainty in computing a single gas velocity vector by cross-correlation was estimated as 0.01 pixels [5]. The uncertainty in computing a particle velocity vector by tracking was estimated as 0.03 pixels. Thus, at the typical pixel displacement of 24 pixels (for a mean velocity of 10 m/s), the uncertainty in calculating single vector measurements was 0.03% and 0.12%, respectively. The two major sources of uncertainty in the present measurements thus were the low-frequency flow fluctuations and the timing jitter, both estimated as less than 1%.

Examples of single-point statistics computed for two-phase flow are plotted in Figure 11. The white bars refer to the first case (30-μm particles), and the dark bars refer to the second case (60-μm particles). The first four plotted quantities, all normalized by the corresponding global means are, from left to right: gas axial RMS velocity, gas spanwise RMS velocity, solids axial RMS velocity, solids spanwise RMS velocity. Particle slip velocities,

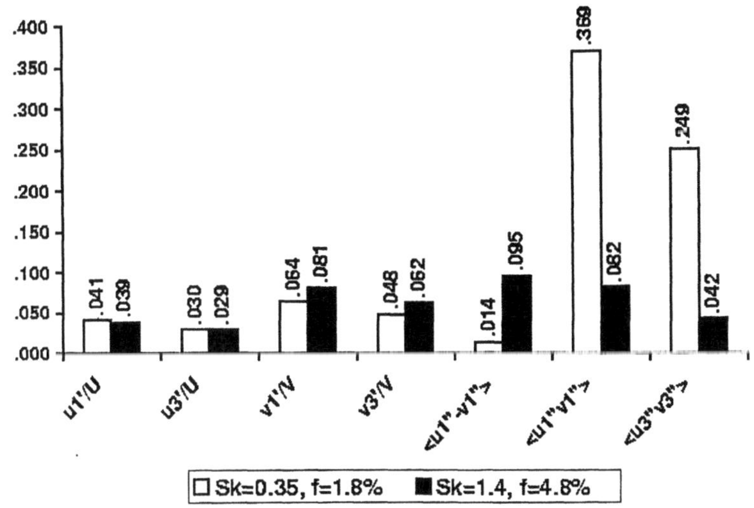

FIG. 11. Normalized ensemble-averaged velocity statistics computed using image separation software in combination with particle tracking and cross-correlation routines. Gas and solid velocities are designated by u and v, respectively; streamwise and spanwise directions are designated by *1* and *3*, respectively

shown in the fifth column, are normalized by the global gas mean. The two principal gas particle covariances, shown in the last two columns, are normalized by the corresponding global variances.

First, examining the gas RMS velocities, one can see that both particle cases yielded similar values. Therefore, the different particle mass loading and sizes appear to have a negligible effect on the magnitude of velocity fluctuations in the gas. This is expected, because at these low mass loadings the particles should have no significant effect on the turbulence. Examination of the particle velocity statistics shows that the larger particles yielded higher RMS values and slip velocities and lower values of the gas particle covariance. As we expect that the motion of larger particles is more independent of eddy motions in the gas, the lower covariance values are also expected.

These results demonstrate the capability of the method described earlier to determine accurate quantitative values for single-point ensemble-averaged statistics. The longer-term goal of this study is to obtain statistics similar to those shown for a range of flow planes, particle sizes, and particle mass loadings. In addition, we plan to obtain two-point velocity correlation statistics based on the streamwise-spanwise measurement planes. All these statistics can be applied to testing two-phase flow models and numerical simulations.

Flow Control Experiments

Actuator Performance

The actuator performance was characterized by measuring its displacement and velocity under a variety of forcing conditions. Displacement and velocity measurements are plotted in Fig. 12. In both graphs, the solid lines represent linear fits to the individual

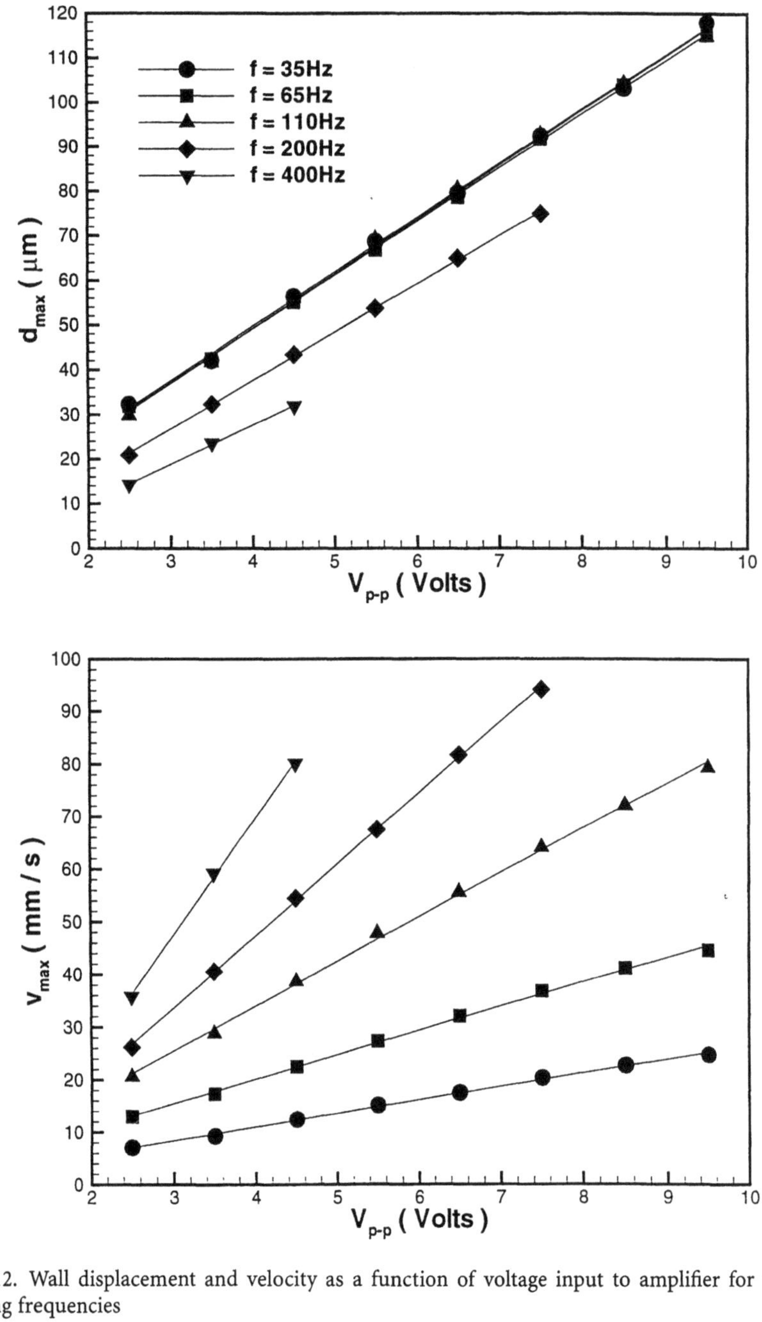

FIG. 12. Wall displacement and velocity as a function of voltage input to amplifier for various forcing frequencies

data points plotted. The upper plot shows the maximum displacement (d_{max}) of the moving plate as a function of the input frequency and peak-to-peak voltage input to the amplifier. In the low-frequency range ($35 < f < 110$ Hz), the curves associated with individual frequencies coincide. The maximum displacement of 120 μm can be achieved when $9.5 < V_{p\text{-}p} < 10$ V. As the forcing frequency is increased ($200 < f < 400$ Hz), however, the maximum achievable displacement decreases because of the maximum current limitation on the linear amplifier. Because of this limitation, the maximum input voltage has to be lowered, resulting in a narrower range of possible displacements.

In the lower plot of Fig. 12, the variation of the maximum velocity (v_{max}) of the moving plate is presented as a function of input frequency and amplitude. For low frequencies ($f \leq 110$ Hz), input frequency and amplitude increase directly with increasing velocity. At higher frequencies ($200 < f < 400$ Hz), the forcing becomes current limited such that the maximum achievable velocity peaks and eventually decreases. One can observe that the maximum value of v_{max} is limited to about 95 mm/s (see $f = 200$ Hz and $V_{p\text{-}p} = 7.5$ V).

Single-Phase Flow Measurements

An (x,y,z) coordinate system is employed to describe the results (see Fig. 2). The x-axis originates at the channel exit and points downstream. The "normal" or y-axis originates at the channel center and points toward the moving plate. The z-axis runs along the span of the channel and originates at the channel center. When the moving plate is not actuated, its inner face is aligned with $y/h = 0.5$. In all the experiments, the mean velocity U_{unf} (0,0) at the center of the channel exit was 4.14 m/s, the turbulence level was 5%, and the Reynolds number based on this velocity and the channel width was 4240. Qualification measurements showed that the mean and RMS velocities were uniform in the z-direction to within 1% over 60% of the channel span. The images and velocity measurements described here were all taken at the midspan of the channel.

In addition to unforced flow, three forcing conditions were investigated. The parameters chosen to characterize each forcing condition are summarized in Table 1. Cases I and II are chosen to be within a lower frequency range, where spectral measurements reveal strong energy content in the flow. Case III is chosen to be in a higher-frequency range where the energy *within the flow* is an order of magnitude smaller than in the other two cases. The maximum displacement of the plate is chosen to be the same for cases I and II, whereas the maximum velocity of the plate in case II is twice that of case I. The maximum velocity in cases II and III is the same. The last column of Table 1 corresponds to the Strouhal number (St) based on the frequency of the input signal, the channel width, and the mean centerline velocity.

Two images of unforced, smoke-marked flow are shown in Fig. 13. The images were obtained by illuminating x–y cross sections with a sheet from a pulsed Nd:YAG laser. In

TABLE 1. Characteristic parameters of the forcing cases

Case	f (Hz)	$V_{p\text{-}p}$ (volts)	d_{max} (μm)	v_{max} (mm/s)	U_{unf} (0,0) (m/s)	St^a
I	65	9.5	115.2	44.6	4.14	0.25
II	110	9.5	114.8	79.5	4.14	0.42
III	385	4.0	31.9	80.1	4.14	1.46

St, Strouhal number

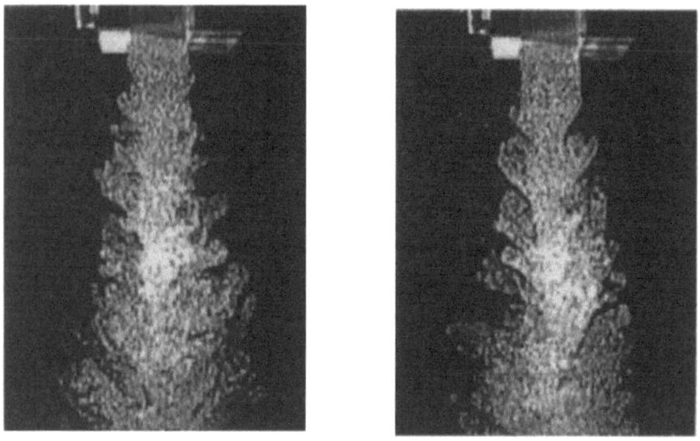

FIG. 13. Instantaneous visualizations of unforced, smoke-marked gas flow exiting turbulent channel

the left-hand image, small-scale structures occupy the near field of the jet. In the right-hand image, both smaller and larger structures are observed. Large eddies with scale similar to the channel width appear in the area close to the jet exit while smaller eddies or structures exist further downstream. Because of the turbulent initial condition, the flow structure is highly complex, and no specific instability dominates the overall structure. The absence of dominant large vortices is also evident in frequency spectra where no strong peaks are present. When the jet is forced, frequency spectra yield strong peaks at the forcing frequency. Nevertheless, the flow structure still appears complex, containing a wide range of eddy scales. Instantaneous images of forced flow appear almost indistinguishable from the unforced case.

Detailed velocity measurements of the gas flow (no particles present) were obtained for a variety of control conditions with a calibrated hot wire probe. The sequence of plots in Fig. 14 shows both forced and unforced mean and RMS velocity profiles taken in the x–y plane at three streamwise locations ($x/h = 1, 4, 10$). Both mean $U(i,j)$ and RMS $u'(i,j)$ velocities [where i denotes the downstream location (x/h) and j denotes the cross-stream location (y/h)] are normalized with respect to the local mean centerline velocity of the unforced case $U_{unf}(i,0)$.

The mean profile of the unforced case at $x/h = 1$ is symmetrical with respect to the centerline, and its shape is typical of those found in fully developed turbulent channel flows. Symmetry persists further downstream, and the jet spreads with increasing streamwise distance (see the mean velocity plots at $x/h = 4$ and 10). At each streamwise location, the mean velocity profiles of all three forcing cases are symmetrical with respect to the centerline. At $x/h = 1$, the forced profiles lie on top of the unforced one except for $St = 0.42$, where the velocity is larger in the outer region of both shear layers. At $x/h = 4$, lower values of velocity with respect to the unforced ones are found close to the centerline for both $St = 0.25$ and $St = 0.42$, whereas no difference is observed between $St = 1.46$ and the unforced case. Further downstream at $x/h = 10$, the profile corresponding with the lowest forcing frequency ($St = 0.25$) is broader, and the velocity decay is larger than in the other cases. In general, one can conclude that although the shape of the mean velocity profiles is not altered substantially, asymmetrical excitation affects the downstream evolution of the centerline velocity.

The different forcing cases did alter the shapes of RMS profiles. At $x/h = 1$, the unforced case yields a symmetrical profile with a low turbulence level of 5% in the core surrounded by higher levels of 19% in the shear layers. The difference in core and shear layer

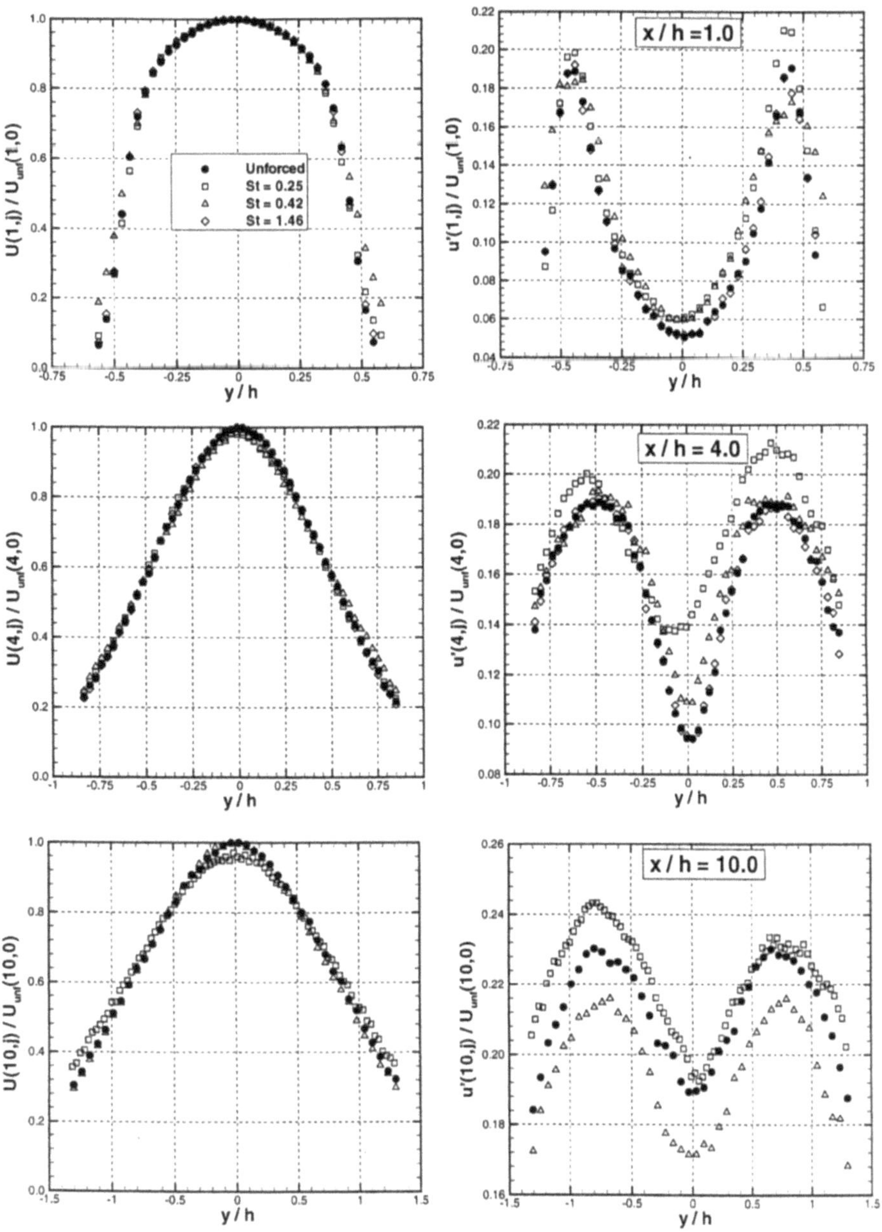

FIG. 14. Mean (*left column*) and RMS (*right column*) gas velocity profiles at three streamwise locations. Moving wall is located at y/h = 0.5

turbulence levels decreases with downstream distance. At $x/h = 10$, the difference in turbulence levels has decreased to 4%. Therefore, at further downstream locations the mixing initiated at the jet boundaries has permeated the entire flowfield. The unforced RMS profiles were symmetrical at all measurement locations.

In the following paragraphs, the three forcing cases are discussed. First, we consider the lowest forcing frequency, $St = 0.25$. At $x/h = 1$, the RMS profile is asymmetrical with respect to the centerline, and turbulence intensities are enhanced compared with the unforced ones. The enhancement is most obvious within the range $0 < y/h < 0.5$. At $x/h = 4$, increases in turbulence intensity have spread across the jet which has become more asymmetric. Specifically, the region of lower RMS velocity has shifted off center, and the shear layer on the actuator side is broader as well as more energetic. The peak RMS value of 0.21 is located at $y/h = 0.5$. At $x/h = 10$, the profile is also asymmetrical, but the peak RMS value of 0.24 has shifted to the opposite side of the jet ($y/h = -0.8$). The actuator-side shear layer at this location has a shape and intensity similar to the unforced one. From this series of plots, it appears that the disturbances caused by the forcing propagate from one shear layer to the other. Within the range $0 < x/h < 10$, no suppression of turbulence intensities compared with the unforced values was detected. Further downstream, however, suppression occurs at all cross-stream locations.

Next, we consider the normalized RMS profiles for $St = 0.42$. In this case, the average kinetic energy of the actuator is approximately four times that for case I ($St = 0.25$). At $x/h = 1$, the RMS profile is asymmetrical with respect to the centerline, and turbulence intensities are enhanced across the core region. The RMS values are comparable with those for $St = 0.25$ except at the peak locations. The maximum RMS values are smaller than those observed for unforced flow and $St = 0.25$. Further downstream at $x/h = 4$, intensities are increased only within $-0.35 < y/h < 0.35$, and the increases are smaller than those observed when $St = 0.25$. At $x/h = 4$ and $x/h = 10$, the RMS profiles appear symmetrical. In fact, symmetrical profiles were found for all $x/h > 2$. Hence, asymmetrical disturbances induced at this frequency appear to permeate the entire flow by $x/h = 2$, and no evidence of a disturbance "crossover" is observed. At $x/h = 10$, the RMS values are smaller than in the unforced case over the entire jet width. Therefore, turbulence reduction or suppression for $St = 0.42$ occurs sooner than for $St = 0.25$.

The third forcing case ($St = 1.46$) is similar to the second ($St = 0.42$) in that the average kinetic energy associated with the actuator was equivalent. Nevertheless, the higher frequency and smaller displacement corresponding with the third case was much less effective at altering the RMS flow pattern. In all locations examined, no significant differences in RMS velocity were found when comparing $St = 1.46$ with unforced flow. Additional velocity measurements as well as frequency spectra are reported in Pothos and Longmire [6].

Both unforced and forced particle-laden flow have been visualized with laser sheet illumination, and differences in the spatial concentration patterns are evident, but, thus far, difficult to characterize. Figure 15, which shows close-up long-exposure images of flow laden with 30-μm and 60-μm particles, indicates that initial particle distributions in the fully developed channel vary with particle Stokes number. A sample image of particle-laden flow downstream of the actuator is given in Fig. 16.

Our next step with this project is to quantify local particle number density and velocity for the four cases just described. This aim will be accomplished using two diagnostic techniques. Particle velocities will be measured with a two-component laser Doppler anemometer system, and local particle number density will be measured by analysis of

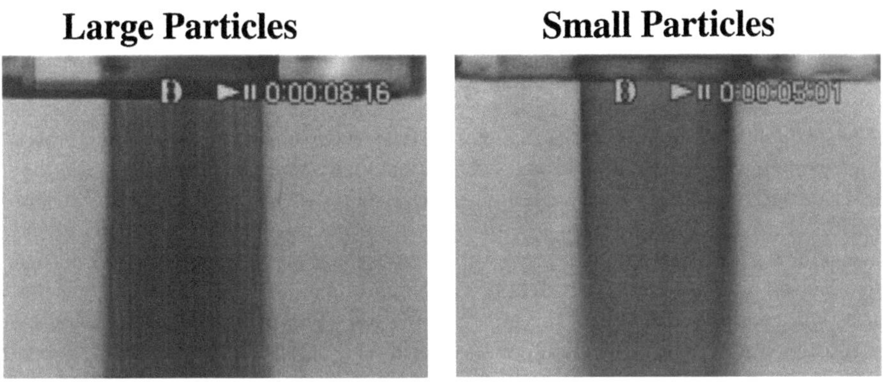

FIG. 15. Close-up time-averaged images of particle-laden flow exiting turbulent channel. Particle diameters are 60 μm (*left*) and 30 μm (*right*)

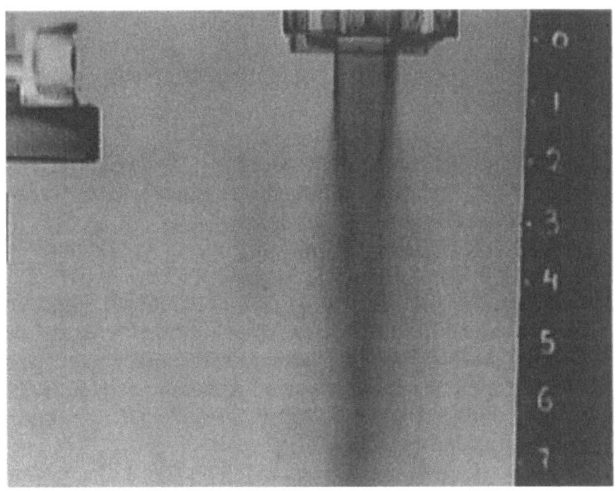

FIG. 16. Visualization of particle-laden flow exiting turbulent channel

instantaneous flow images. We hypothesize that forcing cases II and III, because they have equal energy inputs, should have similar effects on particle behavior for particles of larger Stokes number and possibly different effects for particles of smaller Stokes number. The comparative behavior of particles subject to forcing cases I and II is as yet relatively unknown.

Summary and Conclusions

Particle-laden gas flows occur in many reaction and transport processes related to energy and the environment. Examples of industrial flows include coal combustors, fluidized catalytic cracking systems, and sorbent injection and transport streams. Examples of environmental flows include sandstorms, precipitation in the atmosphere, and contaminant removal systems. In industrial flows, the particle distribution and flow characteristics are extremely important to reaction and process efficiency. In environmental flows, the particle distribution characteristics are important in determining resulting effects on human and other physical surroundings.

Two types of experiments on turbulent particle-laden flows are in progress. In one set of experiments, we attempt to characterize particle–turbulence interactions in a generic flow geometry by obtaining one- and two-point velocity statistics. These data will then be employed to test high-level and engineering-level numerical models needed for prediction of practical flows. In the second set of experiments, we attempt to demonstrate a tunable method, based on a moving wall section, that can be used to control particle distribution and velocity in turbulent flows. Depending on the control parameters, particles will respond either to perturbations of the gas phase or to changes in boundary conditions only. Specific objectives are to determine if actuation can be used to "clear" particles from a wall region and to focus or redirect particles exiting an opening. The longer-term goal of the work is to determine effective control strategies based on an improved understanding of how the wall motions affect particle behavior.

Acknowledgments. The authors gratefully acknowledge support from the ACS Petroleum Research Fund (32534-AC9) and the National Science Foundation (CTS 945-7014).

References

1. McCluskey DR, Easson WJ, Greated CA, Glass DH (1989) The use of particle image velocimetry to study roping in pneumatic conveyance. Particle and Particle System Characterization 6:129–132
2. Brooke JW, Hanratty TJ, McLaughlin JB (1994) Free flight mixing and deposition of aerosols. Physics of Fluids 6:3404–3415
3. Anderson SL, Khalitov DA, Longmire EK (1999) High spatial resolution stereoscopic PIV measurements in a fully-developed turbulent channel flow. In: Proceedings, 3rd international workshop on particle image velocimetry, Santa Barbara, CA
4. Longmire EK, Khalitov D, Borée J, Anderson SL (1999) Planar field velocity measurements in gas-solids flows. In: ASME FEDSM99-7926, 3rd ASME/JSME joint fluids engineering conference, San Francisco, CA
5. Westerweel J, Dabiri D, Gharib M (1997) Effect of a discrete window offset on the accuracy of cross-correlation analysis of digital PIV recordings. Experiments in Fluids 23:20–28
6. Pothos S, Longmire EK (1999) Control of a slot jet with a piezoelectric actuator. Report 1999-1, Dept. of Aerospace Engineering & Mechanics, University of Minnesota, Minneapolis

Part V: Pollutants and Radioactive Wastes in the Earth

The Atmospheric Environment Problem in East Asia for the Twenty-first Century

SHIGERU TANAKA* and YASUSHI NARITA

Summary. Large amounts of air pollutants have recently been emitted from East Asian countries. The emission amount of anthropogenic sulfur in East Asia is especially enormous, estimated to be 30 million tons/year. This amount accounts for about 30% of the anthropogenic sulfur in the world. Most of the anthropogenic sulfur in the atmosphere is washed out by rain and then deposited on the land. However, part of the anthropogenic sulfur emitted from East Asia is transported to the sea near Japan by westerly winds. SO_2 emitted by burning fossil fuel is oxidized to H_2SO_4 in the atmosphere. On land, the oxidized H_2SO_4 in generally neutralized with NH_3, which is generated from the soil by microorganism activity. On the other hand, in the marine atmosphere, SO_2 transported from the land is oxidized to H_2SO_4, but H_2SO_4 is not neutralized because few NH_3 exists in the marine atmosphere. Therefore, H_2SO_4 still exists as a strong acid in the marine atmosphere without neutralization. Furthermore, the transported SO_2 reacts with sea salt particles, which are composed of sodium chloride, and hydrogen chloride as a strong acid is then released from sea salt particles. Thus, the SO_2 transported from East Asia causes increased acidification of the marine atmosphere near Japan. In this study, observations were carried out at several remote islands near Japan and on the routes of vessels over the ocean, from 1991 to 1999, to understand the transport process of anthropogenic sulfur from East Asia and its influence on the marine atmosphere over the sea near Japan.

Key words. Anthropogenic sulfur—Air pollutant transport—Marine atmosphere—The Sea near Japan—East Asia

Introduction

Enormous amounts of air pollutants such as sulfur oxides (SO_x) and nitrogen oxides (NO_x) have been emitted from East Asian areas, mainly from China, which has recently achieved remarkable economic growth. In particular, the amount of emitted sulfur oxides, estimated to be 30 million tons/year, has been increasing along with the expansion of consumption of fossil fuels in recent years. According to Akimoto and Narita [1], the SO_x emission amount from East Asian areas is extremely large, accounting for nearly

Department of Applied Chemistry, Faculty of Science and Technology, Keio University, 3-14-1 Hiyoshi, Kohoku-ku, Yokohama 223-8522, Japan
* Corresponding author

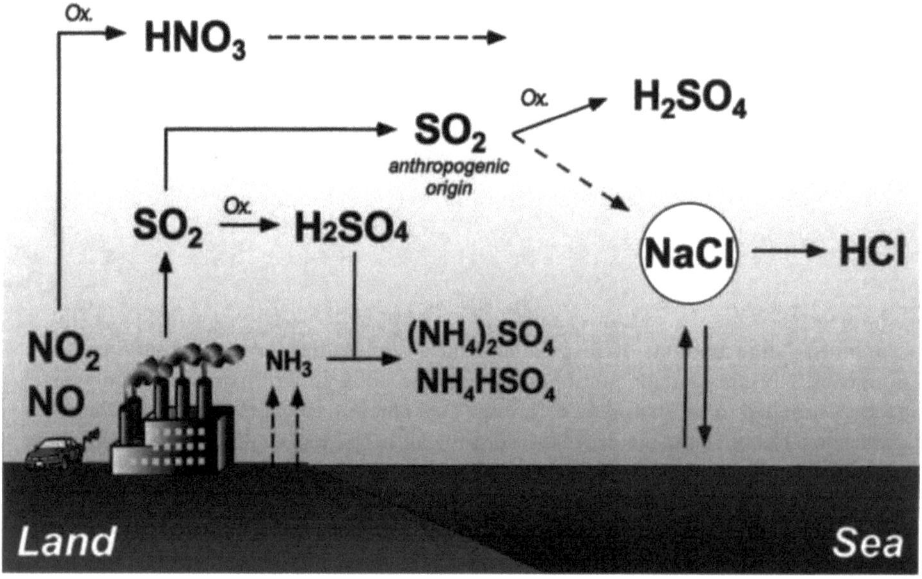

FIG. 1. Transport of air pollutants from East Asia to the marine atmosphere

30% of the world's total emission. Thus, the tremendous amount of air pollutants that are emitted from East Asian areas, mainly from China, are transported over a long distance by westerly winds to the sea near Japan (Fig. 1). In this case, because of transport not over the land but over the ocean, the air pollutants are not neutralized with NH_3; therefore, they are transported as a form of acid. It is accordingly feared that there will be strong influences on the atmospheric environment in the periphery of Japan.

To understand the actual situation of the border-transgressing pollution by such air pollutants from East Asia and to clarify the influence to the atmospheric environment of the periphery of Japan, the authors have been carrying out observations on the atmosphere on isolated islands around the sea near Japan such as Oki Island, Tsushima Island, Yakushima Island, the Okinawa Islands, and on the marine atmosphere in the East China Sea and the West Pacific Ocean by ships and airplanes. In this chapter, the authors mainly report on air pollution in East Asia by research in the literatures and the actual state of the environmental atmosphere around the periphery of Japan from the results of observations on the atmosphere.

Situation of Air Pollution in East Asian Areas

SO_2 Emission Volume in East Asian Areas

Table 1 shows the SO_2 emission amounts that are estimated on the basis of energy consumption as reported in the literatures. According to Arndt et al. [2], the SO_2 emitted from the whole of Asia in 1988 amounted to 31.6 million tons/year. In comparison, total SO_2 emission by anthropogenic causes from all over the world is estimated to be 134.8

TABLE 1. Estimated anthropogenic emissions of SO_2 in Asia

Country	Year	Emissions (in millions of tonnes of SO_2 per year)	Reference
China	1980	13.37	1
	1987	18.87	
	2000	34.04	2
	2010	48.80	
	1988	19.87	3
Taiwan	1980	1.04	1
	1987	0.85	
	2000	1.74	2
	2010	2.22	
Japan	1980	1.44	1
	1987	0.90	
	2000	0.70	2
	2010	0.50	
	1988	0.98	3
North Korea	1980	0.27	1
	1987	0.59	
	2000	0.92	2
	2010	1.28	
	1988	0.41	3
South Korea	1980	1.92	1
	1987	1.22	
	2000	2.72	2
	2010	3.31	
	1988	1.30	3
India	1975	0.83	1
	1980	1.01	
	1987	1.53	
	1988	5.06	3
Bangladesh	1988	0.37	3
Indonesia	1988	0.40	3
Pakistan	1988	0.35	3
Philippines	1988	0.51	3
Thailand	1988	0.70	3
Total Asia	1988	31.59	3
Total Southeast Asia	1988	2.02	4
Total Indian Subcontinent	1988	6.00	4
Total East Asia	1988	23.34	4

1. Akimoto and Nartita, Atmospheric Environment 28 (2): 213–225 (1994) [1]
2. Bhatti et al. Proceeding of the second annual workshop on Acid Roin and Emission in Asia: 5–33 (1999) [6]
3. Arndt et al. Atmospheric Environment 31 (10): 1553–1572 (1997) [2]
4. Streets et al. Atmospheric Environment 31 (10): 1573–1582 (1997) [5]

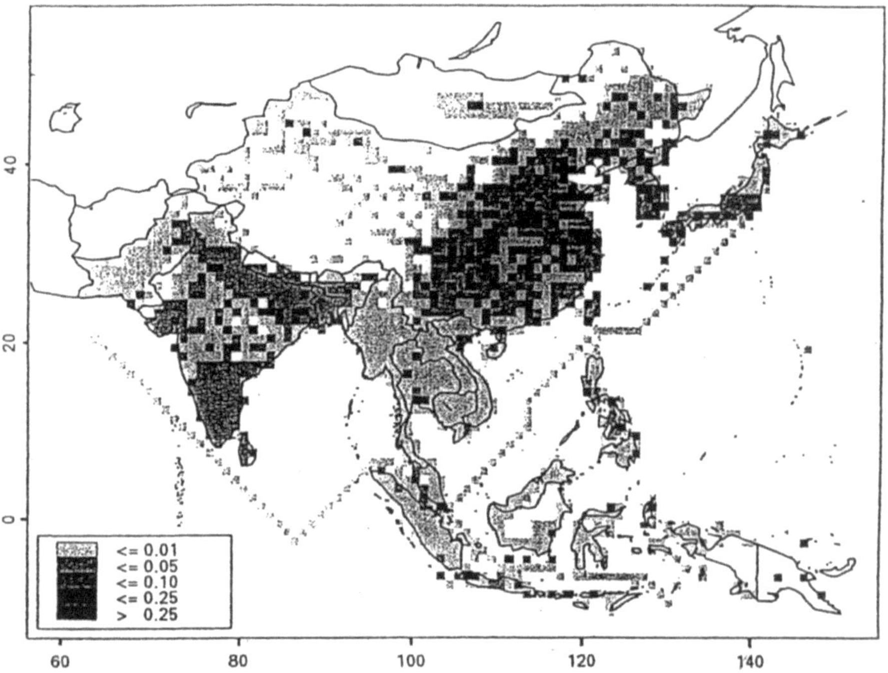

FIG. 2. Spatial distribution of SO_2 emissions for Asia (1988) in 10^3kg SO_2 yr^{-1} (from [2], with permission)

million tons/year [3] or 155.2 million tons/year [4]. These data mean that the SO_2 emitted from Asia accounts for a high percentage, 20% to 25%, of the total SO_2 emission in 1988. In particular, SO_2 emissions from the East Asian areas centered on China and Korea are large, estimated to be 23.3 million tons/year [5], revealing that SO_2 from East Asian areas exceeds 70% of the total amount from Asia. Figure 2 shows the distribution of SO_2 in the Asian areas in 1988. As it is apparent from this Figure, the emission SO_2 shows a marked increase in China along the coastal areas facing the East China Sea. Therefore, it has been feared that there will be effects caused by pollution crossing to Japan, which is located to the leeward of the westerly wind belt in the East Asian areas.

Pollution by SO_2 in East Asian areas can be said to be very serious in 1988. Furthermore, it is anticipated that the amount of SO_2 emitted will continue to increase further, considering the economic growth in these areas. The Asian NIES and ASEAN countries have been continuing their economic growths at an exceptionally high rate for the past 30 years. In Asian areas, the population as of 1995, accounts for 60% of the world, while their substantial GDP (gross domestic product) accounts for 27% (1994), and their energy consumption accounts for 31%, which amounts are not proportional to their population. Therefore, there is a strong possibility that their economic growth will continue and their energy consumption will expand also in the future. In China, especially, not only the consumption of fossil fuels but also the consumption of coal, including much sulfur, is increasing. From such circumstances, it is assumed that SO_2 emission will increase. According to one estimate, the annual growth of SO_2 emissions in this area of

China is 5.3%, which is a very high growing percentage, doubling the amount only in 14 years.

Moreover, as shown in Table 1, because the volume of SO_2 emissions in China as of 2010 is predicted to amount to approximately 50 million tons/year, which is more than 2.5 fold that of the 1990s, it is accordingly anticipated that the problem of air pollution crossing borders from the East Asian areas will increase further in seriousness in the future.

Actual Situation of Air Pollution in East Asian Areas

In East Asian areas, a tremendous amount of air pollutants included sulfur oxides has been emitted because of economic growth and the expansion of energy consumption accompanying this economic growth. It is, therefore, extremely important to understand the present situation of air pollution in the East Asian areas as the source of pollution. Nevertheless, in China, which is the main source of generation of air pollutants, the number of reports on measurements of NO_x and SO_x has been very small. This chapter, based on the small number of reports, compares the results of measurements of the urban atmosphere in Yokohama, Japan, with those in China and Korea so as to summarize air pollution in China and Korea. Table 2 shows the atmospheric concentration of nitrate (NO_3^-) and the atmospheric concentration of the non-sea-salt sulfate (nss-SO_4^{2-}, anthropogenic salt sulfate) compared to observations in Beijing [7] and Seoul [8]. In addition, atmospheric concentrations in Yokohama are also shown.

First of all, regarding the seasonal variation in the atmospheric concentration of nss-SO_4^{2-}, it is known that the concentration increases in winter in both Beijing and Seoul. It is supposed that this increase is caused by coal, which is the main energy resource in China and Korea, and accordingly nss-SO_4^{2-} increases in winter when heaters are used. Moreover, the average atmospheric concentrations of nss-SO_4^{2-} are 18.1 $\mu g/m^3$ in Beijing and 15.2 $\mu g/m^3$ in Seoul, respectively. In either case, the results of measurements exceed the result of measurement of the urban atmosphere 5.09 $\mu g/m^3$ in Yokohama by three- to fourfold. In consequence, it is known that the atmospheric concentration of nss-SO_4^{2-} is high in Beijing and Seoul on account of very large amounts of emitted SO_2. Furthermore, to compare the ratio of concentrations of nss-SO_4^{2-} versus NO_3^-, although the ratio is 1.0 on average in Yokohama, it is 3.0 in Beijing and 2.0 in Seoul as well. Thus, measurement of aerosol content has suggested that air pollution, mainly by SO_x, has been increasing in China and Korea.

Table 3 shows the atmospheric concentrations of NO_2 and SO_2 observed in Beijing [9]. For the sake of comparison, the atmospheric concentrations of NO_2 and SO_2 in Yokohama are also shown [10]. On the other hand, Fig. 3 shows the seasonal variation of the atmospheric concentrations of SO_2 in each city of Korea [11]. To examine these results, the observed average atmospheric concentration of SO_2 emitted directly from combustion of fossil fuel is 43.6 ppbv for Beijing and a value exceeding 50 ppbv for Seoul. Meanwhile, because the atmospheric concentration of SO_2 in Yokohama is at a level of several ppbv, the comparison of SO_2 in the atmospheric concentration confirms that air pollution mainly by SO_x is serious in Beijing and Seoul. Next, Fig. 4 shows an example of measurement of the SO_2 atmospheric concentration in Chongquig where most of the heavy industry in China flourishes [12]. To examine this case, it is known that the SO_2 atmospheric concentrations are high in winter, as in Beijing and Seoul. What is more, regarding their atmospheric concentration levels, it is known that both SO_2 and nss-SO_4^{2-}

TABLE 2. Comparison of NO_3^- and nss-SO_4^{2-} concentrations in Beijing, Seoul and Yokohama

Sampling site	Period		Concentration [$\mu g/m^3$] NO_3^-	Concentration [$\mu g/m^3$] nss-SO_4^{2-}	Ratio [wt/wt] nss-SO_4^{2-}/NO_3^-	Reference
Beijing	1988	Dec.	6.0	17.4	2.9	Sekine et al. [7]
	1989	Jan.	10.1	25.6	2.5	
		Feb.	8.8	17.5	2.0	
		Mar.	5.2	16.2	3.1	
		Apr.	13.2	17.1	1.3	
		May	5.1	10.0	2.0	
		Jun.				
		Jul.	9.7	14.3	1.5	
		Aug.	11.8	15.9	1.3	
		Sep.	4.2	7.0	1.7	
		Oct.	4.4	7.1	1.6	
		Nov.	7.1	11.9	1.7	
		Dec.	9.5	21.0	2.2	
	1990	Jan.	7.4	17.3	2.3	
		Feb.	23.0	54.4	2.4	
	Avg.		9.0	18.1	2.0	
Seoul	1988	Apr.	7.4	14.7	2.0	Hashimoto et al. [8]
		May	5.5	10.1	1.8	
		Jun.	9.0	16.4	1.8	
		Jul.	2.4	11.1	4.6	
		Aug.	3.9	9.6	2.5	
		Sep.	3.2	6.5	2.0	
		Oct.	6.6	17.5	2.7	
		Nov.	5.4	19.3	3.6	
		Dec.	6.9	31.6	4.6	
	1989	Jan.	4.2	16.1	3.8	
		Feb.	4.9	14.8	3.0	
		Mar.	3.9	14.3	3.7	
	Avg.		5.3	15.2	3.0	
Yokohama	Jul. 1992–Nov. 1994		5.24 ($n = 761$)	5.09 ($n = 761$)	1.0	This work

nss, non-sea-salt

concentrations are at extremely high levels, higher than values in Beijing by twofold (SO_2, 250 $\mu g/m^3$ or 100 ppbv), revealing a very serious degree of air pollution.

Thus, SO_2 and nss-SO_4^{2-} atmospheric concentrations more than tenfold those in the urban atmosphere of Yokohama have been observed in the respective cities of China and Korea. The standard for the atmospheric environment of each country is not achieved in some cities. To this extent, air pollution mainly by SO_x has been serious in China and Korea.

Amount of Air Pollutants Transported to Japan

As suggested by the amounts of energy consumption in East Asian areas, it has become clear that an enormous amount of SO_2 has been emitted, that the SO_2 and nss-SO_4^{2-} atmos-

TABLE 3. Comparison of NO_2 and SO_2 concentrations in Beijing and Yokohama

Sampling Site	Period	Concentration [ppbv]		Reference
		NO_2	SO_2	
Beijing	Sep., 1990–Nov., 1990		24.1	1
	Nov., 1990–Jan., 1991		73.9	
	Feb., 1991–Apr., 1991		38.1	
	Apr., 1991–Jun., 1991	13.6		
	Jul., 1991–Sep., 1991		18.7	
	Oct., 1991–Dec., 1991		52.5	
	Dec., 1991–Feb., 1992		54.0	
	Avg.	13.6	43.6	
Yokohama	Apr., 1995	34	8	2
	May, 1995	31	8	
	Jun., 1995	34	7	
	Jul., 1995	40	7	
	Aug., 1995	34	8	
	Sep., 1995	34	9	
	Oct., 1995	39	9	
	Nov., 1995	38	9	
	Dec., 1995	42	10	
	Jan., 1996	35	8	
	Feb., 1996	43	9	
	Mar., 1996	35	8	
	Avg.	36.6	8.3	

1. Ferm et al. Journal of Atmospheric Chemistry 27: 17–29 (1997) [9]
2. Environmental Information Center, Data book of monthly averaged concentrations in the air pollution monitoring station (1995) [10]

pheric concentrations are at very high levels, and that air pollution, mainly by SO_x, is serious in every city of China and Korea. It has been believed that the air pollutants emitted from the East Asian areas are confined to the generating areas, but they are transported by westerly winds for a long distance to the periphery of Japan and the North Pacific Ocean. Therefore, the authors estimated the transport amount of sulfur from East Asia.

Figure 5 shows the distribution of sulfur deposition in each area of Asia calculated on the basis of the predicted SO_2 emission amount reported by Arndt et al. [2]. This figure shows that the sulfur deposition to Japan is at a level of 1.0 to 2.0 g S/m^2·year. Using this value to calculate the annual deposition amount over all Japan, the following equation gives this answer:

$$1.0\text{–}2.0\,\mathrm{g\,S/m^2 \cdot year} \times 380000\,\mathrm{km^2} = 0.38\text{–}0.76\,\mathrm{Tg\,S/year}$$

Moreover, Ichikawa and Fujita [13] estimated that sulfur depositions on the whole of Japan are 0.62 Tg S/year for wet deposition and 0.43 Tg S/year for wet deposition, amounting to a total of 1.05 Tg S/year. Meanwhile, the SO_2 emission amount in Japan is, as shown in Table 1, at a level of 1 million tons per year (0.5 Tg S/year). Accordingly, comparison with the estimate by Ichikawa and Fujita [13] shows that sulfur deposition from East Asia is twofold higher than the amount of anthropogenic sulfur generated in Japan.

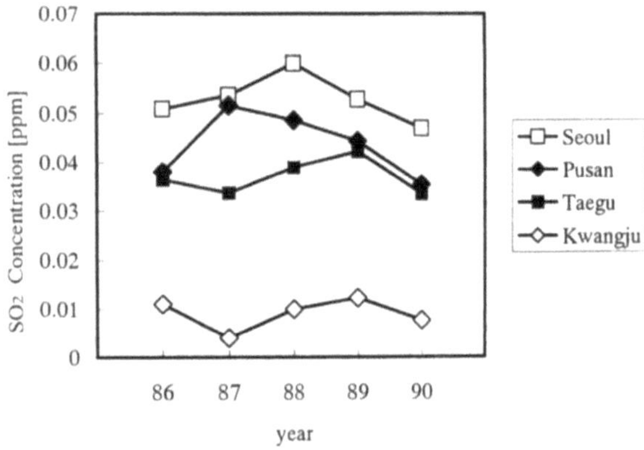

FIG. 3. Yearly concentration of SO_2 in the atmosphere at each city in Korea (from [11], with permission)

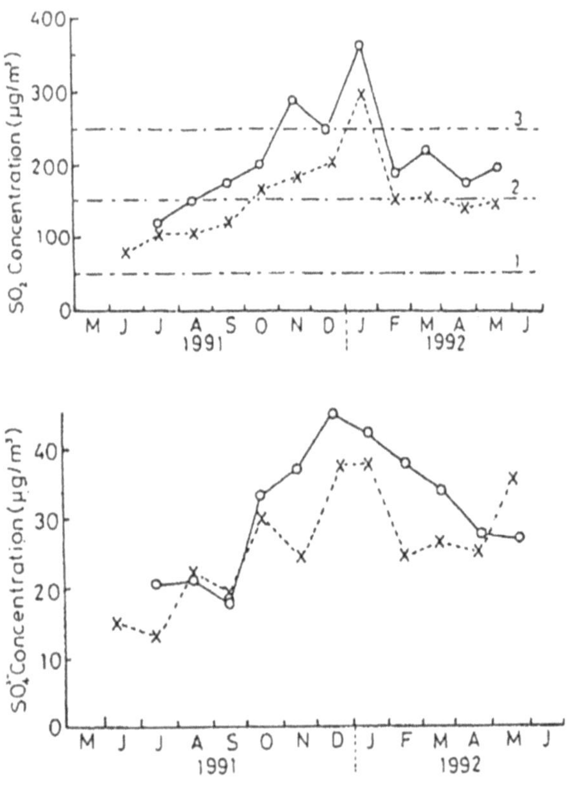

FIG. 4. Seasonal variations of SO_2 and SO_4^{2-} concentration in Chongquing (from [12], with permission)

Even if the contribution of natural sulfur arising from volcanic and biogenic origins is taken into consideration, deposition of a huge amount of sulfur not emitted from Japan is estimated. This estimate indicates that the deposition is strongly affected by sulfur transported over a long distance from the East Asian areas. Also, the amount of sulfur transported over a long distance is estimated to be almost the same amount as the amount emitted from Japan. According to Fig. 6, it is estimated that the percentage of

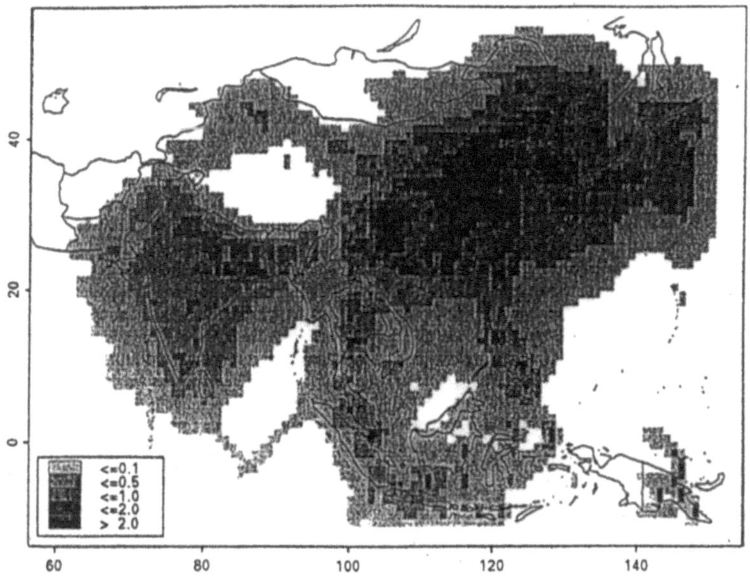

FIG. 5. Calculated annual sulfur deposition (in g S·m^{-2}yr^{-1}) (from [2], with permission)

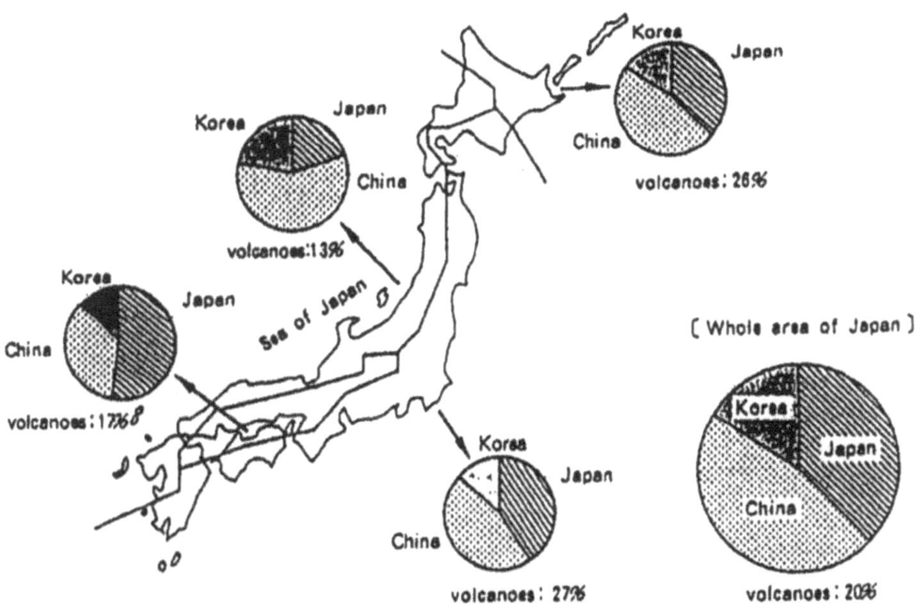

FIG. 6. Sources contributing to wet deposition of sulfate in Japan (from [13], with permission)

sulfur transported from China and Korea accounts for more than 60% of the total wet deposition of sulfur in Japan. The percentage on the side of Sea of Japan is greater than 75%.

In China and Korea, it is anticipated that SO_2 emission amounts will continue to increase as the result of economic growth. It is, therefore, considered that the sulfur transported from East Asia further increases and makes the problem of border-transgressing pollution more serious.

Marine Atmosphere Observation over the Sea near Japan

To understand the actual situation of long-range transport of anthropogenic sulfur from East Asian areas to the sea near Japan, the authors have been carrying out observations on the marine atmosphere from 1981 to the present, mainly on isolated islands in the sea near Japan such as the islands of Okinawa, Oki, Yakushima, Sadogashima, Chichijima, and Hachijo, and also on the routes of ships, such as the R/V *Hakuho-Maru* of the Ocean Research Institute of The University of Tokyo and M/V *Skaugran*, a container ship, navigating between Japan and Canada over the West Pacific Ocean and East China Sea. Table 4 outlines the marine observations.

Atmospheric observation was implemented several times a year on isolated islands of the sea near Japan and on the navigation routes of ships. Nevertheless, the marine atmosphere observations made hitherto were carried out only for 1 or 2 weeks in most cases. It is difficult to accurately clarify the actual state of transported pollution by atmospheric observations for such a short period, and accordingly it can be said that it is essential to conduct observations for a long period of time on a continuous basis. Thus, as a part of the Core Research for Evolutional Science Technology (CREST) program supported by

TABLE 4. Observation projects in the marine atmosphere in East Asia

Project	Sea area	Period
Remote Islands		
Okinawa	East China Sea	August 1996–June 1999 March 1996 October 1991
Sado	Sea of Japan	Nov.–Dec. 1995
Yaku	East China Sea	Nov.–Dec. 1994
Oki	Sea of Japan	August 1997–June 1999
Tsushima	Sea of Japan	October 1991
Research Vessels		
Hakuhou Maru		
KH95-3	North West Pacific Ocean	November 1995
KH94-3	North West Pacific Ocean	October 1994
KH93-4	Central Pacific Ocean	Oct.–Dec. 1993
KH92-4	West Pacific Ocean	Sept.–Dec. 1992
Skaugran	North Pacific Ocean (Japan ↔ Canada)	March 1995–October 1996
Shinanogawa	West Pacific Ocean (Japan ↔ The Persian Gulf)	Feb.–April 1994
Japan Violet	West Pacific Ocean (Japan ↔ The Persian Gulf)	July–Sept. 1991

Japan Science and Technology Corporation (JST), marine atmospheric observations have been carried out on a continuous basis throughout the year since March 1995 on the route of the M/V *Skaugran*, since August 1997 in Okinawa, and since July 1998 in Oki. By means of such continuous atmosphere observations for a long period of time, it can be expected that new findings on seasonal fluctuations will be clarified with respect to the long-range transport of air pollutants from East Asia.

In marine atmosphere observations, aerosols were collected on filters using a high-volume air sampler (air flow rate, 1000 l/min) or a low-volume air sampler (air flow rate, 30 l/min). Aerosols collected on the filters were extracted with deionized water at the laboratory, the sample solution was then analyzed by ion chromatography, and the ion spaces (Cl^-, NO_3^-, SO_4^{2-}, Na^+, NH_4^+, K^+, Ca^{2+}, and Mg^{2+}) was measured. Moreover, as regards the acidic and basic gases (HCl, HNO_3, SO_2, and NH_3), a diffusion scrubber (effective length, 50 cm; air flow rate, 1.0 l/min) was used to collect these gases with deionized water, and an automatic continuous analyzer (DS-IC equipment), which analyzed the gases by ion chromatograph, was used to perform the measurements [14].

In carrying out atmospheric observations on the route of the ship, observation is affected by exhaust gas from the ship itself, depending on such factors as wind direction and wind velocity. Thus, on board the container ship, M/V *Skaugran*, navigating between Japan and Canada, a new system was employed, called the "wind selector," which collects air only when the wind blows from the direction of the bow so that the observation is not affected by the flue gas from the ship.

Anthropogenic Sulfur in the Marine Atmosphere over the Sea near Japan

Concentration Distribution of Anthropogenic Sulfur in the Marine Atmosphere over the Sea near Japan

Sulfate aerosols generated in the atmosphere by oxidation of SO_2 discharged from factories and power plants have a comparatively long life of 2 to 3 days in atmosphere [15]; therefore, there is a possibility that the sulfate aerosol may be transported over a long distance from the generating sources to remote areas 1000 to 1500 km away. For such reasons, it can be considered that anthropogenic sulfur emitted from the East Asian areas is transported by the westerly wind over a long distance to the sea near Japan.

Meanwhile, in evaluating the transport of such anthropogenic sulfur, it becomes necessary to understand the background concentration of sulfur that exists originally in the atmosphere of the Earth. As regards the factors affecting the background concentration of sulfur in the marine atmosphere, it can be said that the one and only predominant factor is the contribution of biogenic sulfur deriving from phytoplankton existing in seawater. Based upon atmosphere observations in oceans and on isolated lands, Saltzman et al. [16] reported that the background concentration of non-sea-salt sulfate (nss-SO_4^{2-}) is 0.3 $\mu g/m^3$ in areas of high latitude and about 0.5 $\mu g/m^3$ in areas of low latitude.

Judging from the results of atmosphere observation in the West Pacific Ocean along the sea route of the R/V *Hakuho-Maru* (KH-95-3) carried out by the authors, it is clear that the concentration decreases as the ship goes away from Japan, and the concentration reaches the background concentration level that was reported by Saltzman et al. [16] in the marine area 2000 km away from Japan (around 165°E).

Figure 7 shows the concentrations of air pollutants on each longitude obtained from atmosphere observation along the route of the container ship M/V *Skaugran* navigating between Japan and Canada over the North Pacific Ocean. It is evident that each content shows a decrease in concentration until the vicinity of 165°E and remains constant in the eastern areas thereof. From this result, it can be said that the long-range transport of air

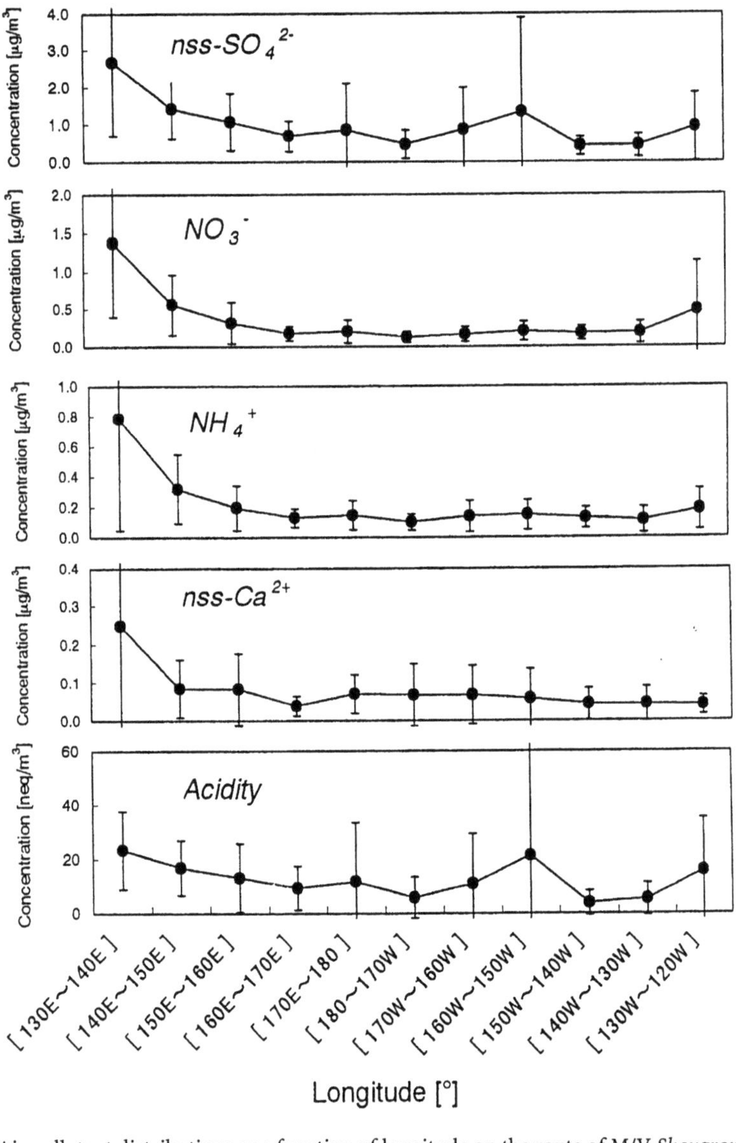

FIG. 7. Air pollutant distributions as a function of longitude on the route of M/V *Skaugran* (March 1995–January 1997) over the north Pacific Ocean

pollutants has an effect as far east as 165°E and a strong effect as far east as 145°E. It is estimated, therefore, that the long-distance transport of air pollutants is as great as 2000 to 3000 km from Japan.

Figure 8 shows the distribution of atmospheric concentrations of the non-sea-salt sulfate (nss-SO_4^{2-}) in the sea near Japan that have been obtained from the results carried out up to the present. It is understood that, while the atmospheric concentration is 1 $\mu g/m^3$ or less and similar to the background concentration on the Pacific Ocean side, the concentration on oceans centered around East China Sea exceeds the background

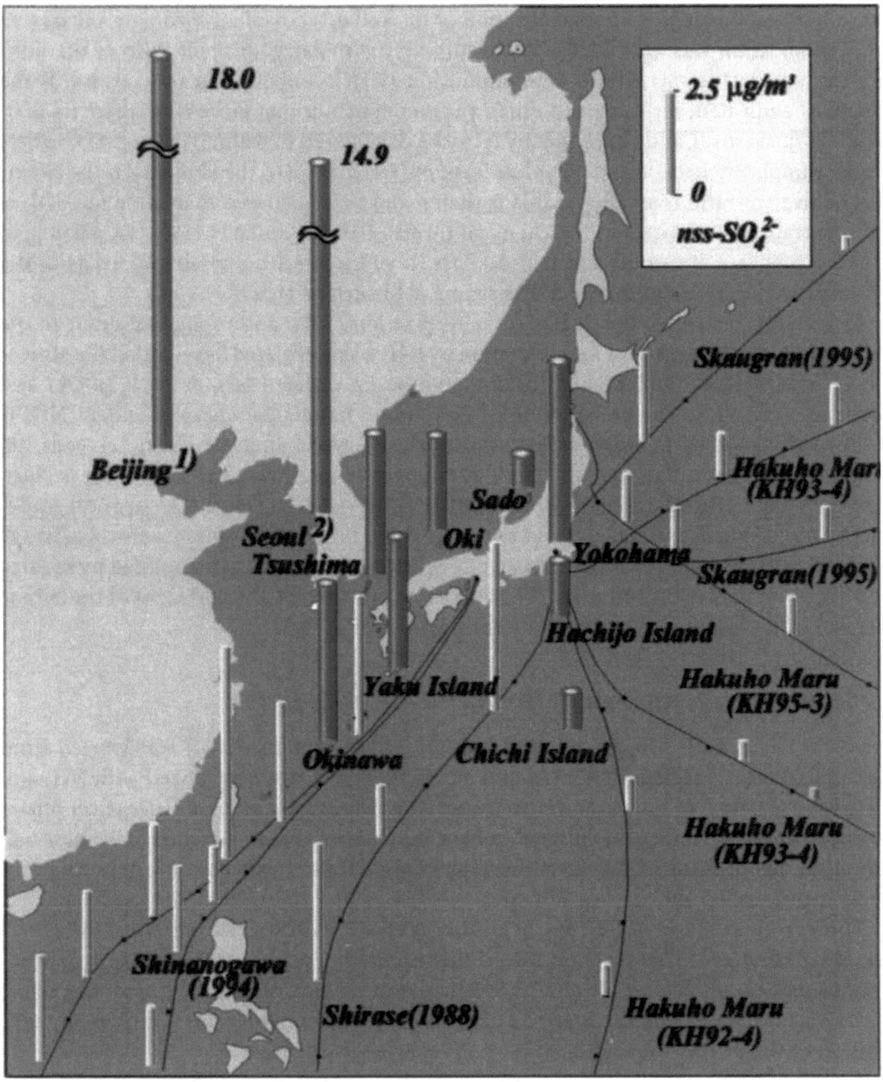

FIG. 8. Concentrations of non-sea-salt sulfate (nss-SO_4^{2-}) in the marine atmosphere in East Asia (Beijing and Seoul from [7] and [8], respectively, with permission)

concentration by a large margin, and that the concentration is at a level equivalent to the annual average value of 5.09 $\mu g/m^3$ (July 1992–November 1994) in Yokohama, which has the typical urban atmosphere. These observation results strongly demonstrated that a large amount of anthropogenic sulfur generated in the East Asian areas has been transported by westerly winds to the sea near Japan.

Degree of Neutralization of Transported Anthropogenic Sulfur

When considering the influence of the anthropogenic sulfur transported from East Asia on the environmental atmosphere around Japan, it becomes very important to take into consideration the degree of neutralization of the sulfate aerosol. Accordingly, the degree of neutralization was examined by determining the molar equivalent ratio of the non-sea-salt sulfate (nss-SO_4^{2-}) and ammonium ion (NH_4^+) contained in the aerosol. If the chemical equivalent ratio is greater than 1.0, then it means that more NH_4^+ than nss-SO_4^{2-} exists in the aerosol, and consequently it can be considered that almost all nss-SO_4^{2-} exists in the completely neutralized chemical form of $(NH_4)_2SO_4$. On the contrary, if the chemical equivalent ratio is smaller than 1.0, then it can be considered that some nss-SO_4^{2-} is not neutralized and exists in the chemical forms of H_2SO_4 and NH_4HSO_4. In particular, if it is 0.5 or less, it can be said that nss-SO_4^{2-} is not neutralized at all and exists in the atmosphere in the chemical form of a strong acid such as H_2SO_4.

To actually determine the degree of neutralization of the anthropogenic sulfur in the sea near Japan, the degree of neutralization by NH_3 was calculated based upon the atmosphere observations implemented by the authors on isolated islands such as Oki and Okinawa (Fig. 9). In the atmosphere of the city of Yokohama where abundant NH_3 is emitted from soil by microorganism activity, the chemical equivalent ratio exceeds 1.0, and consequently sulfate aerosol exists as completely neutralized $(NH_4)_2SO_4$. It is clear, however, that in the marine atmosphere where supply of NH_3 is low, the chemical equivalent ratio is much less than 1.0 and as low as about 0.5, and the sulfate aerosol exists as the acid of H_2SO_4 and NH_4HSO_4. Especially in Okinawa, which is surrounded by oceans, the degree of neutralization with NH_3 is very low, at a level of 1/3, and most of the sulfate aerosol is in the chemical form of a strong acid such as H_2SO_4.

Chlorine Loss Reaction in the Marine Atmosphere

As described, a high concentration of non-sea-salt sulfate (nss-SO_4^{2-}) transported from East Asian areas has been observed and a part thereof is not neutralized with NH_3 and exists as a strong acid such as H_2SO_4 in the marine atmosphere around the sea near Japan. It can be anticipated that the influence of such pollutants on the marine atmosphere will cause the acidification of marine atmosphere directly, and furthermore will cause problems such as acid rain.

Further, the reaction of a strong acid such as H_2SO_4 with sea salt particles (NaCl) will cause chlorine loss. This reaction means that the strong acid, such as H_2SO_4, is neutralized by the sea salt particulate (NaCl) and that at the same time the volatile strong acidic gas, HCl, evaporates into the marine atmosphere. Therefore, this does not offer any mitigation of the process of acidification of the marine atmosphere.

$$H_2SO_4(P) + NaCl(P) \rightarrow NaHSO_4(P) + HCl(g)\uparrow$$

$$NaHSO_4(P) + NaCl(P) \rightarrow Na_2SO_4(P) + HCl(g)\uparrow$$

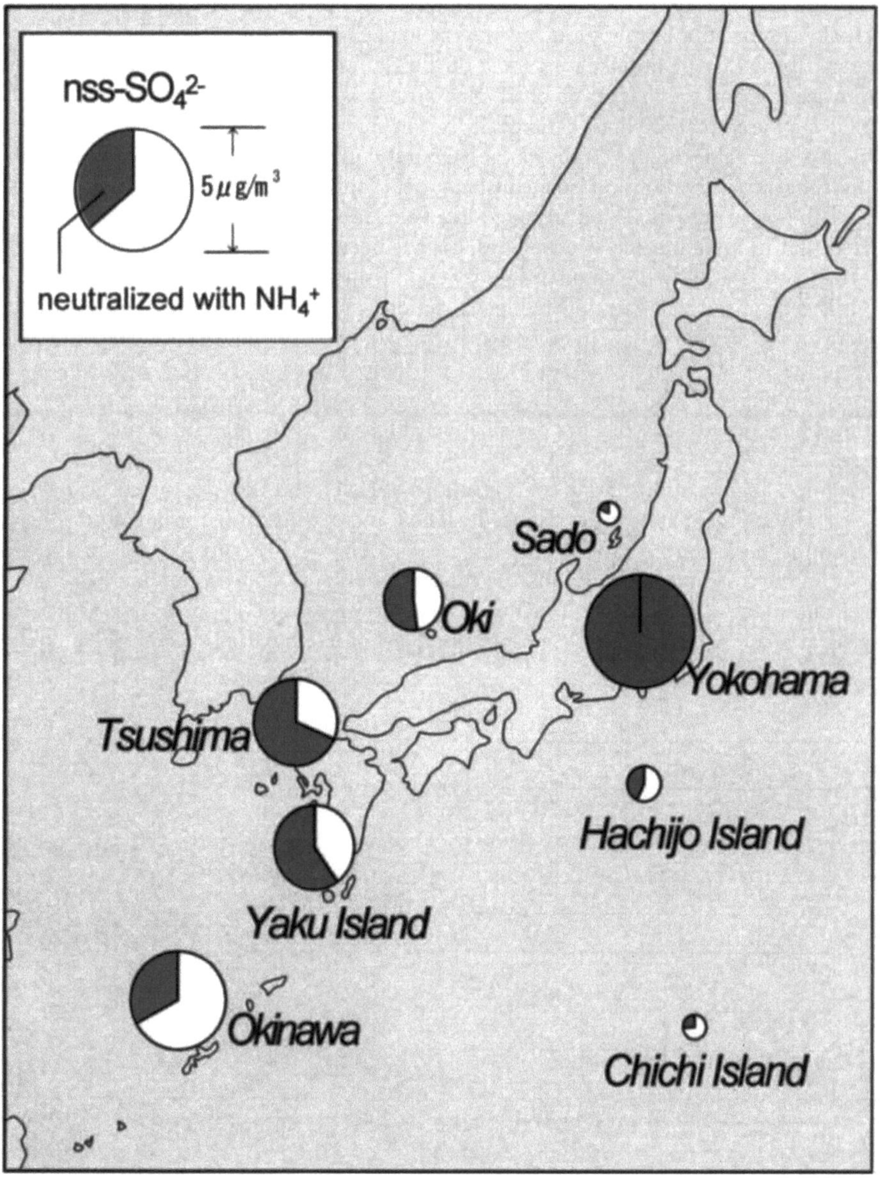

FIG. 9. nss-SO_4^{2-} concentration and NH_4^+/nss-SO_4^{2-} ratio in the marine atmosphere in East Asia

By the reactions given above, Cl^- evaporates as HCl from the sea salt particles, and to actually confirm this evaporation, the mass ratio of Cl^-/Na^+ in the aerosol should be compared. In the marine atmosphere, most of the Na^+ and Cl^- contained in aerosol arises from sea salt, and the mass ratio of Cl^-/Na^+ is 1.8 in seawater. From these facts, it is naturally supposed that the mass ratio of Cl^-/Na^+ in aerosol existing in marine atmosphere will take this value. Figure 10 shows the mass ratio of Cl^-/Na^+ in aerosol obtained from atmospheric observations on isolated islands near Japan and sea routes of ships. In most cases, the mass ratio of Cl^-/Na^+ resulted in a value less than 1.8, and it is thereby recognized that in these marine atmospheres, Cl^- has been evaporated as HCl from sea salt particulates (NaCl) by the chlorine loss reaction with a strong acid such as H_2SO_4.

It is recognized that the chlorine loss reaction is taking place in the marine atmosphere on the sea near Japan. In other words, this can be understand as occurrence of HCl in the marine atmosphere caused by acid substances transported from East Asia. As a result, comparison with other acid gases having no source of generation in the marine atmosphere reveals the existence of a relatively high concentration of HCl in these ocean areas.

Figure 11 shows atmospheric concentrations of HNO_3 and HCl, the typical acid gases, by classifying them as to cities, isolated islands, and oceans. In the urban atmosphere, the atmospheric concentrations of both gases of HNO_3 and HCl are at levels of about 1 ppbv. In the case of isolated islands, the concentration of HNO_3 is about one-half that of HCl, and in the case of oceans, HNO_3 is at a very low level, of 0.1 ppbv or less, while HCl exists at a comparatively high concentration level of 0.1 to 0.5 ppbv. From these obser-

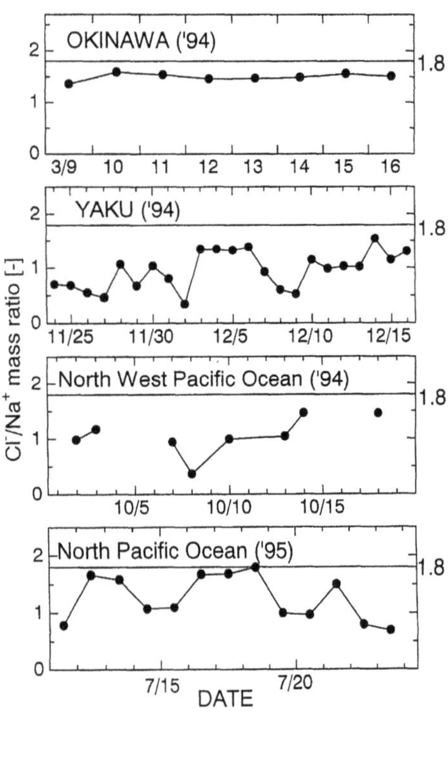

FIG. 10. Cl^-/Na^+ mass ratio in the marine atmosphere

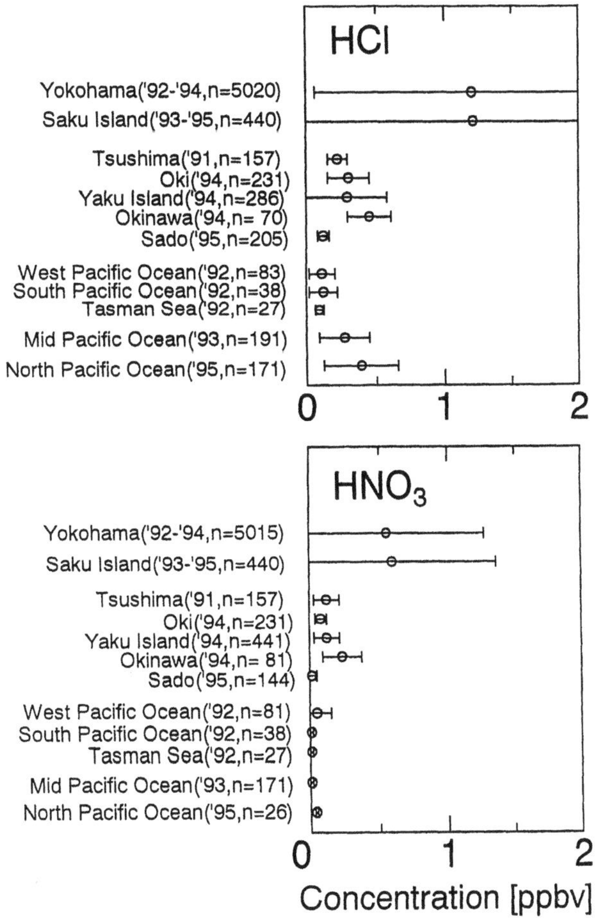

FIG. 11. Concentrations of acidic gases in the marine atmosphere

vations, the generation of HCl due to chlorine loss in the marine atmosphere is strongly suggested.

Conclusion

From the results of observations showing that the concentration of the non-sea-salt sulfate (nss-SO_4^{2-}) being generated secondarily by oxidation from SO_2 in the marine atmosphere in the sea near Japan centered in East China Sea is a high concentration level equivalent to that of the urban atmosphere in Japan, it has become clear that the anthropogenic sulfur being emitted in a large amount from the East Asian areas has been transported by westerly winds to the periphery of Japan. Moreover, from the results obtained from the routes of ships, the influence is seen from transport of such air pollutants spread as far as the West Pacific Ocean 2000 to 3000 km away from the sources of generation, clarifying the strong influence of long-range transport of air pollutants.

Because the sulfate aerosol in the atmosphere is neutralized quickly by NH_3 emitted from soil by microorganisms and human activities, it exists in an almost neutralized state on land. However, in the case of oceans where there are no sources of NH_3 emission, the sulfate aerosol exists in the chemical form of strong acids such as H_2SO_4. Therefore, the anthropogenic sulfur transported from East Asian areas is not neutralized in the marine atmosphere, and a part thereof exists as a strong acid such as H_2SO_4, causing acidification of the marine atmosphere in the sea near Japan.

References

1. Akimoto H, Narita H (1994) Distribution of SO_2, NOx and CO_2 emission from fuel combustion and industrial activities in Asia with 1° × 1° resolution. Atmospheric Environment 28(2):213–225
2. Arndt RL, Carmichael GR, Streets DG, Bhatti N (1997) Sulfur dioxide emissions and sectorial contributions to sulfur deposition in Asia. Atmospheric Environment 31(10):1553–1572
3. Chin MA, Jacob DJ (1996) Anthropogenic and natural contributions to tropospheric sulfate: A global model analysis. Journal of Geophysical Research 101(D13):18691–18699
4. Spiro PA, Jacob DJ, Logan JA (1992) Global inventory of sulfur emissions with 1 degrees × 1 degrees resolution. Journal of Geophysical Research 97(D5):6023–6036
5. Streets DG, Carmichael GR, Arndt RL (1997) Sulfur dioxide emissions and sulfur deposition from international shipping in Asian waters. Atmospheric Environment 31(10): 1573–1582
6. Bhatti N, Street DG, Foell WK (1990) Acid rain in Asia. Proceedings of the Second Annual Workshop on Acid Rain and Emissions in Asia. Foell WK, Sharma (Eds.):5–33, AIT, Bangkok
7. Sekine Y, Hashimoto Y, Nakamura T, Chen Z, Mitsuzawa S (1992) Characterization of Atmospheric Aerosol Components at Beijing, China. Journal of Japan Society Air Pollution 27(5):237–245
8. Hashimoto Y, Kim HK, Otoshi T, Sekine Y (1991) Air quality monitoring at Seoul, Korea, May 1986–March 1989. Journal of Japan Society Air Pollution 26(1):51–58
9. Ferm M, Rodhe H (1997) Measurements of air concentrations of SO_2, NO_2 and NH_3 at rural and remote sites in Asia. Journal of Atmospheric Chemistry 27(1):17–29
10. Environmental Information Center (1995) Data book of monthly averaged concentrations in the air pollution monitoring station (a MO publication), Tokyo, Japan
11. Japan Society for Air Pollution (1997) The global atmospheric environment problem and its environmental conservation policies a point of view from Asia (in Japanese). Ohmsha, Tokyo, p 181
12. Shen J, Zhao Q, Tang H, Zhang F, Feng Z, Okita T, Ogura N, Totsuka T (1995) Concentrations and depositions of SO_2, SO_4^{2-} etc. in a Chongqing suburban forested area. Water Air and Soil Pollution 85(3):1299–1304
13. Ichikawa Y, Fujita S (1995) An analysis of wet deposition of sulfate using a trajectory model for East Asia. Water Air and Soil Pollution 85(4):1927–1932
14. Komazaki Y, Hamada Y, Hashimoto S, Fujita T, Tanaka S (1999) Development of an automated, simultaneous and continuous measurement system by using a diffusion scrubber coupled to ion chromatography for monitoring trace acidic and basic gases (HCl, HNO_3, SO_2 and NH_3) in the atmosphere. Analyst 124(8):1151–1157
15. Pszenny AAP, Keene WC, Jacob DJ, Fan S, Maben JR, Zetwo MP, Springeryoung M, Galloway JN (1993) Evidence of inorganic chlorine gases other than hydrogen-chloride in marine surface air. Geophysical Research Letters 20(8):699–702
16. Saltzman ES, Savoie DL, Prospero JM, Zika RG (1986) Methanesulfonic Acid and Non-Sea-Salt Sulfate in Pacific Air: Regional and Seasonal Variations. Journal of Atmospheric Chemistry 4:227–240

Demonstrating the Safety of Repositories for the Final Disposal of Radioactive Wastes

William M. Miller

Summary. Radioactive wastes are produced by civilian nuclear power generation programs as well as by some military, medical, industrial, and scientific research activities. The preferred option for the disposal of these wastes is burial in purpose-built repositories. Near-surface, engineered facilities are already in operation in a number of countries for the disposal of the least radioactive wastes. The most radioactive wastes are planned to be buried as much as 1 km underground in mined repositories containing a series of engineered barriers designed to provide long-term containment of the waste. It is anticipated that groundwater will eventually degrade these barriers and potentially may transport a small proportion of the waste back to the surface. However, by the time this process could occur, natural radioactive decay will have reduced the activity of the waste to levels equivalent to those in the surrounding rock. To demonstrate quantitatively the future safety of radioactive waste repositories, mathematical models are used to simulate these degradation, transport, and decay processes and to calculate the radiological hazard to future populations in the vicinity of the repository. Observations from geologic systems containing natural radionuclides, such as uranium ore bodies, are combined with laboratory experiments to condition and validate these models. Extensive modeling studies undertaken in a number of countries indicate that the releases to the surface which might occur in the future should pose no significant health or environmental risk.

Key words. Radioactive waste disposal—Natural analogues—Performance assessment

Introduction

Radioactive (nuclear) wastes are unwanted by-products of civilian nuclear power generation, military nuclear programs, and, to a lesser extent, medical, industrial, and scientific research activities. The safe disposal of these wastes presents a unique technological problem for our generation because they will remain dangerously radioactive and, therefore, must be isolated from the human environment, for periods of up to 10^4 to 10^6 years in the future.

QuantiSci Ltd., 47 Burton Street, Melton Mowbray, Leicestershire, LE13 1AF, UK

Most industrialized countries have some radioactive wastes that require disposal, although the volumes and types of these wastes vary considerably from country to country. The preferred disposal option for these wastes is underground disposal in purpose-built radioactive waste repositories, containing a series of engineered barriers to provide long-term containment of the waste. However, other disposal options have been considered and a few have been practiced to some extent. The location and design of a repository depend partly on the level of radioactivity and partly on the physical and chemical properties of the waste it is intended to contain. Several repositories around the world have already been built to contain wastes with low levels of radioactivity, while repositories for wastes with the highest levels of radioactivity are now being planned.

It is anticipated that some residual waste radionuclides might migrate away from a repository and return to the surface in very low concentrations at some time in the distant future as the result of degradation of the waste and of the engineered barriers and subsequent transport by groundwater. However, all repositories are designed to contain the majority of wastes for so long that, by the time such releases might occur, natural radioactive decay will have reduced their activity to levels similar to those of the surrounding rocks or of natural ore bodies. To indicate this process, a containment period equal to 10 half-lives leads to a reduction in radionuclide inventory by three orders of magnitude. The optimal containment period depends on the type of waste and its radionuclide content. For most wastes, containment periods from a few hundred years to many thousands of years are required. Eventually, radioactive decay will convert the wastes to stable, nonradioactive materials.

A primary objective in radioactive waste disposal is to evaluate the safety of a repository in a quantitative and robust manner, and to demonstrate that any releases which might take place in the future will pose no significant health or environmental risk. These evaluations are based on complex mathematical models that simulate all the most important processes which control the evolution of the repository, such as degradation of the waste and the engineered barriers, radioactive decay, groundwater flow, and contaminant transport. To condition and validate these models, a considerable amount of both qualitative and quantitative information is obtained from laboratory studies. However, standard laboratory experiments cannot replicate either the complexity of the repository physicochemical environment or the long time periods involved in waste disposal. There are considerable uncertainties associated with extrapolating laboratory measurements made over days or weeks to the thousands of years required for radioactive waste disposal.

To circumvent these problems, laboratory experiments are supplemented by studies of natural systems that have some similarity with a repository, so-called natural analogue studies [1]. Examples of natural analogue studies are investigations of uranium ore bodies to obtain information on the geochemical processes that affect the transport of uranium and its daughter nuclides, and investigations of the long-term stability of engineering materials such as steel and concrete by examination of archaeological artifacts made from similar materials.

The disposal of radioactive wastes is in contrast to the waste management practices often applied to other types of industrial waste such as toxic chemicals and heavy metals. Many of these wastes will remain hazardous forever, but unfortunately the controls on the disposal of these wastes are usually far less strict than those applied to radioactive wastes and, as a consequence, the immediate environmental risk posed by these other wastes is often far greater. One future benefit of radioactive waste disposal research could be the transfer of safety standards and practices from radioactive wastes to other hazardous wastes.

Origin of Radioactive Wastes

The majority of the radioactive wastes created around the world are the by-products of electricity generation using nuclear power [2]. Large amounts are also generated in some countries by military activities (e.g., nuclear weapons manufacturing and decommissioning, and nuclear-powered warships), although these military wastes are often managed separately from civilian wastes. In addition, various radioactive wastes are also created in relatively small amounts by a large number of industrial, medical, and scientific research activities that use radioactive materials. These last uses mean that most industrialized countries generate some radioactive waste, even if they do not have active nuclear power or nuclear weapons programs.

The nuclear fuel cycle (Fig. 1) is the term used to encompass all activities associated with the production of nuclear power for electricity generation. Wastes are generated at each stage, and these wastes are variously radioactive. At the front end of the fuel cycle (the part before fuel reaches the nuclear reactor), uranium mining and milling operations generate large masses of natural material, such as displaced rock, and discarded waste, mainly in the form of mill tailings, which are the residual materials from ore processing. On average, 86 000 tonnes of uranium-depleted tailings are produced every year for each reactor [2]. These wastes are not normally considered for disposal in a repository and are treated separately. However, they emit levels of radiation above average natural background, and care is required to ensure safe management of this processed material.

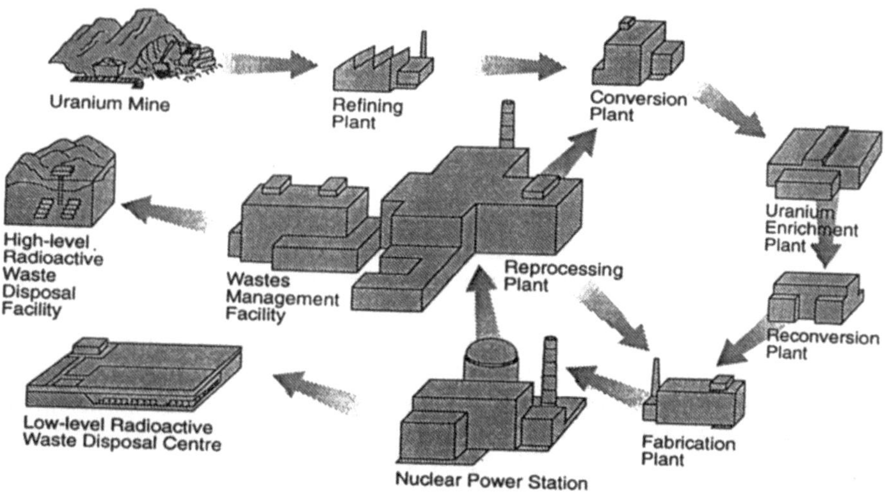

FIG. 1. The nuclear fuel cycle. Radioactive wastes are generated at each stage, and many of these will be sent for disposal in a repository. Spent fuel may be sent for direct disposal to a repository or may be reprocessed to extract unburnt ^{235}U for inclusion in new fuel elements

Most nuclear reactors burn uranium dioxide fuel (UO_2). The ^{235}U content is enriched from the natural isotopic abundance of about 0.7% to about 3% in the fuel. Wastes are produced during the uranium conversion and enrichment processes and during fuel fabrication, and some of these wastes may be scheduled for disposal in a repository, depending on their levels of radioactivity.

The routine operation of a nuclear power plant generates many different types of waste in both liquid and solid form. The most significant waste, in terms of both volume and activity, are ion-exchange resins that are used to recover radionuclides from liquid wastes. These liquid wastes are generated during routine decontamination procedures, either from chemical processes (e.g., primary coolant and fuel storage pond cleanup) or from detergent waste (e.g., laundry and personnel decontamination). Additional radioactive liquid streams are produced during solid waste treatment and volume reduction processes. The raw solid wastes are classed either as "wet" (e.g., sludges, ion-exchange resins) or "dry" (e.g., rubber gloves, paper tissue). The ion-exchange resins are generally dehydrated and powdered, and the final powder is solidified in either a bitumen or cement matrix. Other solid wastes are compacted to reduce their volume and are then variously packaged in steel or cement containers before being solidified in cement or bitumen. The low-activity, combustible, solid wastes may be incinerated to reduce their volume, creating an ash that is then immobilized in a cement matrix. Incineration itself generates secondary wastes such as the air filters that are used to clean the incinerator flue gases. These secondary wastes must also be processed and become part of the total inventory of wastes to be disposed.

The UO_2 fuel burned in a reactor becomes "poisoned" over time with nuclear fission products and transuranic elements produced as a result of the nuclear reactions. These poison elements reduce the efficiency of the fuel, and consequently it must be replaced every few years. This spent (used) fuel may be disposed of directly to a repository or it can be reprocessed to extract the unburnt ^{235}U for incorporation in new fuel elements. Plutonium may also be extracted for use as a fuel in certain reactors or for military applications. Reprocessing is not undertaken in all countries. Some countries such as Sweden and Spain plan to dispose of spent fuel directly to a repository. Other countries such as the United Kingdom, France, and Japan currently operate, or are building, industrial-scale reprocessing facilities.

Reprocessing produces large volumes of waste with a broad spectrum of physical and chemical compositions and varying levels of radioactivity. The most radioactive wastes initially are in liquid form, produced after dissolution of the spent fuel and chemical extraction of the uranium and plutonium. These highly radioactive liquids are subsequently solidified by mixing into a glass matrix. Less radioactive reprocessing wastes comprise a range of materials from ion-exchange resins used in the extraction process to swabs used for cleaning laboratory benches and disposable clothing. These lower-activity wastes thus contain a high proportion of solid, organic material. To facilitate disposal, these wastes are compacted and immobilized in either a cement or bitumen matrix in the same manner as the operational wastes from the reactor. Organic wastes present a particular problem in a repository because they can generate large volumes of gas, principally CO_2, H_2, and CH_4, which may incorporate some radionuclides directly (e.g., as $^{14}CO_2$). These gases are relatively mobile and may migrate rapidly away from the repository. In addition, the degradation of organic wastes can generate populations of colloids and organic complexants that enhance radionuclide solubility in the groundwaters.

At the end of the operational life of a nuclear facility, it must be decommissioned. These facilities include nuclear power plants and fuel fabrication and reprocessing systems. Various decommissioning options are possible, but all involve partial or complete dismantlement of the facility and associated buildings. Some of the materials generated by this process are radioactive. For example, in a nuclear power plant, the reactor vessel and the surrounding reinforced concrete structure (the biological shield) become radioactive because of neutron irradiation by the reactor core, while other components, such as primary coolant piping, may contain residual radioactive liquids or solids. These decommissioning wastes may be scheduled for disposal in a repository, depending on their levels of radioactivity.

Most radioactive wastes created in the nuclear fuel cycle exist in solid form (or are solidified), but a small proportion of very low activity wastes arise in liquid or gaseous waste streams and may be discharged to the environment directly from nuclear installations. Solid radioactive wastes may be categorized in several different ways, such as by the manner in which they were created, or by their physical or chemical form. However, in most cases, these wastes are categorized by level of radioactivity. Usually, three categories of radioactive waste are recognized [2]:

1. High-level waste (HLW) includes the spent fuel and the solidified (glass) waste streams generated during reprocessing of spent fuel. These wastes are characterized by high levels of radioactivity, contain a component of very long lived radionuclides, including heavier-than-uranium (actinide) elements, and generate a considerable amount of heat. The radioactivity of these wastes generally ranges from 10^{16} to 10^{18} Bq/t.
2. Intermediate-level waste (ILW) includes a diverse range of materials such as the ion-exchange resins and wastes from normal reactor operations and from spent fuel reprocessing. These wastes are generally solidified in either a cement or bitumen matrix. They are characterized by significant levels of radioactivity and have a component of long-lived waste nuclides but are not heat generating. The radioactivity of these wastes is generally higher than 10^9 Bq/t but below the limit for HLW.
3. Low-level waste (LLW) essentially includes all other wastes that have a radioactivity content below the threshold for ILW but above normal background or exemption levels. These wastes generally come from normal reactor operations, fuel fabrication, and reprocessing. These typically include the "dry" operational wastes (e.g., paper tissues and disposable clothing) that are compacted and packaged in steel or cement containers.

As indicated, all radioactive wastes are conditioned to create a stable, solid wasteform, such as the spent fuel itself or a glass, cement, or bitumen matrix. These wasteforms are designed to be both physically and chemically inert in a repository environment. The wasteforms are variously packaged in metal or reinforced concrete containers before being emplaced in a repository. However, the heat-producing HLW will be stored on the surface for about 50 years to allow its radiogenic heat output to decay before being placed in a repository.

Disposal Options

Many different options for the disposal of radioactive wastes have been proposed over the past few decades and a few have been practiced to some extent in the past [2]. The most commonly discussed alternatives include these:

- Transmutation
- Disposal into space
- Disposal in the polar ice caps
- Disposal on or beneath the seabed
- Surface storage until radioactivity levels decay to safe limits
- Shallow land burial
- Deep geologic disposal

Transmutation, the conversion of radioactive materials to stable elements using controlled nuclear reactions, theoretically can reduce the proportion of longer-lived radionuclides in the wastes but currently no practical process has been developed that can deal with the industrial volumes of radioactive wastes which currently exist. However, research into this option is still in progress.

Disposal into space (or firing the waste into the sun) has not been investigated in any great detail because the technique, using current rocket technology, is extremely expensive and potentially dangerous, as highlighted by the explosion during take-off of the *Challenger* space shuttle in 1986.

Suggestions to dispose of radioactive wastes in the polar ice caps have not been pursued because of political opposition and difficulties in demonstrating long-term safety. In contrast, disposal on the seabed has been carried out in the past in deep waters under international agreement by some countries, such as the United Kingdom. Furthermore, comprehensive international research programs to develop and assess methods for disposal under the seabed (subseabed disposal) were undertaken in the 1970s and 1980s [3]. Seabed disposal by any method is now prohibited by international agreement [4]; however, several studies have shown that subseabed disposal may actually be, in radiological terms, a very safe option [5].

Surface storage is frequently promoted by environmental organizations opposed to the burial of radioactive wastes. However, this is a practical option only for the short-lived wastes because the necessary institutional and management control cannot be guaranteed for the very long (thousands of years) isolation period required for the longer-lived wastes [6]. Furthermore, in many parts of the world, future environmental changes (such as sea level rise or glaciation) will mean that it is impractical to leave the wastes on the surface.

Given the current situation, the majority of effort in most countries is focused on disposal in repositories. Although it is possible that other alternative disposal technologies may be developed in the future, ethical considerations suggest that disposal should be undertaken now by the generation that created the waste and which benefited from the nuclear power, rather than leaving the problem to future generations to solve.

Repository Designs and Siting

There are basically two repository designs for the disposal of radioactive wastes that are actively being considered, although other designs and variants have been suggested. These are shallow (to a few meters deep) repositories for LLW comprising surface or near-surface bunkers covered with engineered barriers, and deep (50–1000 m) underground burial in mined repository excavations for HLW and ILW. To be feasible, a repository must meet a number of fundamental design requirements, the most basic being that the repository achieves the following aims:

- Ensures the waste will not be released to the surface environment in concentrations that would represent an unacceptable hazard
- Effectively isolates the waste from the effects of human activity or catastrophic natural events
- Is capable of being built using available technology and achievable at a reasonable cost
- Is sufficiently simple such that its future behaviour and safety can be assessed quantitatively and reliably

In addition, certain other requirements might be placed on some disposal system designs, such as to ensure that retrieval of the waste is both technologically and economically feasible if so desired by future generations.

The exact design details of any repository depend on the level of radioactivity and on the physical and chemical properties of the waste, the geologic environment at the disposal site, and any engineering constraints imposed by the host rock. However, all modern repository designs are based on the multibarrier concept whereby the wastes are emplaced inside a series of nested engineered structures and natural barriers that act in concert to contain the waste for long periods of time. These barriers are generally devised such that they have no common failure mechanism. Meaning that if some event or process were to cause one barrier to fail, this should not lead to the other barriers also failing. Because of the long-lived radionuclide inventory in HLW and ILW, repository designs for these wastes tend to be deeper and more robust than those for LLW.

Repository Designs for LLW

The disposal of LLW in near-surface or shallow repositories is practiced routinely in many countries, including Japan, Spain, France, the United Kingdom, and the United States. These repositories are located at or close to the land surface because the low activity and short half-lives of the waste they contain mean that they do not require the very long isolation times necessary for HLW or ILW. The basic design of a modern LLW repository involves a series of reinforced concrete bunkers or trenches that may be located on the ground surface or set fully or partially into the ground (Fig. 2). The waste itself is compacted and solidified in cement and contained in large concrete boxes or steel containers in most cases. These waste packages are stacked in the bunkers and trenches, and all residual spaces between them are backfilled with earth, clay, or cement. After each trench or bunker is filled, it may be closed with an "anti-intrusion" slab of reinforced concrete designed to prevent inadvertent excavation into the wastes in the future. When all disposal activities are completed, the entire repository system will be covered with a low permeability earth cap to provide an outer protective barrier that may form a mound or hill. This cap often comprises layers of low-permeability clay or an impermeable membrane to reduce water percolation into the repository. Specially chosen vegetation is often planted on top of the earth cap to further limit water percolation. In some designs, drainage systems below the bunkers are included.

Repository Designs for ILW

A number of ILW repositories have now been built and are currently receiving waste in Sweden, Finland, and the United States, and several others are in the planning stage. Some of these are designed to contain both ILW and LLW. Typically, ILW repositories are located at depths about 50 to 100 m below the ground surface, although some designs are

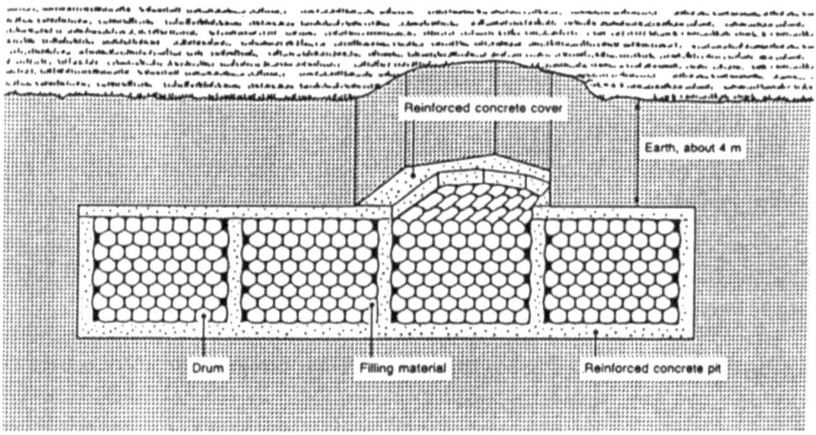

FIG. 2. A generic low-level waste (LLW) repository in which the waste packages are contained in reinforced concrete bunkers and covered with a concrete cap and several meters of earth. Some LLW repository designs are above ground, and the earth cover will form a hill. LLW repositories are already in operation in a number of countries

deeper. The excavations for an ILW repository generally consist of a number of large disposal caverns (sometimes referred to as vaults), and certain designs also include a reinforced concrete silo structure set into a cavern to contain the higher-activity wastes, such as the Swedish ILW repository at Forsmark (Fig. 3). The ILW is generally solidified in a cement matrix, although bitumen or resins are used for certain types of waste, and contained in steel drums or large concrete boxes, depending on the waste type and repository design, to provide a stable waste package. In a typical repository, these waste packages are stacked in the caverns and silo, and all residual spaces between them will be backfilled with concrete. The large volumes of concrete in a typical ILW repository have an important safety role. As groundwaters infiltrate the repository after closure, they will slowly dissolve the cement minerals, creating hyperalkaline conditions with a pH around 13, and will also corrode the steel canisters, creating chemically reducing conditions. High-pH, chemically reducing environments are favorable because many radionuclides are poorly soluble under such conditions.

Repository Designs for HLW

No repository for HLW has yet been built, although a number of conceptual designs have been developed and several countries are involved in site selection activities. Most concepts for HLW repositories have a great deal of similarity, and the designs for the high-activity glass-reprocessing wastes and unreprocessed spent fuel are essentially the same. The basic design features for a HLW repository include a series of tunnels in which the waste will be emplaced (sometimes referred to as galleries), which are excavated at depths typically between 500 and 1000 m, and a number of vertical or inclined shafts that connect the repository to the surface (Fig. 4). The glass or spent fuel waste will be placed in massive, thick-walled metal canisters to isolate the waste from groundwaters for long periods of time. The metal used may be either steel or an unreactive metal such as copper

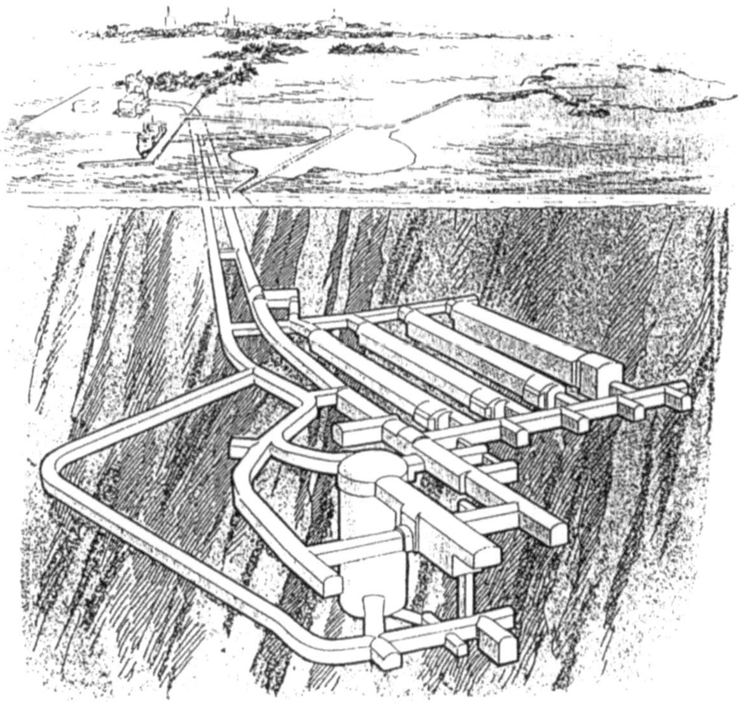

FIG. 3. Artist's impression of the Swedish ILW (intermediate-level waste) repository at Forsmark. This repository consists of several caverns to take the lower-activity wastes and a reinforced concrete silo for the most radioactive wastes, such as ion-exchange resins. The repository is located 60 m below the Baltic Sea

or titanium. Steel has the disadvantage of corroding faster than the other metals but, in so doing, will buffer the redox conditions by scavenging free oxygen from the groundwaters, thus maintaining strongly reducing conditions.

Chemically reducing environments are again favorable because many radionuclides are poorly soluble under such conditions. Copper and titanium corrode much more slowly than steel, providing a potentially longer canister lifetime but less capacity for controlling the redox conditions. Several geometric options for locating the canisters in the tunnel are under consideration. The simplest has the canisters arranged axially at intervals along the tunnel. Alternative designs locate the canisters in individual disposal holes drilled into either the floor or the walls of the tunnels. All the spaces between the canisters and the rock will be filled with a buffer of compacted bentonite clay. This clay material has the unusual property of swelling, when water saturated, to about double its dry volume. In the disposal tunnel, where room for swelling will be restricted, the bentonite will become very compressed and effectively impermeable to water flow. Transport of water from the rock to the canister, and subsequently of dissolved radionuclides from the canister to the rock, will thus occur by diffusion rather than advection.

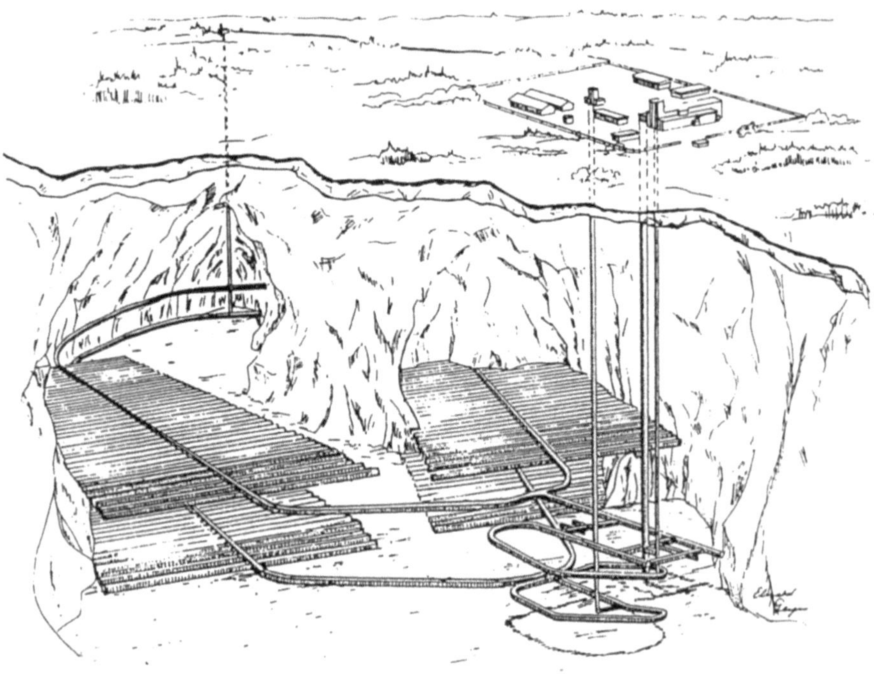

FIG. 4. Artist's impression of the proposed Swedish HLW (high-level waste) repository for spent fuel. The spent fuel is contained in massive metal canisters that are emplaced in the many parallel disposal tunnels, arranged in "panels". Alternative tunnel geometries are possible, controlled partly by the engineering properties of the rock

Repository Siting

Choosing a suitable site for a repository requires consideration of both geologic and environmental or social criteria. The host rock should be relatively stable, with a low hydraulic permeability to limit the groundwater flow around the repository. The basic function of the rock is to provide a physically and chemically stable cocoon for the repository, protecting the engineered barrier system and the wastes. The selection of suitable host rocks for deep repositories has concentrated on three principal rock types:

- Crystalline (igneous and metamorphic) basement rocks such as granites and gabbros
- Low-permeability sedimentary sequences such as clays
- Thick evaporite (salt) deposits

The groundwater flow conditions are important when selecting sites because they provide an important control on the transport of radionuclides away from the repository. Consequently, considerable effort is made to characterize the groundwater system at potential repository sites. In rocks with plastic behavior, such as clays and evaporites, groundwater and contaminant movement occurs predominantly by diffusion, which is a much slower mechanism than groundwater flow. In contrast, hard crystalline rocks are often fractured, and groundwater flow will occur through the fracture networks. However, repository construction in crystalline rocks is often easier than in plastic rocks because they are stronger and can be self-supporting. The mineralogy and geochemistry of the host rock are also important because the mineral surfaces exposed to groundwaters containing radionuclides provide sites for radionuclide precipitation and sorption (electrostatic attraction between charged mineral surfaces and radionuclides in oppositely charged ionic form). The potential for precipitation and sorption is strongly controlled by the geochemistry of the groundwaters, which is itself largely determined by the composition of the minerals in the rock.

Various environmental and social criteria may be used in the site selection process. If possible, a site may be chosen close to the existing nuclear facilities that generate the waste and to transport networks to optimize movement of the waste to the repository. Consideration is also given to proximity to urban populations and to current land use.

Evaluating Future Repository Safety

The future safety of a repository will be evaluated using a series of computer codes that simulate repository behavior in a *performance assessment*. These computer codes are derived from conceptual models that describe all the most important processes which act on the repository system, such as degradation of the waste and the engineered barriers, radioactive decay, groundwater flow, and contaminant transport. The total safety of the repository will be judged by the calculated hazard to humans and the environment over a range of future time periods. A license to build and to operate a repository will be granted only after detailed performance assessment modeling has provided adequate demonstrations that the hazard from the repository is below strict legally defined safety (or licensing) criteria. Normally such criteria are defined in terms of radiological dose or risk to humans, where risk is defined as risk of death due to cancer arising from exposure to repository releases. Typical quantitative criteria for a specific target individual or group of individuals in the vicinity of the repository are as follows [2]:

- A radiological dose <1 mSv/year (often <0.1 mSv/yr)
- A risk target $<10^{-6}$/year

Depending on the type of waste, quantitative assessments of dose and risk may be required for some specified future postclosure time period, which is generally up to 10^4 or 10^6 years.

Because the repository system is very complex, the conceptual models used to describe its behavior are, by necessity, simplifications or abstractions. Furthermore, due to the computational limitations of current computers, the mathematical models (computer codes) derived from the conceptual models may be simplified still further. Although this may seem, at first sight, to be an inappropriate way to handle such an important issue, a large safety margin is built into the assessment by the use of "robust" or "conservative" assumptions when making abstractions of the repository system. In so doing, it is acceptable and common practice to exclude from the computer code representations of processes that may act to increase repository safety but which are hard to quantify. For example, a process known to occur that acts to retard or hinder radionuclide transport can be safely omitted from the models because the consequence is only to underestimate repository safety in the final analysis.

Clearly, the primary objective when making abstractions for the computer codes is to ensure that all processes and their couplings that act to reduce repository safety are accurately represented in the code. However, a pragmatic approach is usually adopted and codes are designed to be conservatively realistic, meaning that they include all the most important or first-order processes, those that act to increase or decrease safety, but with an emphasis on the latter. At the most basic, performance assessments require models to simulate the following:

- Degradation of the engineered barriers that contain the waste
- Dissolution of the solid wasteform
- Solubility (concentration) of the radionuclides in the groundwater
- Transport of groundwaters and dissolved radionuclides through the engineered barriers and the surrounding rock formations to the surface
- The various processes that may retard radionuclide transport in the engineered barriers and the rock, such as sorption and precipitation
- The incorporation of radionuclides into the food chain and eventual uptake by humans

A comprehensive performance assessment requires a chain of codes, with a separate code constructed to represent each of these processes. These codes are then linked together, with the output from one model fed as input to the next (Fig. 5). The output from a performance assessment is usually expressed graphically, showing calculated radiological dose (or risk) and its variation with time after repository closure. An example is given in Fig. 6, which shows the results from a number of performance assessments for HLW

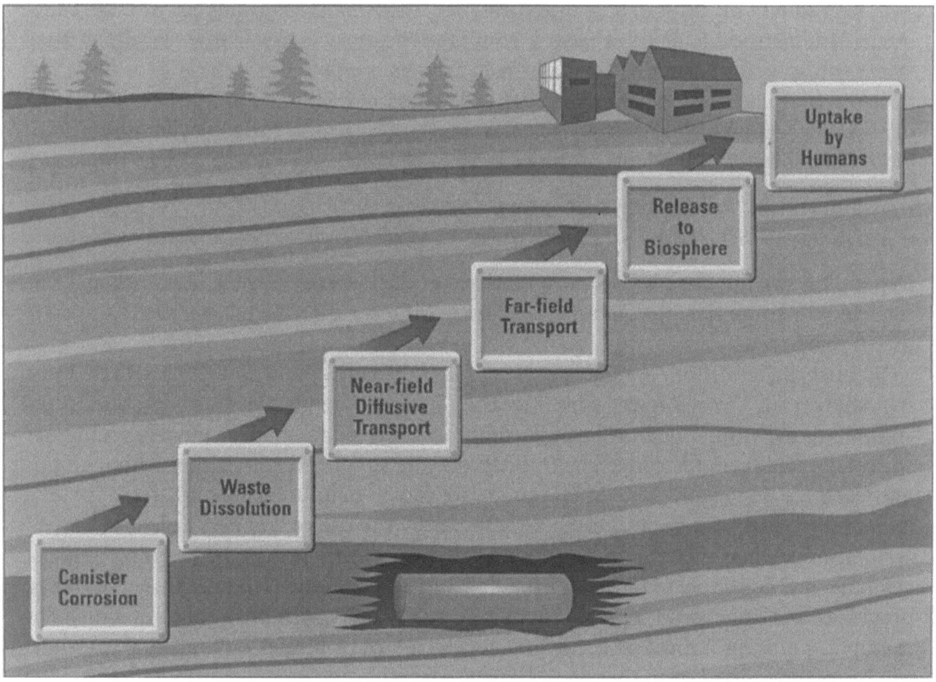

FIG. 5. Diagrammatic representation of the model (code) chain employed in a comprehensive performance assessment. Different models simulating the behavior of each part of the repository system are linked together to evaluate the overall safety of the repository

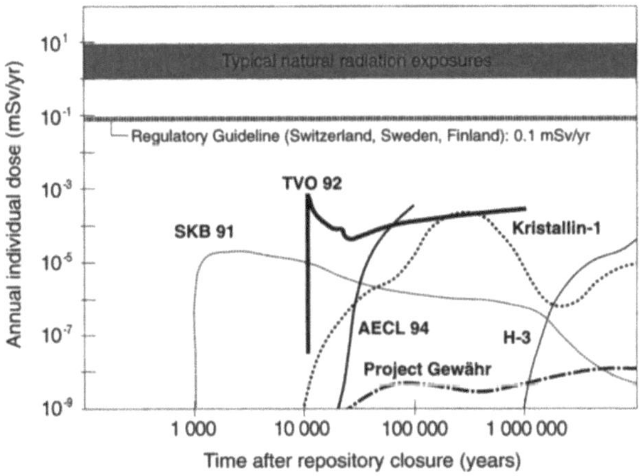

FIG. 6. Example outputs from a number of performance assessments for different HLW repositories. In this graph, the outputs are expressed as calculated radiological dose to target individuals in the vicinity of the repository. The different release curves reflect differences in the wastes and the repository designs, as well as variable assumptions made in the performance assessments. In all cases, the releases are several orders of magnitude below the regulatory limits. The assessments are from different countries: *SKB 91*, Sweden; *TVO 92*, Finland; *AECL 94*, Canada; *H-3*, Japan; *Kristallin-I* and *Project Gewähr*, Switzerland

repositories [7]. It can be seen that, for all cases, the calculated doses are below the safety criteria limit (1 mSv/year, in this case) for all times. Furthermore, the graph also shows that the calculated repository releases are at least three orders of magnitude less than the radiation exposure from natural background sources, providing some indication of the relative hazard presented by a HLW repository.

Natural Analogue Studies

The codes and models used in a performance assessment must be conditioned and validated to ensure that they are capable of describing future repository evolution. Laboratory studies provide a large amount of the information required for these purposes, but they are limited because standard laboratory experiments are unable to replicate either the complexity of the repository physicochemical environment or the long time periods involved in waste disposal. For this reason, a considerable amount of interest has been shown in natural analogue investigations of geologic or archaeological systems that exhibit some similarity with a repository system [1]. Essentially, natural analogues provide very long term experiments in which many processes relevant to the repository can be observed and measured, such as migration of naturally occurring radionuclides or the degradation of materials such as steel or concrete. However, natural analogues also have their limitations. Often the boundary conditions of geologic systems cannot be accurately defined, such as the age of the system, and their very complexity often makes it difficult to isolate and quantify the effects of a single process. Nonetheless, natural analogues provide the only way to examine the long-term operation of repository relevant processes in the geosphere, avoiding uncertainties caused by the extrapolation of short-term laboratory experimental data.

Many natural analogue studies have examined the geochemical systems in and around uranium ore bodies. A common uranium ore mineral is uraninite, which has the same bulk composition and mineralogical structure as nuclear fuel, that is, UO_2. Uraninites from many ore bodies in a number of locations have been examined as natural analogues. Some of the most detailed studies have taken place at Cigar Lake, Canada [8]; Poços de Caldas, Brazil [9]; Alligator Rivers, Australia [10]; Tono, Japan [11]; and El Berrocal, Spain [12]. A number of these uranium ore bodies are hundreds of millions of years old and thus demonstrate the potential long-term stability of UO_2 in geologic systems.

The degree of analogy between a uranium ore body and a repository can be demonstrated by comparison of the Cigar Lake ore body in Canada with the Canadian concept for a HLW repository for spent fuel (Fig. 7). The Cigar Lake ore body is one of the largest and richest uranium deposits yet found. It is located at a depth of 430 m, at the contact between metamorphic rocks and overlying younger sandstones. The ore body formed around 1300 million years ago, when uranium-rich, oxidizing groundwaters flowing through the permeable sandstone encountered strongly reducing groundwaters flowing upward through fractures in the metamorphic rock. The rapid change in redox

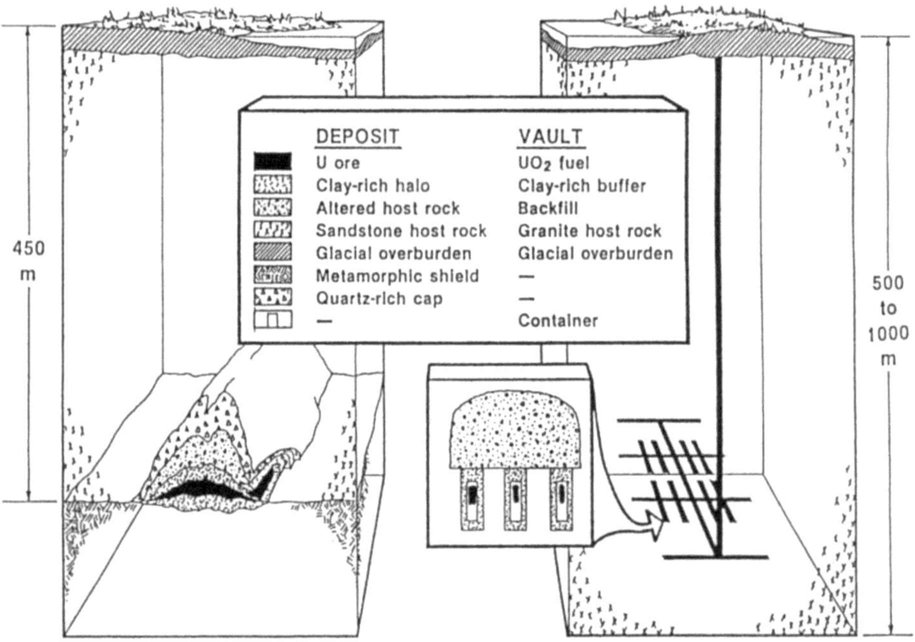

FIG. 7. Demonstration of the analogy between the Cigar Lake uranium ore body in Canada and the proposed design for a Canadian repository for spent fuel. Both the ore body and the repository are located hundreds of meters below ground in stable rock; both comprise a UO_2 matrix surrounded by a clay envelope. The main difference is that there is no metal canister analogue at Cigar Lake. Despite this, there is no geochemical evidence for any radionuclide migration having taken place from the ore to the surface

conditions caused the uranium to precipitate as uraninite and so form the ore body. The high-temperature conditions at the time of ore body formation caused alteration of the surrounding sandstone to form a clay rich halo around the ore [8]. In this respect, the Cigar Lake ore body demonstrates all the main components of a HLW repository apart from the metal canister. An important lesson from Cigar Lake is that there is no geochemical evidence at the ground surface for the presence of the ore body. The ore body was only detected using remote geophysical techniques that measure differences in the magnetic and gravitational fields between the ore and the rock. There are no measureable increases in the concentrations of any elements from the ore body, including uranium, or any increase in radioactivity in the soils, lake waters, or plants at the surface. The implication is that no significant radionuclide transport has occurred between the ore body and the surface over the past 1 billion years.

Despite the obvious similarity in bulk composition and mineralogical structure between spent fuel and uraninites from ore bodies such as Cigar Lake, there are also significant differences. Most importantly, spent fuel is enriched in ^{235}U compared to the present-day natural isotopic composition of uranium in normal uraninites and, due to the fission processes in the reactor, spent fuel also contains a range of fission products, actinides, and their daughters that do not occur in most natural uraninites. However, there is one particularly interesting uranium ore body at Oklo, Gabon, where some of these features can be investigated, because this ore body experienced nuclear criticality.

Criticality at Oklo occurred approximately 2 billion years ago, at a time when natural uranium had a high isotopic abundance of ^{235}U (which has since decayed to the current low levels). However, at that time, uranium was widely dispersed in throughout the rocks and sediments in mineral form, and critical masses could not accumulate because surface conditions were chemically reducing, meaning that uranium remained as a solid and so could not easily be mobilized. At the onset of the evolution of plants capable of photosynthesis, surface conditions gradually became chemically oxidizing, allowing uranium in near-surface environments to go into solution. At Oklo, this newly soluble uranium was mobilized from a primary source and accumulated in a sequence of organic-rich sediments in sufficient mass to achieve criticality. The fission reactions were moderated by water in the sediments and operated intermittently for between 10^5 and 10^6 years [13]. Estimates suggest up to 12 tonnes of ^{235}U underwent fission, producing almost 4 tonnes of plutonium and other fission products. The heat from the reactions (approximately 600°C) was sufficient to cause mineralogical alteration of the surrounding sandy sediments, to form a halo of clay around the reactor cores (Fig. 8). The criticality processes could only have occurred in the relatively short period (thought to be a few million years) between the time the atmosphere became oxidizing and the time radioactive decay reduced natural ^{235}U isotopic abundances to below concentrations where criticality would be possible. As a consequence, the formation of the natural fission reactors at Oklo is an event that it is believed may be unique in Earth's history.

At present-day Oklo and in the surrounding area, a total of 16 separate reactor cores have been identified since 1972, but unfortunately all but 1 have largely been removed by mining. Nonetheless, during the past 25 years, the Oklo natural fission reactors and their surrounding clay aureoles have been studied in great detail. Recent studies [13] suggest that limited dissolution of the uraninite occurred during or soon after criticality but that about 90% of uraninite material has remained in the same spatial configuration since criticality ceased. The actinides formed within the uraninites (Np, Pu, and

FIG. 8. Photograph of one of the exposed natural fission reactor zones at Oklo, Gabon (*the darker rock area*): note the geologist (*in the foreground*) for scale. The reactors went critical approximately 2 billion years ago and produced plutonium and other fission products. The majority of these radionuclides remained in place within the uraninite minerals and did not migrate away from the reactor zones

Am) were also largely retained, as indicated by the presence of stable daughter nuclides, although some fission products that were less compatible with the uraninite matrix (Re, Ru, Te, Nb, Zr, Sb) were partially lost from the matrix but retained within the clay halo. More relevant to repository conditions is the conclusion that virtually no observable uraninite alteration occurred after high-temperature conditions ceased, except when uraninites were exposed to surface conditions, allowing weathering and erosion of the rock to take place.

The studies at Oklo and Cigar Lake, and other natural analogue investigations of uranium ore bodies, provide some of the best evidence to support performance assessment results, which indicate that only very limited releases of radionuclides will occur to the surface from a HLW repository.

Natural analogue studies have also investigated the long-term stability of some of the materials (e.g., metal, cement, and clay) that will be used in the engineered barrier systems in different repository concepts. The corrosion of two metals (copper and iron) in a variety of burial environments has been investigated in a number of archaeological analogue studies. Most studies [14] have been, in the main, qualitative and focused on the different sources of ore and methods of smelting, and how these differences affected the durability of the artefacts and contributed to their survival. Generally, however, the metal compositions and burial conditions were too dissimilar to the repository system to prove very useful. However, one study in particular did provide quantitative values for corrosion rates. In this study [15], both copper, copper alloy, and iron artifacts were examined. Average corrosion rates thus determined were 0.03 to 1.3 μm/year for copper and 0.1 to 10 μm/year for iron. This uniformity is surprising, considering the various environmental conditions from which the artifacts originated, and these values may be considered upper limits to those expected in the repository because of the harsh burial environments experienced by some of the artifacts. To put these values into perspective, the authors suggest that a 10-cm-thick steel HLW canister should survive for at least 10 000 years in a repository environment.

Although these corrosion values are encouraging, they provide only bounding limits to the likely rate of corrosion in a repository because of differences in the chemistries of the burial and repository environments. Unfortunately, it is quite common for there to be little or no information on the chemistry of the burial environment from which an archaeological artifact is recovered because this information has not often been of interest to traditional archaeologists. For studies of this type to progress further, it is essential that soil, sediment, and porewaters are sampled at the time of the excavation to obtain representative chemical analyses. Only then can the measured corrosion rates be put into meaningful context.

Average corrosion rates are not necessarily the best indicator of the lifetime of a disposal canister because it takes only one small hole in a canister to allow radionuclides to escape. Consequently, attempts have been made to quantify the average pitting factor (the ratio between pit depths and mean corrosion rate) for some metal artifacts from repository-relevant environments [16]. This study involved the examination of Roman coins and vases, objects from the Swedish Bronze Age, seventeenth-century coins all made from copper and copper alloy, and one lump of native (metallic) copper. The objects were all of different age, came from a wide range of environments, and had differing compositions, but all (excluding the native metal) indicated pitting factors of less than 3; the native metal had a pitting factor in the range 2 to 6. In a Swedish performance assessment of a generic repository concept, pitting factors of 25 were used: the archaeological evidence shows this to be safely conservative.

One widely quoted study of copper corrosion is that of cannon recovered from the wreck of the sixteenth-century Swedish warship *Kronan* which sank in 1676 in the Baltic Sea during the Battle of Öland [17]. When the *Kronan* sank, copper cannon from the ship became embedded in the marine sediments. The cannon remained partly buried in a vertical position, muzzle down, in clay sediments until recovered in the 1980s. This is an analogous situation to copper canisters surrounded by bentonite clay in a HLW repository concept. The analogy is fairly close because the cannon had a high copper content and the clay was tightly packed and water saturated, although the seawater has a composition somewhat different than deep groundwaters. The sediment porewaters around the Kronan cannon had neutral pH with variable oxidation conditions; porewaters nearest the top of the sediments were more strongly oxidizing due to ingress of seawater. The change in redox condition with depth was evident from a change in sediment color, resulting from oxidation of iron minerals, from brown in the top 10 cm to gray at depth. Identified corrosion products included Cu_2O and Fe_3O_4, suggesting a generally oxidizing environment. A corrosion rate of 0.15 μm/year was calculated, and this was constant over the surface of the cannon. However, inclusions of CuO slag in the bronze have weathered more rapidly. The fact that the cannon has suffered only minor corrosion increases confidence in the suitability of copper as a HLW canister material. Indeed, the corrosion of a copper disposal canister should be slower than that of the cannon because of the less harsh repository environment (reducing rather than oxidizing) and higher-quality manufacturing process (no inclusions).

An interesting archaeological study that demonstrates low iron corrosion rates is that of a hoard of iron nails discovered at the most northerly legionary fortress in the Roman Empire, at Inchtuthil, Scotland. At Inchtuthil, more than 1 million nails were buried in a 5-m-deep pit, which was then covered with compacted earth in a successful attempt to prevent the iron from falling into the hands of Pictish tribes when the fortress was abandoned in A.D. 87. The fortress was excavated, and the nails unearthed, in the 1950s. The nails vary in length from 6 to 40 cm, with smaller ones being most common. Although all the nails were composed predominantly of iron, they were heterogeneous in composition, with regions of high and low carbon content. The surfaces of all the nails exhibited some corrosion but the degree of corrosion was greatly controlled by location in the hoard. The nails on the inside of the hoard show minimal corrosion, limited to the formation of a thin corroded layer, while those on the outside of the hoard, and in particular those at the top, were corroded to such an extent that they formed a solid crust [18]. This crust would presumably have had a low hydraulic permeability and this, combined with the oxygen consumption by the outer nails, ensured that anaerobic conditions were maintained at the center of the hoard, despite its position in the floodplain of the River Tay.

It was noticed that, on a few limited areas on some larger nails, appreciable localized corrosion (pitting) had occurred. Unfortunately, no information is given as to the chemistry of the burial environment or of the location within the hoard of those nails exhibiting pitting, and no quantification of the pitting factor (the excavation was undertaken for archaeological purposes and not as a natural analogue study). It may be that the pitting corrosion is influenced by iron composition. It has been noted that the larger nails have higher carbon contents, but without further investigation this must remain speculation. It is clear that the corrosion rate of the outer nails was fast in the oxidizing conditions but that the large mass of iron probably ensured a negligible corrosion rate at the center of the hoard for almost 2000 years. This situation can be considered as an analogue for the iron canisters planned to be used in a number of HLW disposal concepts [1]. The analogue is not complete, however, because the conditions at Inchtuthil, being strongly oxidizing with a high water flux, are much more aggressive in comparison to the reducing conditions, with a very low water flux, expected in a repository. Nonetheless, qualitatively it may be concluded from this analogue that, where large volumes of iron (or steel) are present in a repository, much of the iron may be largely unaffected by corrosion for a duration in excess of that at Inchtuthil.

Archaeological concretes have also been examined as natural analogues in a number of studies. Concrete and cements have been used since at least the Neolithic period in the Middle East [19], although early cements are chemically dissimilar to the modern Portland cements that will be used in an ILW repository. Portland cements largely comprise calcium silicate hydrate (CSH) compounds, together with other substances to control porosity, permeability, etc. The oldest cements with compositions and mineralogy that closely simulate modern Portland cements were manufactured by the Romans, who sometimes used siliceous materials (pozzuolans) in the cement mixtures, such as crushed vitreous fireclay, siliceous limestones, or cherts. These materials sometimes reacted with free lime to form CSH compounds, although this reaction may be more by accident than by design. Numerous Roman structures are still standing, in part because of the durability of these CSH-bearing cements. Examples include Hadrian's Wall in northern England and the Pantheon in Rome.

The 1700-year-old Roman mortars used in Hadrian's Wall in northern England contain substantial amounts of CSH compounds and have been studied specifically with the behavior of a proposed ILW repository in the UK in mind [20,21]. The origin of the CSH compounds in Hadrian's Wall is thought to be from calcining of siliceous limestones to produce lime or by the inclusion in the mortars of metamorphosed cherts found locally in limestones. Several other studies have been performed to determine the chemistry and mineralogy and to test the physical properties of archaeological concretes and cements with and without pozzuolans. As with the Hadrian's Wall mortars, some other archaeological concretes contained CSH compounds with compositions and structures similar to those found in modern cements. Those concretes with CSH compounds all contained crushed vitreous fireclay as a pozzuolan, as opposed to the usual volcanic material. The CSH compounds are thought to have formed by reaction between the vitreous pozzuolans and free lime: the vitreous fireclay is more reactive than normal pozzuolans. The formation of the CSH compounds reduced the porosity and permeability of the cement, which helped to ensure its preservation for up to 2000 years. The durability of ancient cements was also enhanced by careful selection of materials and good workmanship. The writings of Vitruvius (27 B.C.) make clear that effective quality control was exercised by the Romans.

Portland cement itself was first manufactured in 1824, so that the oldest concretes manufactured specifically to include CSH compounds are only around 150 years old. Several studies have examined the durability of Portland cement-based concretes. Two interesting cases from the UK are those of a 60-year-old concrete from a seawall and a concrete formed from a cement paste consignment found in a shipwreck from 1848 [22]. In the latter case, the paste hardened in its barrels and was later retrieved and the barrel-shaped concrete blocks used to form a sea defense. When examined, the CSH compounds in both cements were found to be indistinguishable from those found in modern Portland cements, despite several decades of exposure to harsh oxidizing aqueous conditions. In another investigation [23], samples of concrete were taken from Britain's first multistory reinforced concrete framed building, Weaver's Mill, Swansea (1897–1898) and its first reinforced concrete marine structure, Woolston Quay, Southampton (1899). The CSH compounds in concrete from these two structures were found not to exhibit any form of degradation, although hydration of the cement was incomplete. It might be expected that further curing and strength gain could occur if hydration proceeds to completion.

The archaeological analogue studies performed to date on concretes have shown that a compositionally wide variety of ancient (more than 1500 years old) and more recent industrial cements show similar results for their durability under quite varied conditions, and that these cements may be expected to survive for as long as thousands of years. However, there is scope for expanding these studies further to examine other issues of relevance to the behavior of cement in a repository environment. In particular, it would be interesting to examine the foundations of piers, dams, thermal baths, etc., where the cement has matured in a water-saturated, chemically reducing, alkaline environment. The latter case, thermal baths, may offer the possibility to investigate the uptake of radionuclides by the cement minerals and their degradation products, because hydrothermal waters often contain elevated concentrations of uranium and its daughter products plus other trace metals of interest.

Cement minerals also occur naturally, but rarely, in a few unusual geologic environments. The most detailed analogue investigation of natural cement minerals was undertaken in Maqarin, Jordan. At Maqarin, the rocks of interest include bituminous marls and limestones [24,25] that contain a rare assemblage of cement minerals, including portlandite. These minerals formed when the bituminous rocks underwent spontaneous combustion caused by the rapid introduction of oxygen after a landslide followed by hydration by infiltrating groundwaters. The present groundwater chemistry is controlled by the dissolution of the cement minerals and is hyperalkaline, with a pH in excess of 12. This process and the resulting geochemical conditions are directly analogous to those predicted to control the evolution of the groundwater chemistry in a cementitious ILW repository.

Natural analogues have also provided information on the long-term stability of clay minerals similar to the bentonite clay planned to be used in HLW repositories. One particularly appealing study was undertaken at Dunarobba, Italy. At Dunarobba, several 2 million-year-old tree trunks were unearthed during quarrying operations for sand. The bases of the tree trunks were still in their original upright, growing positions. Visually, the wood can be seen to have been well preserved, as indicated by the fact that the wood can be sawn, polished, and burned. Subsequent chemical and isotopic studies confirmed that only limited degradation of the natural cellulose had occurred [26,27]. The preservation of the tree trunks has been attributed to a low-permeability clay layer that enveloped the trees, isolating the wood from oxygenating groundwaters. In the sediment sequence at Dunarobba, groundwaters preferentially flowed through the high-permeability sandy horizons. This analogue provides only qualitative evidence for the behavior of bentonite in a repository because the clays at Dunarobba are somewhat different in terms of both mineralogy and structure from the compacted bentonite that will be used in a repository. Generally, it might be expected that the hydraulic conductivity of compacted bentonite would be even lower than in the clays at Dunarobba, meaning that the preservation of the trees might represent a minimum level of the preservation compared to that provided by bentonite in a HLW repository.

These examples represent just some of the many natural analogue studies that have been undertaken during the past 25 years. Combined, these studies provide powerful evidence that the future behavior of the repository can be robustly represented using the currently available performance assessment models, although further work to validate these models by applying them to other natural geologic systems is still required.

Conclusions

The disposal of radioactive wastes underground in specially designed repositories is the disposal option favored in all countries currently investigating the issue. Shallow or near-surface repositories are operating at the present time in several countries and are accepting a variety of LLW and ILW. Conceptual designs and site selection programs for HLW and spent fuel repositories are also well advanced in a number of countries.

Demonstrating the future safety of these repositories using performance assessment modeling approaches has been significantly aided by natural analogue studies of relevant geologic and archaeological systems that simulate some aspect of the repository system. Such studies have been able to provide a range of qualitative and quantitative information on issues as diverse as the stability of UO_2 and the hydraulic properties of clays. Combining natural analogue information with data obtained from laboratory studies is an essential aspect of conditioning and validating the performance assessment models.

Helped by information from these natural analogue studies, numerous performance assessment modeling exercises have been undertaken in many countries during the past two decades for a range of repository designs and for different wastes. Overall, these performance assessments have pointed to the fact that any radionuclide releases to the surface environment that might take place in the future will pose no significant health or environmental risk to future populations in the vicinity of the repository. However, this is no cause for complacency, and further efforts are essential to provide additional proofs of safety. The widespread feelings of unease and mistrust held by the public need to be addressed. In this regard, natural analogues can be used to provide nontechnical demonstrations of safety [28] to all interested parties, including the general public, so as to demonstrate that geologic disposal is a sound and credible technology.

Acknowledgments. The Tokyo Electric Power Company (TEPCO) and Keio University are gratefully acknowledged for inviting the author to participate in the symposium, and for their financial and organizational support. This paper benefited from Phil Richardson's careful review of an early draft.

References

1. Miller WM, Alexander WR, Chapman NA, McKinley IG, Smellie JAT (1994) Natural analogue studies in the geological disposal of radioactive wastes. Studies in environmental science, vol 57. Elsevier, Amsterdam
2. Savage D (ed) (1995) The scientific and regulatory basis for the geological disposal of radioactive wastes. Wiley, New York
3. NEA (1988) Feasibility of disposal of high-level radioactive waste into the seabed. Vol 1. Overview of research and conclusions. Nuclear Energy Agency of the Organization for Economic Co-operation and Development (NEA-OECD), Paris, France
4. Sjoeblom KL, Linsley GS (1994) Sea disposal of radioactive wastes: the London Dumping Convention 1972. Bulletin 2. International Atomic Energy Agency, Vienna, Austria
5. de Marsily G, Berhendt V, Ensminger D, Flebus C, Hutchinson B, Kane P, Karpf A, Klett R, Mobbs S, Poulin M, Stanner D (1988) Feasibility of disposal of high-level radioactive waste into the seabed, Vol 2. Radiological protection. NEA-OECD, Paris, France
6. IAEA (1992) Storage of radioactive waste. TECDOC-653. IAEA, Vienna, Austria
7. Neal F, Smith P, Sumerling T, Umeki H (1995) Putting HLW performance assessment results in perspective. Nagra Bulletin 25:47–55, Wettingen, Switzerland
8. Cramer JJ, Smellie JAT (1994) Final report of the AECL/SKB Cigar Lake analog study. AECL-10851 TR 94-04. Atomic Energy Agency Canada (AECL), Whiteshell, Canada; Swedish Nuclear Fuel and Waste Management Company (SKB), Stockholm, Sweden
9. Chapman NA, McKinley IG, Shea ME, Smellie JAT (1990) The Poços de Caldas Project: summary and implications for radioactive waste management. SKB-TR 90-24. Stockholm, Sweden Nagra; NTB 90-33, Wettingen, Switzerland; UK DoE WR 90-055. United Kingdom Development of the Environment (UKDoE), London, UK
10. Duerden P, Lever DA, Sverjensky DA, Townley LR (1992) Alligator Rivers analogue project. Vol 1. Summary of findings. Swedish Nuclear Inspectorate (SKI) Stockholm, Sweden RR/92/071. Her Magesty's Inspectorate of Pollution (HMIP), London, UK
11. Yoshida H, Sakuma H, Yusa Y (1995) The Tono natural analogue study program. In: von Maravic H, Smellie J (eds) Natural analogue working group, sixth meeting, Santa Fe,

September 1994. EUR 16761, 315–318. Commission of the European Communities (CEC) Nuclear Science and Technology Report, Luxemburg
12. Rivas P, Hernán P, Bruno J, Carrera J, Gómez P, Guimerà J, Marín C, Pérez del Villar L (1997) El Berrocal project. Characterisation and validation of natural radionuclide migration processes under real conditions on the fissured granitic environment. EUR 17478. Commission of the European Communities (CEC) Nuclear Science and Technology Report, Luxembourg
13. Blanc PL (1996) Oklo—natural analogue for a radioactive waste repository (phase 1). EUR 16857/1. European Commission (EC) Nuclear Science and Technology Report, Luxembourg
14. Tylecote RF (1979) The effect of soil conditions on the long-term corrosion of buried tin-bronzes and copper. Journal of Archaeological Science 6:345–368
15. Johnson AB, Francis B (1980) Durability of metals from archaeological objects metal meteorites and native metals. RNL-3198. Battelle, Pacific Northwest Laboratories, Richland, WA
16. Bresle A, Saers J, Arrhenius B (1983) Studies in pitting corrosion on archaeological bronzes. TR 83-05. Swedish Nuclear Fuel and Waste Management Company (SKB), Stockholm, Sweden
17. Hallberg RO, Östlund P, Wadsten T (1987) A 17th century cannon as analogue for radioactive waste disposal. In: Côme B, Chapman NA (eds) Natural analogues in radioactive waste disposal. EUR 11037. CEC Radioactive Waste Management Series, Commission of the European Communities, Luxembourg, pp 135–139
18. Angus NS, Brown GT, Cleere HF (1962) The iron nails from the Roman legionary fortress at Inchtuthil, Perthshire. Journal of Iron and Steel Institute 200:956–968
19. Malinowski R, Garfinkel Y (1991) Prehistory of concrete. Concrete International 13:62–68
20. Rayment DL, Pettifer K (1987) Examination of durable mortar from Hadrian's Wall. Materials Science and Technology 3:997–1004
21. Jull SP, Lees TP (1990) Studies of historic concrete. EUR 12972. CEC Nuclear Science and Technology Report, Commission of the European Communities, Luxembourg
22. Steadman JA (1986) Archaeological concretes as analogues. In: Côme B, Chapman NA (eds) Natural analogue working group, second meeting, Interlaken, June 1986. EUR 10671. CEC Nuclear Science and Technology Report, Commission of the European Communities, Luxembourg, pp 165–171
23. Mallinson LG, Davies (1987) A historical examination of concrete. EUR 10937. CEC Nuclear Science and Technology Report, Commission of the European Communities, Luxembourg
24. Linklater CM (ed) (1998) A natural analogue study of cement buffered hyperalkaline groundwaters and their interaction with a repository host rock: phase II. S/98/003. Nirex Science Report, Harwell, UK
25. Smellie JAT (ed) (1998) Maqarin natural analogue study: phase III. TR 98-04. Swedish Nuclear Fuel and Waste Management Company (SKB), Stockholm, Sweden
26. Ambrosetti P, Basilici G, Gentili S, Biondi E, Cerquaglia Z, Girotti O (1992) La Foresta Fossile di Dunarobba. Ediart, Rome
27. Lombardi S, Valentini G (1996) The Dunarobba forest as natural analogue: analysis of the geoenvironmental factors controlling the wood preservation. In: von Maravic H, Smellie J (eds) Natural analogue working group, sixth meeting, Santa Fe, September 1994. EUR 16761. CEC Nuclear Science and Technology Report, Commission of the European Communities, Luxembourg, pp 127–133
28. Miller WM (2000) Communication with natural analogues. In: von Maravic H, Alexander WR (eds) Natural analogue working group, eighth meeting, Strasbourg, March 1999. EUR 19118. EC Nuclear Science and Technology Report, European Commission, Luxembourg (in press)

Part VI: Education on Energy and Environmental Concerns

On the Educational Issues and the Environmental–Energy Concern

Azusa Tomiura

Key words. Education—Environment—Energy—University—Civilization

Prologue

Civilization based on materials and energy culminated in the twentieth century, and the development of modern economic and technological systems has undoubtedly brought us material affluence. However, the social system of mass production, mass consumption, and mass disposal has altered the consciousness of individuals, and traditional ideas of old standing such as reverence toward nature, the spirits of virtue and veneration, a broad mind for accepting different values, and so on have been lost.

As a consequence, detestable social phenomena have become apparent, such as the loss of consideration and politeness toward others, the collapse of families, the increase of juvenile delinquency, and the deterioration of the customs of keeping the immediate natural environment clean and conserving material. These social problems are by no means unrelated to environmental and energy concerns, and are strongly related to problems regarding the education of students, parents, and citizens.

Although problems in education and concern regarding the environment and energy seem to be different phenomena, they are the same at the root. The purpose of the panel discussion will be how education can be reformed and how we shall take part in solving the environmental and energy concerns and educational issues simultaneously. Before starting the panel discussion, the chairman would like to describe the current circumstances of education for energy and environment in Japanese universities.

Environmental Education and Studies at Japanese Universities

The history of environmental education and studies in Japan is long. In 1957, the Faculty of Sanitary Engineering was founded at Hokkaido University. In 1971, the Environment Agency was started, and in 1996, the Basic Law for Science and Technology was legis-

Nippon Steel Corporation, Sogo-Kojimachi No. 3 Bldg. 1-6 Kojimachi, Chiyoda-Ku, Tokyo 102-0083, Japan

TABLE 1. Courses of environmental education and studies

Faculty	Number of courses
Engineering	27
Agriculture	17
Science	11
Economics	4
Others	25

lated. As the Environment Agency has advocated the importance of environmental education, and the Basic Law for Science and Technology has emphasized the promotion of research and development that contributes to the solution of global issues, new courses of environmental education and studies at universities have been successively inaugurated.

As of 1997, 66 of 567 universities have courses of environmental education/studies. There are 84 courses, which can be classified as shown in Table 1. Of the 25 others in Table 1, 16 courses are related to human life science and the remaining 9 are miscellaneous courses.

Current Issues of Environmental Education and Studies

As seen from Table 1, about 30% of the courses of environmental education/studies cannot be defined by the conventional makeup of university faculty. Even in the conventional faculty, the environmental courses are environmental subjects hybridized with other conventional subjects. This fact means that environmental education/studies are multidisciplinary or transdisciplinary, integrating natural science and sociocultural studies.

As an example, although the mechanism of the composition and decomposition of dioxin is said to require chemical studies, integrated studies from disposal and collection to incineration of waste, which are obviously matters of social and personal concern, are indispensable for the reduction of dioxin. Frankly speaking, present environmental education/studies in Japanese universities can hardly be conducted covering such a holistic approach. Herein lies the first issue of environmental education and studies.

The second issue is this: While the role of the university is to conduct advanced scientific studies and to educate students for higher professions, current studies in universities have further dispersed more advanced science and technologies and have neglected to synthesize elements of knowledge thus discovered into effective technologies that are applicable for solving problems. As a result, although students learn many subjects and enrich their knowledge, they cannot learn how this knowledge converts to wisdom for problem solving. It is, therefore, doubtful if graduates of environmental faculties can contribute to society or enterprise as higher-level professionals.

These two issues concern not only the area of environmental studies but also contemporary education in general. Numerous problems have emerged in the new fields that cannot be covered by conventional education/studies based on nomothetic sciences.

Engineering ethics, e-commerce, financial engineering, optimization of multifarious entities, virtual reality, and, of course, environmental and energy issues are good examples. Because all these new studies are related to conventional studies, graduates, regardless of natural science or social and cultural science basis, cannot discharge their role as highly educated professionals unless they have knowledge of these new fields simultaneously with knowledge of their own conventional fields. How the university educates all students on environmental and energy concerns is the third issue.

In summary, current issues of environmental education and studies are these:

1. Lack of holistic education/studies for solving environmental issues
2. Lack of education/studies to synthesize individual scientific knowledge to wisdom that is applicable for environmental concerns
3. Lack of global scope of education for environmental and energy issues for all students in the university

Relationship Between Environmental and Educational Issues and Their Simultaneous Resolution

The development of modern economic and technological systems has undoubtedly brought us material affluence. However, the social system of mass production, mass consumption, and mass disposal has changed the consciousness of individuals, and ideas of long standing such as reverence toward nature, respect for the elderly, the spirits of virtue and veneration, a broad mind for accepting different values, and so on have been lost. As a consequence, undesirable social phenomena have successively appeared, such as the loss of consideration and politeness to others, the collapse of families, the increase of juvenile delinquencies, the adherence to DaeWoo-ism (which means big-I-ism in Chinese), and loss of the customs of keeping the immediate natural environment clean.

To improve such social phenomena, the reform of education starting from the elementary school level is now being seriously discussed. These social problems are by no means unrelated to environmental issues. Although the problems of educational as well as environmental issues seem to be phenomenally different, they are the same at the root. The devastation of the environment as well as education can be considered to progress spirally, keeping the mutual linkage. Environmental issues, therefore, cannot be solved only by university education and studies, but should be solved by the change of individual consciousness from the pursuit of material affluence to the achievement of spiritual affluence.

The Science Council of Japan set up a special committee on this matter and is now making earnest studies. The committee has the concept that the devastation of the environment as well as education originates from the same roots, and that they cannot be cured by symptomatic treatment as the roots are derived from the humanities. The committee is going to discuss the following proposals:

1. Correct scientific understanding of issues and accurate comprehension of the humanities
2. Integration of science and technology with ethics
3. Creation of new studies and technologies
4. Founding of an international center of excellence

Epilogue

Because our predecessors developed a civilization that was founded on the harmony of humans with nature, we have always tried to keep this tradition. Much of this thought and culture is being lost in our time under the influence of modern culture and the rapid progress of science and technology. We, however, now must play an effective role in solving the present global question by syncretizing (synthesizing different principles and tenets) of material civilization with spiritual civilization. In this sense, it is expected that tasks in which we can all cooperate mutually could be discovered at this panel discussion.

Panel Discussion

A. Tomiura: It is my great honor to moderate this panel discussion. First of all, I would like to deeply thank all professors and doctors who willingly volunteered to be panelists. I also thank all the participants for joining us this morning. We should not forget the enormous efforts devoted by the organizing committee members and their staff.

In my experience, whenever I participated in such an international panel discussion, I always had some frustration because of the time restriction for the discussion and the inconvenience of speaking a foreign language. Therefore, I would like to ask all panelists to make not-too-long but full comments in easy English, and simultaneously to solicit all participants for your active contribution to the panel discussion. You are allowed to speak "Internet" English by just uttering key words.

Now, I would like to introduce the panelists in alphabetical order: Professor Assael from Greece, Professor Bose from the United States, Dr. Englezos from Canada, Dr. Kalman from Israel, and Dr. Miller from the United Kingdom. To save time I do not want to introduce the professional career of each panelist, but suggest that you refer to the brochure of the panel discussion instead. I am very happy to have such distinguished panelists representing various nations.

As you well understand, education and studies on environment and energy are intrinsically multidisciplinary, integrating natural sciences with social and cultural sciences. This complicated character of education and studies on environment and energy brings us such difficulties as those mentioned in my note for the panel discussion.[1] The first is the lack of holistic education and studies, and the second is the lack of education and studies to synthesize individual scientific knowledge with wisdom that is effective for solving environment and energy concerns. The third is the lack of global scope of environment and energy education for all university students.

In the meantime, you notice that material affluence in modern society has led us to a wasteful culture, which strongly relates to environment and energy concerns, while spiritual satisfaction has gradually been lost. As a result, detestable social phenomena such as irreverence toward nature, lack of respect for the elderly and for the spirits of virtue and veneration, lack of a broad mind for accepting different values (especially among different races), and many others can be observed. Current concerns such as juvenile delinquency, like violence in schools, accidents caused by negligence of the operational

[1] See the chapter by A. Tomiura, this volume.

standard in industries, like the accident in the nuclear fuel recycling plant in Japan, and regional conflicts in Eastern Europe and the Middle East are by no means unrelated to these issues.

To improve such social phenomena, I firmly believe that reform of education, starting from elementary school and continuing to university, should be required, because environment and energy concerns are strongly related to education. We, therefore, do not need symptomatic treatment but remedies by completely tackling the source of the disease. This is the background with which we took up the issues on education.

In this context, I would like to invite various comments from our panelists as well as participants. I would be very happy if we are able to find a panacea to heal our concerns.

Now, I would like to ask each panelist to comment on this matter. We will start in alphabetical order. The first speaker will be Dr. Assael from Greece. Dr. Assael prepared a talk about education in elementary schools.

M. J. Assael: Thank you, Mr. Chairman. Ladies and gentlemen, I think the Chairman has put everything in the right perspective. It has always been my belief, that if we are going to educate people in environmental issues, we must start from elementary schools. We must start from the beginning. I will use my country as an example of this. Hence, the extent to which we have achieved this goal, that is, to make elementary school pupils aware of environmental issues in Greece, and how, will be the subject of my short talk.

Let's start by stating some facts. Greece is a very small country of about 10 million people. It has a very small old industry, which used to be rather dirty. However, due to recent antipollution legislation the situation is becoming better. It is also fair to add that, since Greece has become a member of the European Union, a lot of antipollution legislation has been introduced. Furthermore, we should keep in mind that because we have always considered Greece to be an old country with a beautiful countryside, beaches, and islands, environmental issues have been important to us.

Until 1970, universities more or less monopolized environmental concerns and research. Anything concerning the environment was the work of universities. In the early 1980s, environmental concerns started to be connected to energy issues following the oil crisis in 1972.

The first changes in this situation started in 1987, when the Ministry of Education appointed Environmental Education Advisors in 20 municipalities with the task of promoting environment issues in secondary schools. In 1990, Environmental Educational Advisors were appointed in most municipalities around Greece. They now have the task, in cooperation with secondary and elementary schoolteachers, to promote new environmental ideas through lessons and projects. So in that sense, after 1990, the teaching of environmental issues started in elementary schools—maybe more in private schools than in public ones.

In 1992, in parallel to the appointment of Environmental Education Advisors, another new idea started in Greece—the idea of Environmental Education Centers. The first such center was founded in 1992, and was followed by seven more in 1995 all over Greece.

So until today, the three basic social services that deal with the teaching and awareness of environment issues are the teachers of elementary and secondary schools, the municipal Environmental Education Advisors, and the Environmental Educational Centers. This goal was achieved as follows:

a. In most elementary or secondary schools, teachers teach special courses on environmental issues, while at the same time environmental ideas are taught indirectly through most subjects. Many environmental projects also are carried out in schools locally or in cooperation with the Environmental Educational Centers.

b. The Environmental Education Advisors' task is to advise and help. They organize, in cooperation with schools and the Environmental Education Centers, events and exhibitions and different projects that aim to stimulate pupils' ideas, such as why the sea and the beaches are contaminated and how we can save energy.

c. The Environmental Education Centers are independent of schools. They provide schools with materials and organize local and international networks. The Centers, of course, are funded from the third European Union framework. I would like to add one more thing about the Centers. Their task is to make people sensitive, aware, and active. These three words are quite important. We have to make people sensitive about educational issues. We have to make them aware of the problems. And then, most importantly, we have to make them active as far as issues of environmental protection are concerned (such as forest and greenery fires, landscape devastation, sea and river pollution) as well as in energy conservation and maintenance of environmental policies, etc.

Nevertheless, it seems that in Greece, with the help of the Ministry of Education legislation, the European Union funding, and the elementary school teachers initiatives, children have started to become aware. I can say that because my son is 8 years old, and he is much more environmentally conscious than I was at his age, and even on certain issues more aware than I am now. So it is very interesting that over the last 8 years such a turning in thinking and mentality was accomplished.

There are, however, questions still to be posed: "Where is industry's part in all this?" What I talked about referred to no industrial involvement. I mentioned previously that I was quite impressed with TEPCO's idea of creating and financing this program. We have nothing like this in Greece, no industrially financed seminars. I believe, in simple terms, that although the environment is strictly an industrial issue the Greek industry has not yet made any attempts to educate pupils in environmental issues.

How do we further educate more children in environmental issues? I think we need to concentrate on indirect ways, which the children might find more interesting. For example, all children in Europe watch animated programs on TV. These are either computer generated or, in some cases, of the old-fashioned animation kind. Nevertheless, the majority of these programs come today from Japan. Maybe producers have to be convinced through strict legislation to include indirect environmental messages in such programs. It would be beneficial if we could pass forward such messages through these TV programs.

Finally, and maybe most importantly, "How do we educate parents?" This might be an almost impossible task, as somebody who is 40 years old or 45, like me, of course knows everything and hence does not want to be educated.

I will stop here by reading to you a sentence taken from UNESCO's constitution that I like very much, and then I will propose a change. UNESCO's constitution states that "Since wars begin in the minds of men, it is in the minds of men that the defenses of peace must be constructed." My proposition is to change this to "Since the destruction of the environment begins in the minds of men, then it is in the minds of men that the defenses of the environment must be constructed." And maybe this could be part of the Earth's constitution instead of the UNESCO's constitution.

Thank you very much for listening to me.

A. Tomiura: Thank you, Dr. Assael. I was impressed by the ambitious educational program that Greece is promoting. Now, although some participants have questions, I will accept them when we have finished the comments from all participants. As next speaker, I would like to invite Professor Bose.

B. K. Bose: First, I would like to thank the TEPCO and Keio University team for holding a symposium on such an important issue as the environment and energy conservation, and this panel discussion. I have never heard of such a symposium and panel discussion before. Therefore, I think this kind of panel discussion is extremely important, and it also shows the concern of Japan on these issues. I only wish that this kind of panel discussion could be held for a wider audience, such as an international conference, so that all the public in the whole world can be exposed to this kind of thing, rather than being confined to a group of specialists.

In my opinion, most of the energy and environment issues are just commonsense kind of things. We can make a lot of improvement, and can do a lot just through public education, and raise public consciousness about energy. It is such a vital issue today. The technical issues are of course important, but technical issues often require funding, but the public issues and public consciousness are the first items we should address in the whole world. And I think, in these aspects, the public in the affluent countries like the United States should be addressed first.

In my presentation, I mentioned that 33% of energy in the United States is simply wasted because of bad public habits. It is the cheap energy and bad public habits that lead to waste. But, of course, it is true that affluence encourages waste. But, if it is harmful for the society, then this wastage should not be permitted. There are so many unwritten laws in the society: "Simply because I'm affluent doesn't mean I should waste something that adversely affects others."

So, what I am saying is that public education and public consciousness constitute the No. 1 issue today, which can be followed by technological issues through funding and research. Technological issues, of course, should be followed by governmental regulations. But the regulations should come toward the end rather than the beginning. Also, widespread publicity is needed everywhere around the world. This publicity can be done by television and newspapers, magazines, posters, seminars, etc. to educate the general public and make them conscious about these energy problems. This can let people see how energy is simply being wasted in the United States, where energy consumption is the maximum and the maximum amount of pollution is created today, and it is much higher than in China and India today. Of course, China and India are moving toward greater pollution in the future.

Because we belong to the older generation, it is somewhat difficult to educate us. But it is very important today that we start with the school and college students some amount of curricula, or maybe seminars and discussions, to make them environmentally conscious of the long-term issues so that at least the next generation and future generations can be educated solidly along these lines.

Of course, saving energy does have the direct incentive of saving money, you can say, but when you are giving a course or seminar to the school and college students, it should be some kind of incentive that will make them follow this way.

This panel is a very historically important event, in my opinion. I would like to suggest that to have a dramatic start, we should not stop here. We should start building momentum and draw up an action plan. I would ask the chairman to draw up some initiatives to follow in the future. Otherwise, this may just end here. It does not do justice to such a vital issue. And I would like to suggest that to make a dramatic start, the year 2000 should be declared "The Save Energy and Prevent Pollution Year," publicized everywhere, with public education and raising public consciousness. It's such a simple and economical way to do it. And I would request of Japan, because Japan has a tremendous amount of regard for a clean environment, saving energy, and very impressive records, that Japan take the lead and, of course in cooperation with the other countries of the world, take the public education initiative.

And finally, I would like to say that this panel discussion and symposium being held today are very historically important events. We should not just stop here. We should go on building momentum for these issues for dramatic actions in the future.

A. Tomiura: Thank you, Professor Bose. I will take up a few issues that you mentioned in your presentation, and discuss them later. The next speaker is Dr. Englezos.

P. Englezos: When I read the e-mail from Professor Mori regarding this panel discussion, the first thing that came to my mind was something that is attributed to Leonardo da Vinci when he was observing how life was evolving in Florence at the time we call the Renaissance. He said, "Oh human misery. Of how many things you make yourself the slave of money." The reason I bring this up is that, in my opinion, when we talk about educational issues we have to distinguish between education for the general public, society in general, and then address academic issues. Because technology is involved, specialists are needed to address these issues appropriately. Starting from society, we have to respond to complex questions. The situation is not easy because we have to reconcile environmental and economic interests that are not easily reconciled.

If we look at the Gross Domestic Product (GDP) per capita for 1989 in U.S. dollars versus the 1991 total energy consumed per capita in equivalent barrels of oil for several countries,[2] we see that there is a correlation between the two quantities. For example, we see that Canada and the United States have a high GDP but they also consume a lot of fossil fuel. And we know that consumption of fossil fuel is associated with environmental pollution. On the other hand, Switzerland has a high GDP, but not as high consumption of fossil fuel. Subsequently, we look at places like India—and I believe China should be somewhere there—very small GDP, very small consumption of fossil fuel.

Why do I bring this up? Yesterday we heard a talk about pollution from China and NO_x, SO_x, etc. coming to Japan following the global air circulation patterns.[3] China is on the way to becoming an industrialized nation and increasing GDP. So, it's following a path that the United States, the U.K., and other countries have followed. In Europe, forests do not cover as much area as they did a few hundred years ago, so people have become sensitive. But we cannot impose on the Chinese people stringent environmental regulations just because we want a cleaner environment.

What do we have to do? In my opinion, because the Chinese also want a piece of the prosperity pie, what we have to do is promote international collaboration to address the global problems and avoid past mistakes. Environmental problems are now becoming

[2] Ristinen RA, Kraushaar JJ (1999) Energy and Environment. Wiley, New York.

[3] See the chapter by S. Tanaka, this volume.

global. Pollution generated in place X appears in place Y, such as we saw yesterday. And I can give you another example. Recently, it was reported that trace PCBs (polychlorinated biphenyls) were found at the top of the mountains in British Columbia.[4] They found their way there because of global air circulation. Pollutants "stick" there and find their way into waterways, etc. Therefore, we must have strong international cooperation to address those complex problems.

Regarding energy issues in the future, we have to think in terms of what is called integrated energy development. This is a concept which recognizes that no single energy source can possibly provide all the energy required by the developing and developed countries. So we are talking about local sustainable energy planning, or integrated energy planning, to provide reliable sources of energy that are not hard on the environment. In terms of energy policy we basically see two approaches. First, the "continue as we are presently" scenario. In my opinion it is sometimes easy for politicians to do that, and for countries as well, because the other path is to find alternatives to fossil fuels, such as the sun, biomass, wind, and geothermal. We also must increase the efficiency of the relevant processes.

When it comes to educating the public, I would like to quote Michael Smith, a Nobel laureate in chemistry, who made a very interesting observation. He said that schools still produce outstanding scientists, but the most profound need to address is the scientifically literate citizen.[5] We need to have a human society that is scientifically literate. And I was glad to hear Professor Assael addressing exactly this point. Still, schools do a great job. You take a class, and 1 or 2 of the 50 students might get a Ph.D. and become an outstanding scientist or engineer. But you also have to look at the rest, the 49 people who will become journalists or accountants, etc. They have to have a basic understanding of scientific issues. In my opinion, several problems are magnified by some in the media, for example, who do not have the required complete understanding of scientific issues when it comes to pollution. We've seen that in Canada with certain industries there. It's easy to say "Not in my backyard." But why? Technology may offer the appropriate solutions, so we have to have the ability to understand the complex issues.

Another thing that I would like to be more specific about regarding the teaching of students at the universities is the concept of project-based learning. And as a chemical engineer, I can appreciate that more because in the final year we have a so-called Design Project course. The idea behind is simple: you learn by doing. You pose a problem, and then you try to learn key concepts through addressing the problem. And I'm not saying to abandon the curriculum and change it radically, but gradually introduce this idea to problems of environmental interest.

Finally, there is the issue of setting examples. And I will look back to the Renaissance. There is a very nice book called "Daily Life in Renaissance Italy".[6] It has a very nice picture in which the caption says "A popular book advised cooking once a day only in the morning for the poor people to save fuel. But the kitchens of the well-off were busy all day with activities like cheese making." So I want to send a message to rich countries to collaborate with the poor who want a piece of the prosperity pie, and also within

---

[4] Ogilvie C. The Globe and Mail, October 4, 1999.
[5] Smith M. Science and Society in the Forthcoming Millenium. The Killam Annual Lectures, Vancouver, B.C., December 10, 1998.
[6] Lee CL, Jr. (1975) Daily Life in Renaissance Italy. American Heritage Publishing, New York.

society, for industries, etc. to set examples for the citizens so that environmental issues are not forced upon us to follow but become a part of our life.

A. Tomiura: Thank you Dr. Englezos. I would also like to continue discussion of North–South issues later. The next speaker is Dr. Kalman from Israel.

H. Kalman: I would like to make a few short comments. And I would like to divide them into two parts. The first part is the general public education. We all spoke about education in schools, but we forgot our part as parents in the education of our children. I think that some of the social issues that were raised by the chairman, like lack of respect for the elderly, are a part of our modern life. In our modern life, actually, as parents we are not rich enough to take our vacations here in Hakone unless someone else is paying, but we are wealthy enough to get almost all the things our children need and wish. So they are raised to think they don't have to struggle for anything. Later, they only ask what the society should and can give to them. They don't ask themselves what they can contribute to the society. How can we change that? I'm not sure we can, because this is human nature, that we will do everything for our children.

But we can do more in formal education, besides the education of individuals for preserving the environment. Most of us are teachers in universities, and we are training professionals in different disciplines. At this part we can and should do many things.

First, I would like to give a very short example, which is a story that sometimes we tell between professors in Israel. In a course in fluid mechanics, not in my university but another university, the students had an assignment to solve a problem to convey some amount of flow in a 3-inch pipe of blood from Haifa to Tel Aviv, which is 100 kilometers away. So they all looked for relevant handbooks, looking for the density and other properties they needed, calculating the Reynolds number, looking into the Moody diagrams, and looking at what power they need to pump this blood from Tel Aviv to Haifa. No one asked, "Hey, wait a minute! Blood? From where? What is the reason?" So, as professionals, we have to ask all the questions together while practicing our professions. And the environmental issues should be connected very strongly to all the professions that could be concerned with environmental issues.

In my university, as in probably many other universities, we have a special master's program of environmental aspects. It is now a good topic. You can get a lot of money from the government to establish such an educational program. But this, in my opinion, is not enough, because the environmental problems should be integrated into all the professional issues. Whenever we are designing something, whatever it may be, the environmental issues should be considered right in the design stage, and we should not as engineers design anything and then bring in the environmental specialist to look at the environmental aspects. It should be integrated right in the beginning.

As a demonstration, I would like to show you the new master's program of environmental technology that is initiated now in my university (Table 1). Some of the courses are mandatory—the first four; and others are selective. But the environmental problems are concerned within these courses. Topics taken up in some "selective courses" like no. 4, no. 5, and no. 11 are mainly environmentally related; and topics taken up in some other "selective courses"—nos. 2, 6–8, 12, and 13—and also in "mandatory course" no. 3 involve environmental problems in their curricula. When the graduate engineers from this program get to the industry and design something, the environmental issues should be taken into consideration right from the design stage. The situation in which one of the

TABLE 1. A new master's program at Ben-Gurion University of the Negev on Particle Technology

Mandatory courses	
1	Topics in Chemical Engineering
2	Topics in Mechanical Engineering
3	Introduction to Particle Technology
4	Characterization of Particulate Solids
Selective courses	
1	Crystallization Processes
2	Fine Particles Technology
3	Two-Phase Flow
4	Separation of Particles from Liquids and Drying
5	Control of Particle Air Pollution
6	Pneumatic Conveying of Solids
7	Storage and Mixing of Bulk Solids
8	Mechanical Conveyors
9	Industrial Measurement and Control
10	Micro and Nano Particles Technology
11	New Technologies for Particle Classification and Separation
12	Size Enlargement
13	Size Reduction

industries I am familiar with, in the south of Israel, releases daily about 15 tons of particles into the atmosphere should not be repeated. It is far away, it is not in an urban area, but in any case it is a lot of pollution.

To conclude, I would like to say that in professional terms we can divide all the issues into four parts. The first part, which is the most important, is awareness. It was mentioned before, and it is the most important. If someone is not aware of something, he cannot even think about how to solve it. The second one, which I think is a necessity (we'll be glad if it is not needed), is regulations, which are mainly governmental regulations. The regulations are necessary because industry doesn't have the awareness to solve the problems. The third one is education (I talked about that professional education, which is very important, in my opinion), and the fourth one is the cost, because you always have to take into account the cost. As professionals in other areas, we should develop new efficient technologies that cost less to make the conflict with industry smaller.

A. Tomiura: Thank you, Dr. Kalman. You emphasized that environmental issues should be an integral part of everything. It is a very significant comment. The next speaker is Dr. Miller.

W. M. Miller: This photograph (Fig. 1) was sent to me by e-mail on the same week that I was sent the background information for this panel discussion. I am sorry it is not very clear. It's actually a picture of a protest in Switzerland about potential sites for locating a repository for radioactive waste. It shows a local group of people very much like Greenpeace in attitude, who immediately started protesting as soon as a particular site was mentioned. And what they have here are made-up cows, large papier-mache cows with

Fig. 1. Demonstration by local people in Switzerland protesting the proposed siting of a repository near their village. They have made a model of a "radioactive cow" to illustrate their fears that radioactivity from a repository could contaminate the milk supply. Such fears are unfounded, but this photograph illustrates the need for better communication between scientists and the public about the potential risks and benefits from modern technological advancements. Improved scientific education would help people make informed decisions about these issues. (Photograph courtesy of Nagra, Switzerland)

radioactivity symbols on them. And the false message that they were trying to put forward is that a radioactive waste repository near the town will pollute the milk. And everybody likes to drink the milk in Switzerland. I show this photograph because in radioactive waste disposal we have a particular problem, and that is education and communication, particularly with adults. I agree with everything that has been said so far about education for younger children, from primary school through to university. But in the present day it is the responsibility of adults to address current environmental problems. It is the current generation of people who can have an impact on decision makers and governments. Maybe in the future, when our children grow up, if they have been educated to understand environmental issues, things might change. But basically we can't wait for future generations to come through.

Obviously, I work in radioactive waste disposal, so what I say has this as a background. But I think a lot of the issues we face in radioactive waste disposal can apply to other environmental issues as well. This is a short summary of what I said yesterday.[7] In radioactive waste disposal we have a problem. The wastes exist today. We must dispose of them. We must do something with them. However, we can only do this with the approval of the public. But the public has a dread of radioactivity and therefore no one wants a repository built near their home, as I said, the NIMBY effect (Not In My Back Yard!). But the same will go for many other waste materials or environmental problems as well.

[7] See the chapter by W. M. Miller, this volume.

The problem we face is that if the public is not keen on progressing with an environmental solution, the government will not be either, because the government only has its mandate from the people. So if a government fears public opposition, the government won't act. And that opposition could be based on a false fear of the waste material, in the case of radioactive waste, or the facility needed to treat the waste. But it could also be a cost issue, as was mentioned just now. For example, cleaning up the atmosphere is always going to cost money. If we try to reduce carbon dioxide or other emissions, it's going to cost. And if that cost becomes a tax to the public, the public will ask, "Do I vote for the government that is potentially green, who is going to tax me more, or do I vote for the government that potentially is going to tax me less, although it may be more polluting?" Very often, when it comes down to putting the check on the ballot paper, people vote with their pocketbooks. And this is, I think, a big issue that we must overcome, because at the moment there are a lot of issues which are not being dealt with, not because there is no solution, but because basically there is no will on the part of the government and the general public to put these solutions into place.

So how do we solve this problem? For a start, the public must be better informed of the issues. This is the broad educational aspect. But as someone asked earlier, "Where is industry in all this?" Industry must take part. And industry must listen to the people. It's not simply a case of "The public must do as they are told." Industry must take on board the concerns of the people and act accordingly.

How do we make this happen? This is what we have learned in radioactive waste disposal. It is not as simple as saying, "Here is a problem. You the general public must take a part in solving it. This is what you must do." People don't necessarily want to do this. They don't believe what governments tell them. So we have to have, I think, quite a fundamental research program investigating communication with the public, communication on environmental issues, communication on risks, and how we should explain risks, because—let's face it—everybody knows it's dangerous to smoke, but people still smoke. They take that risk. They choose to do that. So when you ask people to use their car less or to turn the lights off in their houses to try to save electricity, people do not generally do it. They know there's a risk, but they don't see it as a risk that is personal to themselves. There is a broad problem of many people isolating themselves in their minds from the environmental impact of what they might be involved in. And therefore educating people requires much more than just *telling* people. There is a real problem in actually *communicating*.

In radioactive waste disposal, people have been trying to communicate with the public for several years. Many leaflets, posters, and brochures about radioactive waste disposal have been written for the general public to read, but generally they have no impact. People read them and then they discard them. And still they don't support the building of a radioactive waste repository. Just like when people see a newspaper advertisment that reads "Stop smoking! It's going to give you cancer!" and people turn the page and then light up a cigarette. Telling people is not enough.

So what do we do? How do we solve this problem? First, we have to start proper dialogue among the "stakeholder groups," that is, everybody who is involved in the issue, be it in radioactive waste disposal, or trying to reduce carbon dioxide emissions, or asking people to drive less and use public transport. And there are a number of groups that must be involved in this dialogue: this includes the public of course, and the government because at the end of the day it's going to be governments that drive these things through. The environmental organizations must also be involved. This is another thing we have learned in radioactive waste disposal. There is no point in working behind closed doors and talking about the solution but at the same time keeping the environmental organizations, like Greenpeace, out. We must bring the environmental organizations in and let them be involved in the discussions, and then the schools and the colleges and the universities. We must listen, learn, and then speak. It's not just a case of imposing your views on other people. It must be a proper dialogue.

I have been talking about radioactive waste disposal most of the time. But I think the dialogue should not be compartmentalized into separate environmental problems. It should be a broad discussion, a broad discussion of environmental issues across a range of problems, and should involve the public, industry, environmental organizations, and governments. It should consider all environmental problems: from waste disposal to energy-saving issues and food safety. I mention food because we have particular problems in Britain, related to BSE (bovine spongiform encephalopathy) and genetically modified crops. There is a debate going on, but as somebody said earlier, the debate is generated and tends to be run by the newspapers. It is sometimes very difficult for the general public to obtain unbiased information to read and to make their own assessment of what is safe and what is not safe. It's very difficult to find that sort of information.

So these are a few views about the problem of communicating environmental issues and educating the public. Issues such as waste disposal, health, and food safety are intertwined in the sense that people have to understand the issues. We need a communication strategy to allow people to understand the impact of what they do. And it's this impact that I think is important. Generally, we need better teaching of science and technology across all groups, not just in universities, not just in schools. We must make all groups in public life relate the impact of their actions to an understanding of social responsibility. There is a real problem that people don't understand that by not turning off the lights it increases global warming. People don't seem to see the connection. In terms of science they might understand the issue but they don't seem to think it's real as applied to themselves.

I also think we need a better understanding of where governments come into decision making, because at the end of the day it is governments that will make and enforce legislation to clean up the environment. But people often don't understand that we have a direct link into government to make the government reach the decisions that we want them to.

So I think the objective of this communication exercise is to instill into the public a sense of active responsibility and to help people feel involved and empowered in the decision-making process, because if they are not, they won't take part. If they believe there is no reason for them to say what they feel, if they believe that nobody is going to listen, and nobody is talking to them, then people just feel isolated and walk away from the problem.

That's it from my point of view. This is very much a radioactive-waste-biased talk, but I really feel there are lessons here that apply to many aspects of the environmental and education discussion we are having today.

A. Tomiura: Thank you very much, Dr. Miller. I was impressed by your final phrase regarding the instillation of active responsibility into the public. I would now like to ask comments from participants representing other sectors. How about Professor Park? Can you make some comments on the Korean situation? After that, Professor Watanabe.

J.-K. Park: Just a short comment not included in the panelists' addresses, although I think it was included already yesterday.[8] Around East Asia there are big countries, like China, that consume energy increasingly. So I want to stress the need for international cooperation among the Near East countries. That is one point. And another point is that in university education I propose developing new curricula that can be included in the core curricula, like liberal arts, which are related to environmental education. For example, there can be the technology histories, or engineering ethics, such kinds of new curricula that can be transdisciplinary, or interdisciplinary. And they should be taught in every field of the academic paradigm. For example, the table prepared by Dr. Tomiura[9] shows all the environment-related education and courses that are taught in the natural sciences and engineering only. So I think this kind of environmental curriculum should be taught in every field, including the humanities and the rest. We should develop new environment-related courses in the core curriculum, at the very low level of undergraduate education. Two things: international cooperation and new curriculum in the core curriculum.

A. Tomiura: Thank you, Professor Park. I can share the same thought that you mentioned. How about Professor Watanabe with a perspective from Brazil?

E. H. Watanabe: When we talk about energy and environmental problem in Brazil, we can say that many works are being developed at universities, especially at the graduate level. However, when you talk about education in this field, especially for the common people, in general, we can see that they are not aware and do not understand the problem. So, what is the problem with energy or environmental problems? I believe that one of the problems is that people are not exactly conscious about the necessity of energy and the problems that relates it with the environment. For example, 2 or 3 weeks ago I was talking with a colleague, who, I think, is a member of an organization similar to Greenpeace. He was just saying that he was against hydroelectric power plants, in general, and especially in the Amazon region. And I said, "Okay, but let's first turn off the air conditioning and open the windows to feel the meaning of energy." The idea was to show that it is easy to be against an energy generation plant if you are in a cool room. It is wise to remember that temperatures in some parts of Brazil may go as high as 40° Celsius. Opening the windows and feeling the hot temperature might change the meaning of the importance of energy. Some people are against building dams or power plants, but they do not give up using the energy that comes from these plants.

Here at Hakone, maybe we should be discussing energy and environment problems in this session without the heating system. We would feel the difference, because outside is very cold today.

I think we need to have some way to make people, in general, understand what energy is. Not from physics, but understand it from everyday experience. I have one example that I am used to present to my students: Once I was talking with a friend who complained that he had to charge the batteries of the cellular telephone every day. He said that we, from electrical engineering, should do something so we could just generate the necessary power while we talk at the cellular telephone. At that time, I made some calculations and said, "That's too much energy to generate. You will have to generate some-

[8] See the chapter by S. Tanaka, this volume.
[9] See Table 1 in the chapter by A. Tomiura, this volume.

thing like 1 watt-hour." He suggested: "That's nothing. I use more than 10 000 watt-hours per day at home." In addition, I said, "Let me see if it's too much or not." Suppose you buy 10 kilograms (kg) of food, and you live in the fourth floor of a building without an elevator. This means that you have to go up about 10 m with the 10 kg of food. Therefore, you carry the 10 kg to the fourth floor using the staircase, in about 1 min. You will have to repeat this four times to generate something close to 1 watt-hour. This example shows that if we want to generate energy by ourselves, the effort is enormous, but when we use it we forget this effort. Therefore, I think this kind of example should be given to students to make them understand the real meaning of energy. I feel the necessity to create examples like this one, which is easy for people to understand. The problem is "How much effort one has to make to produce 1 watt-hour and in my home I use about 10 000 times more per day." This is true during winter, in Brazil, because there is no need of air conditioning. However, in summer it may go up three times because of the hot weather and the air conditioning.

In Brazil, I am always saying that we have to change the culture in some aspects. For example, we have the imported culture of going to work and meetings wearing a necktie and a jacket when the temperature in the street is 40° Celsius. This habit is very uncomfortable. I prefer not to wear a necktie and jacket, and set the temperature of the cooling system to 27°. That is the temperature I set on my air conditioning at home. My son always complains, saying it's too hot. Nevertheless, I say "No, it's okay, because for each 100 watts you spend here at home you need about 100 square meters of flooded area of dam somewhere else.

I think we should produce different simple examples to make people more conscious about energy problem. Maybe, as Professor Assael said, using the Japanese animation may be a good way to reach the children. I think possibly this is one way to get to the people and make them understand the meaning of energy and how it is related to the environment. Just a final comment: We have been working with this problem at our university, together with the Technical University of Berlin, with a project for solar energy and energy conservation. Although this project produced some papers, I feel that it is difficult to get information to the public in general.

A. Tomiura: Thank you, Professor Watanabe. Easily understandable dialogue between an amateur and a professional is quite important. I would like to touch on this issue later.

E. K. Longmire: I have a couple of comments, which are related to other peoples' comments. The first has to do with this concept of personal awareness of one's effect on the environment, and use of energy in terms of daily life. And I think in terms of educating adults as well as children this is a very good direction to take, because you can tell people anything, but it will only affect their actions if they think their actions make a difference. So I think that could be an educational approach that is very important.

The second thing, which I don't think has been discussed much, which is certainly extremely important in the United States, is that a lot of the costs of using energy or producing waste are not borne by the producers or the users at that time. And I think for substantial change to occur, those costs have to be incorporated into things, let's say at the producer end, within a business, as well as perhaps by the users of different products. It's a very difficult change to make, because there are a lot of interests that would fight against that. It has to come from the general public.

A. Tomiura: Thank you for your short but significant comments. As for personal awareness, I will take it up later. We still have some guests and participants who did not make comments. I would like to continue asking for comments from those participants and guests. Dr. Celata, please.

G. P. Celata: First of all, I wish to say that I totally agree with Professor Assael, and especially with the first declaration of his presentation, that education should start since the beginning, in the elementary school and so on. But I would also say primarily in the family. Why should we start since the beginning? First, because children can learn very easily and very much, and second because not all the children will become scientists or engineers, and cannot have this kind of information in the university.

I am a Catholic, so I believe that Family—with a capital "F"—is very important for the formation of a person. The information a child gets from the family is very important. I can give you an example. I am very fond of ancient music. And my daughter, who is 10 years old, loves this kind of music, too. Of course she loves the Spice Girls, Backstreet Boys, and other modern music, but she plays the piano, and she loves ancient music too. This demonstrates the importance of education since the beginning. If we do not properly address our children they may believe that TV, computer, mobile telephone, etc. always existed, and have no feeling of their impact on the environment.

We have a problem, already stated by Professor Assael: education of parents. And what is the situation in Italy? I have to be honest. It's very bad. We have a green movement, which is pretty demagogic, aiming to increase their presence in the parliament rather than to educate people. Just as an example, they say, "If you use the washing machine, you will pollute the rivers." Okay, we can also agree to change our shirts once a week, but this is not the solution. If we do that we must go back in the standard of living. We should maintain our standard of living, improving it, if possible, but have to have concern about the environment. Saving energy means to respect nature and economy, but we cannot reverse or change completely the standard of life. We must learn how to respect the environment and how to minimize the impact of our presence in the environment.

I have to say also that very few attempts are made in the elementary schools (I am always referring to Italy). They are made on a voluntary basis only. Next Saturday I will be at my daughter's school, and I will try to talk with the students about energy production and the impact on the environment, because another important thing was said yesterday by Dr. Miller,[10] that energy production produces waste. And I would add that almost every human action produces waste. I believe information given to the public, information and information, drop by drop, is the most important thing. We have to learn soon how to minimize the impact of these wastes on the environment. I remember when I studied nuclear power plants at the University, they taught me that water temperature out of the condenser was fixed by the law. Nobody said that the water was discharged in a river with its own habitat, and that we should limit our impact on it, although we will impact anyway.

Finally, getting back to the family, I think another problem is our modern life, which doesn't leave much time for energy education for children. For example, I have been called a weekend father, because on weekdays I am at the office. And sometimes I skip

[10] See the chapter by W. M. Miller, this volume.

my weekend parental duty, like this weekend, as very often I travel abroad. And we tend to spoil our children during the weekend, because we feel guilty about our absence during the week. Of course modern society doesn't help to teach the children more.

Another thing about the Italian government, and this is not a joke. The first action of the Italian parliament, after the Kyoto Conference a few years ago, was the introduction of a new tax. They called it "carbon tax." It means that you if you pay you may pollute. This is really a nonsense, in my opinion.

A. Tomiura: I am interested in the voluntary teacher and a story of the weekend father. I would also like to touch on those issues later. Thank you, Dr. Celata. Professor Mujumdar, do you have any comments?

A. S. Mujumdar: I'll just make a few comments, mostly based on the Canadian experience. I agree with almost everything that has been said by all the panelists. There is really very little I could add to that.

I feel that public awareness of environmental issues, at least in Canada, and maybe the rest of North America and western Europe, is already at quite a high level, mainly through education, but also through the media. And of course often it is biased, so we may not necessarily get the right picture. But certainly the awareness is there. Educational institutions at the university level, I think, are doing extremely well in terms of teaching environmental aspects from the social, scientific, and engineering points of view. And there are all kinds of courses or environmental issues taught at all kinds of levels. Some universities are offering even an M.B.A. in environmental management. So I think a lot is being done so far as education is concerned. I believe that much needs to be done at the primary and secondary school levels, as a number of panelists have said. I think that's very important. I think it is relatively easy to teach or train students in the educational system; it is also easy to legislate all kinds of environmental regulations, which politicians are doing all over the world. But it is much more difficult to implement what we teach or legislate because of the costs and the fact that they are taken more as an expense than as an investment.

We see very large corporations deciding to close operations rather than meet environmental regulations. Several companies, particularly in the United States and even Europe, have actually discontinued production of certain chemicals because they simply cannot economically meet the environmental regulations. So what is happening is that these products are being made in other countries, like India or China, and then they are imported back to North America. So the pollution is essentially transferred to another country where the environmental regulations are not as stringent. I know of cases in Canada where companies have threatened to close operations rather than meet the new environmental regulations. And of course the politicians then relent, because they do not want to cause unemployment in certain areas. They are forced to relax the enforcement of some of these regulations. I'm not quite sure what the solution to this is, but I think that economics are driving this issue quite significantly all over the world.

Another thing I would like to mention as many panelists have said, the level of pollution has a direct relationship to resources consumed. And I read somewhere (I don't know where they got their numbers) that a child born in North America will consume up to 30 times as much of the Earth's resources as one born in India. So if you were to define the "environmental population" of the world, the population distribution maps would look quite different. If you take the population of India to be about 1 billion people, then the "environmental population" of the United States will be about 9 billion. If you compare the resources consumed, the industrialized world is very significantly larger than the developing world. I feel that the developed world owes to the developing world tangible support so that they can increase their living standards without consuming excessive amounts of resources, which would lead to catastrophic environmental pollution on a global scale. I don't know if there is any immediate solution in sight. But this, I think, must be considered in trying to handle this problem.

The developing world will no doubt add to the environmental pollution as they raise their standard of living, which requires greater per capita consumption of energy. This must be complemented by reduction in per capita consumption by the industrial world through conservation and aggressive R&D to improve energy efficiency in industrial, commercial, and transportation sectors. Renewed interest in renewable energy sources will provide a partial answer, I believe.

According to a recent estimate, world energy consumption stood at 350 quads (1 quad $= 10^{15}$ BTUs $= 1.056 \times 10^{18}$ joule) in 1990 with the developing and nonindustrial world taking up 32%. By 2025, these numbers will be 465 quads and 60%, respectively. The world population will balloon to 8.4 billion from 5.3 billion in 1990. The affluent nations have the potential (but may not have the willingness) to cut per capita energy consumption by 30%, but the developing nations will raise their consumption by 60% by 2025. The cost of compliance to Kyoto protocol has been variously estimated to be such that the cost of coal will have to go to some $350 per metric ton carbon from about $30 in 1992. For the United States, the loss in real GDP just due to compliance with Kyoto protocol could be $2800 per household per year. It is anybody's guess whether the general population is prepared to absorb such costs. Reduction in such costs will require changes in lifestyle along with aggressive funding of R&D and technology transfer—especially to the developing countries. Because global warming and the greenhouse effect know no geographic boundaries, the rich countries will have to consider this as their duty to mother earth and not to specific countries.

Further, we need to advance both source conversion technologies and end-use technologies in commercial, industrial, and transportation areas. Because the life span of assets is rather long (30–50 years) in the former two areas but shorter (10 years or so) in the transportation area, it more likely that we will see more efficient transportation reducing the energy demand over the next decade. Special tax incentives are needed to make a dent in the other two areas.

In my area of research interest, industrial drying, there is potential to conserve up to 30% energy in most industries. Because 12%–25% of the national industrial energy consumption goes for some form of thermal dehydration, this is a significant amount that is achievable by industry–university interaction and proactive legislative support of appropriate R&D. It is also important to try and utilize renewable energy technologies to the maximum extent possible. A longer-term vision is needed to make a real difference, however.

Thank you very much.

A. Tomiura: Thank you for your excellent comments representing not only Canada, but also India. Mismatch among regulatory demand, social demand, and industrial demand is always a difficult issue. As for biased public awareness, I would like to discuss this later. How about Professor Taylor?

A. M. K. P. Taylor: First, I think the panelists did an excellent job in handling a difficult soft subject. They focused on education. And in Britain there are organizations similar to the Japan Society of Promotion of Science, called the Royal Society, the Royal Institution, and the British Association for Advancement of Science. And each of these three has set up funds for university staff to promote what is called "public understanding of science." And that involves lecturing. It involves a series of Christmas lectures delivered primarily for young people, which is not to trivialize the issue, but to promote the feeling that people should become scientists and deal with these problems. So I think that sort of promotion is very valuable, because we have to be frank and realize that people really do not want to become engineers anymore.

And there are many reasons for that. But if we do not produce high-quality engineers, then there is very little chance of providing the solution for stemming pollution.

That's the first set of comments I would like to make. The second set is that I think hijacking "Pokemon" for propaganda is a dangerous move. One should clearly separate Pokemon as being entertainment from the message of environmental protection. Because if we could take control of Pokemon for the environment, then perhaps other people can control Pokemon for something that we might not be so happy about, karaoke, for example.

The next point I would like to make is that education is a long-term solution. And there is no question that it is the best solution. But while Professor Assael's son is growing up, what do we do between now and then? I have to say that the global corporations with whom I am in contact in England, particularly Unilever, take their environmental responsibilities about wasting power very seriously. They consider themselves a "multilocal" corporation. Also in India, they take their environmental responsibilities as seriously as the government-legislated marketplace will allow them to do. I think that there are some corporations that are behaving very responsibly, beyond the economic need to do so. And there are others, and I think Professor Mujumdar will know that Union Carbide, as it was called—they changed their name now; they are called something else—were less than wonderful on that.

The final set of comments is that the ability to sit and agonize about the environment is a rich man's pastime. And I'm not quite sure what the Third World and Brazil and Indonesia and China will do. I suspect that they will do nothing, and that we will not in the rich part of the world, the G-7, contribute money to solving their pollution problems. It simply is a case of what Dr. Miller said: "We won't do it for ourselves [we won't stop smoking for ourselves]." It is impossible to believe that we will be generous to other countries. And there, I don't know how to solve the problem. That is really difficult, unless it is done at World Bank level, whereby the self-interest of selling advanced desulfurization plants or advanced PFBC or something like that to China is tied to loans to China. That's the only way. I don't see how else it's going happen. It's not.

And I don't see why the Chinese should be asked to stop. Nobody asked us when we cut down our forests in Europe. And if they had, I suspect that our answer would have been what we call in England "dusty." So I don't know how to solve that. And I think the First World will solve its pollution problems over the next 30 years. But the pollution problems from the Third World will be much bigger, much longer. And I don't see how to solve it.

Sorry to end on a nonoptimistic note.

A. Tomiura: It is not a pessimistic perspective, but an intuitively possible story. I am interested in the fund for lectures on public understanding of science in order to encourage young people to be scientists or engineers. I would very much like to touch on this issue later. Thank you very much. And our final commentator will be Professor Sloan.

E. D. Sloan: Well, it's a very complex problem. And it seems like there is some consensus at the table about what should be done. And I don't think there is reason for pessimism. But with your permission, I would like to turn away from the consensus about what should be done and talk a little bit about the question raised by our chairman, the second question. And his question is "In education how should we connect knowledge to wisdom?"—in particular, wisdom about what should be done with energy and the environment.

It's a Socratic question. And it certainly belongs to the culture of Professor Assael and Professor Englezos and Professor Taylor—if they were 3000 years old. And as a Socratic question, the first answer must be with Socrates. And that is we really don't know what either wisdom or knowledge is with absolute certainty, much less do we have absolute knowledge about how to convert knowledge to wisdom. However, as engineers are frequently asked to give answers in the absence of certainty, we give approximations.

So the first is this question of knowledge. When we think about knowledge, we all think about the tremendous influence of the computer on the traditional knowledge path. For example, when we once wished to know something, some of us turned to the library for information. However, a current poll shows that if a Berkeley engineering student cannot find information on a subject within 2h on the Internet, he or she abandons that search without going to the library. We all recognize the extreme danger of giving the Internet such absolute authority status, particularly when knowledge is available from the Internet with absolutely no guarantee of quality. This has led to the common saying about the computers, that student computer users are drowning in information or knowledge, but they are thirsting for wisdom.

We see, for example, the students' use of computers, and sometimes professors' use of computers, which are blindingly fast machines with computer codes that are hundreds of pages, virtually indecipherable. And the challenge for us in education is, "How do we teach students to stand up to a computer?" "How do we teach a student to say to a computer (as the most fundamental example) 'When you give me a default water density of 52lb. per cubic foot, that's off by 20%.'?"

When we look at the other side of the coin, at wisdom, I think it's also important to be prudent. For example, alumni returning to the university tell us that we should educate students in business skills, in communication skills, and people skills, etc. at a time when it is difficult enough to help young people find their way in a technical area of environment and energy during the early part of their lives. It's important not to ask too much of these young people, who already have so much to concern them. In other words, we cannot ask a 22-year-old to have the wisdom of a 40- or 50-year-old by the time they graduate.

It may be enough to consider a halfway point between knowledge and wisdom. Let me suggest that a halfway point between knowledge and wisdom is judgment. However, we must be very careful regarding what we mean by "judgment." Judgment is not opinion. But it is opinion that is fortified by evidence, and particularly experience-based evidence of knowledge. Judgment relies upon knowledge for the evidence to become something more than opinion. So knowledge forms a necessary base upon which judgment is founded. And at the far side of judgment, when a number of consistently good judgments have been made over a span of time, we say a person is wise. So the appeal is to provide methods for good judgment as a means of achieving wisdom later, in our case wisdom related to energy and the environment.

Methods for judgment must come from the older community to provide wisdom. In particular, they must incorporate the life experience of the older and the wiser people. We call those methods of judgment heuristics, or rules of thumb, which incorporate the evidence gained by life's experience. These heuristics are not absolutes, but just general rules for the basis of action. And with enough heuristics, one can generate the state of the art, or the paradigm. For example, when we think of an analog system to absorb cancer-producing chemicals into the body, we think of Joel Hildebrandt's rule "Like dissolves like," which leads us to iso-octanol to stimulate the body in dissolving cancer-producing chemicals. However, while this heuristic works for most chemicals, it doesn't work for all of them.

As another example, in the medical community doctors sometimes say as a heuristic for action in the emergency room, "When you hear hoofbeats, think horses, not zebras." This is a good heuristic outside of Africa, but it doesn't work very well in Kenya.

So in the end, as a proposal to connect knowledge with wisdom for environmental and energy problems, it is suggested that we consider heuristics as a road to a paradigm for the problem at hand. In the end, this leads to Professor Ohnishi's closing address at the Yagami campus yesterday, namely, that it is very important for us to have a basis for judgment which transcends culture to deal with worldwide environmental and energy problems in the hope of achieving wisdom and good solutions for the world culture.

A. Tomiura: Professor Sloan discussed the relationship among knowledge, wisdom, and judgment. With regard to judgment heuristics incorporated with experience, Professor Englezos pointed out project-based learning. Is there any connection between judgment heuristics and project-based learning?

E. D. Sloan: Certainly. Let's ask Professor Englezos if he would care to address that question.

P. Englezos: Heuristics would be a part of that, but certainly to address a project of environmental nature one has to rely on knowledge that you have up to this point or knowledge that you gain through addressing the problem. And that's the really important issue: that you learn by doing something. So you acquire knowledge by just doing this particular project. And of course a set of heuristics would be part of what you will learn. And that set of heuristics is a product of the distillation process that occurs in the teachers' minds. And it is transferred to the young students.

A. Tomiura: In my observation, there is little of this type of education in Japanese universities. Thank you for your clear explanation. Before going to ask for comments from Japanese participants, I would like to get comments from students. Miss Fujiwara, what kind of desire for education in the classroom do you have?

A. Fujiwara: First, for me and for the students in Keio University, we have many chances to have classes in which we hear advice and comments from many kinds of people from different fields. But those classes are not the regular classes. They are given by visiting lecturers. And at Keio, there are no regular classes that specialize in that. This summer I met a woman who works at a company that gives advice on environmental issues to other companies. She studied that subject in New Zealand and the United States, not in Japan. So maybe she didn't have a chance to study that in Japan. I would like to see that kind of curriculum in Japan in the university but starting in junior high school. We should think more about what kinds of subjects we should have for that specialty. And they should be in the right order. I heard a small story that a girl in junior high school asked her mother about energy produced from human products. And the mother couldn't explain it to her.

A. Tomiura: Mr. Ohmura, do you have any comment?

R. Ohmura: First, the students must be given the state of the art and an overview of the actual situation of the energy and environmental issues around the world. In such a sense, I think the lectures given in the course "Advanced Energy Conversion Engineering I"[11] were nice lectures. I think that such kinds of lectures are needed for the university students.

What I would like to suggest is that first the students learn the state of the art and an overview of the energy and environmental issues. After that, the students should seek some cultural or philosophical knowledge so that they will be able to select the direction in which they should steer their civilization. To make them have such knowledge, I think the students should be taught environmental ethics. The education of environmental ethics is probably established in the United States, but we rarely find courses of environmental ethics in universities in Japan.

For students in the field of science and technology such as mechanical and electrical engineering, I think lectures on life-cycle assessment should be also given.

A. Tomiura: Engineering ethics is a great topic in Japan. In the United States, a large book has already been published describing numerous case studies. In my observation, there is a difference in teaching engineering ethics in Japan and the United States. In the case of the United States, every professor discusses ethical issues associated with his regular lectures, while in the case of Japan, many professors in the engineering faculty are likely to rely on professors from the literature department to give lectures on ethics.

Thank you, Miss Fujiwara and Mr. Ohmura. Are there any comments from the Japanese participants?

M. Maeda: Professors would have to study things of the real world. As I studied at the Technical University of Aachen in Germany, but a long time ago, I learned students had to have practical experience in industries for more than 10 weeks, and even the full professor is still required to have experience in an industry related to the dissertation title.

I would say, stressing on that, that one should have an experience anyway before building up a concrete and effective concept in education. I am somewhat confused as to how to bring our concept of education to solving the pollution problem. Anyone who has no experience of being burnt by fire does not really know the danger. Most of the develop-

[11] One of the TEPCO chair courses taught in the Graduate School of Science and Technology, Keio University, 1995–1999.

ing countries do not really want to understand pollution because of the lure of getting energy in an easy way. We have already experienced injury by fire so that we should probably distribute the knowledge on our undesirable experiences to them.

As Dr. Celata mentioned, communication is easier now and effective. I agree, but in another side a devil is hiding there, because they look at just the happy side for enjoying a higher living standard by using energy without considering how it has been gotten. It's quite natural behavior for humans.

I think, then, it would be very important to transfer the true and effective information to them.

G. P. Celata: I would like to make a comment to Professor Maeda and Professor Taylor about the education of developing countries, to teach them that fire is dangerous. But I have a question. Why do we sell very old technology sometimes to developing countries like China? I was recently in Beijing, and you can see very, very old generation cars, Volkswagen or Mercedes cars, of three or four model changes ago. And they pollute much more than the most recent cars. And this is even worse than not giving education. We are selling very old technologies, because they are exhausted in our country, so we can no longer get money from them. But we can sell them to developing countries because we can get more money. So we don't seem to care about education in developing countries, or at least in some cases.

A. Tomiura: I think that your comment seems to be similar to that of Dr. Taylor's. I have some comments as to the desulfurization plant delivered to China and old Mercedes. We should understand that the environmental and energy concerns depend strongly on the cases such as nation, locality, economic conditions, and so forth. In the case of Mercedes, they have no alternative but to buy second-hand cars because of the economic reasons. In the case of desulfurization technologies, a professor of Keio University and chemical and civil engineering professors of other universities are now carrying out an interesting project cooperating with Chinese municipal authorities for developing an appropriate desulfurization system for a specific local area. What they are doing is very suggestive for solving local issues. In China, agglomerated coal is still used in individual homes. Because this agglomerated coal generates a lot of SO_x, the Japanese professors tried to produce unique briquettes by mixing pulverized lime with the coal. This briquette coal is extremely effective to reduce the generation of sulfur dioxide, and the ash after burning was spread in a desert area for neutralization of alkali soil, which is unsuitable for agriculture as well as the plantation of trees. This case is similar to the story of killing two birds with one stone. This type of technology, even though it seems old-fashioned, is very useful. We need not provide the most advanced technologies to developing countries. However, it's a very controversial story that high-level government officials always want to have the most modern technologies from advanced countries. Even so, we dare to recommend to them appropriate technologies depending on local conditions, national economics, and so on. This story is my real experience.

Are there any other Japanese participants who wish to make a comment? Mr. Ichihara, please.

Y. Ichihara: Let me talk about the issue of education from a different point of view, that is, social education about industry. In reality, today's world is supported by industry. All the sectors in the society, including government officials, lawyers, journalists, schoolteachers, farmers, etc., are dependent on industry, maybe indirectly, for their life. And even the developing countries are deriving benefit from the industry of industrially advanced countries through mutual trade. And the fact is that people are not informed about that kind of social relationship or interdependency. Environment problems are very complicated and controversial, but no doubt the existence of human beings themselves is the biggest source of damage to the environment. And modern urban life itself is a cause of environmental problems. Of course many industries are responsible for the direct effects of environmental hazards. Industrialists and engineers are responsible for that. But in a sense they are bearing their share of responsibility by necessity or demand of the society. The actual measure they use is to be questioned if it is implemented with the best use of knowledge or not. But the objectives that respond to social necessity should be understood by people who receive the resultant benefit, and responsibility for the deed should be shared by the whole society. Particularly, this kind of common base is necessary to discuss environment issues such as nuclear waste.

So far various efforts have been made to increase people's understanding on industry and environment from the industry's side. Various communications have been tried. For example, TEPCO is inviting many schoolteachers and students and ordinary citizens to our facilities to look at things for themselves, asking them to understand what the real industry is by their own observation. This tour has been successfully accepted by people, but it doesn't solve the whole issue of people's understanding of the industry. Although today's society depends on the industry, the industry does not get the appreciation it deserves. Rather, engineers are humiliated in today's consumer-oriented public cultural trend. They are just criticized in case of fault or accident, but they are not respected for their achievement or contribution to the society. As a result many young people hesitate to participate in the industry. All these phenomena come from the same root. Probably there is a lack of common base of understanding of industry-related things in this society. This issue has a close relationship with education in many respects, particularly with social education and elementary education. The industrial sector should feel its responsibility to establish its trust, and should do many things to improve the situation. But also here I would like to ask people of the academic society to be reminded of your responsibility.

Today's situation in industry reflects a failure of science and technological society in a very general sense, for example, a failure to build the appropriate education system and a failure in establishing the value of technology and science in the society. Technology and science themselves are still in progress, but in a general sense academia has failed to establish the respectful position of the community of people engaged in technology and science in this society, most of whom were educated by them. Those people working in the industry have lost the respect of the people. There must be some way to make people understand how they are themselves in today's industrially interwoven social structure. And I think it starts with understanding of basic logic and sequence of things. Young people don't know anything about how their family earns a living and what functions of society they depend on. This kind of understanding of social relationship should be shared by the society and by all the people. And in order to realize that, today's methodology to educate or transfer technology and science to people seems to have some defects.

Let me talk about the system of units for quantitative expression, for example. At the time of using "horsepower" as a unit, it reflected some idea of force people could understand. But at present, many units used in science and technology are very difficult for people to understand. When we talk about the incident related to radioactivity, people always have difficulty in understanding the issue. In the newspapers, we frequently find confusion in the use of units. For example, sometimes "micro" and "milli" are confused. That kind of funny but serious mistake happens easily. But there is no effort among academia to make an easier system of units for people to understand those things in a quantitative way. Academia seems to be content with today's situation as an isolated sector in the community. They don't seem to feel the necessity of new methodologies to transfer information and knowledge to the public, to cultivate and increase the wisdom of the human race as a whole. Simply, among academia there is a lack of responsibility to maintain social intelligence at the level it should keep. That is a cause of the whole issue. I think there should be some movement and proposals for industries and academia to collaborate worldwide on a new approach for social education.

A. Tomiura: Mr. Ichihara made a critical comment about academia and academics. I would like to come back again to this issue later. Now, Dr. Kalman, do you have any comments?

H. Kalman: I would like only to make a few comments on the things that were said, like the "wisdom" raised by Professor Sloan, and "touching of fire" by Professor Maeda. We have an old Hebrew saying from the ancient wise people, that goes in free translation as "A clever person can learn from his own experience, but a wise person will learn from the experience of others." This is the way that we should transfer information and make others wiser so that they will be able to learn from the experience of others.

I would like also maybe to comment on the pessimism of Professor Taylor, and also what others said before about international cooperation. I would like to do this by starting with a joke: It's about heaven and hell. The air in hell was very polluted, very smelly and not pleasant. But the people in heaven didn't care about that, because they had a very thick and long wall to separate them from hell. But after a time, when some holes started to form in the wall, and the bad smell from hell came to heaven, they started to argue about that. And the heaven people said, "Your bad smell is not good for us. We don't like it. Please fix the wall." And the people in hell said, "If it bothers you, fix it yourselves." They couldn't come to an agreement, so they went to court. And the hell people won the case, because the best lawyers were there.

But, if we leave the joke part out of it, the example of the SO_x or other pollutants coming from China to Japan, and some industries, as Professor Mujumdar said, that are closed in developed countries because of the environmental aspects and opened in developing countries, emphasize the case. All this will come back to us sometime, somewhere in the future.

The real solving of problems should be by cooperation between the developed and developing countries. And actually the cost of that should be borne by the developed countries, because they are the only ones that can bear the cost of these global problems. It shouldn't be a national or a local problem. This issue should be dealt with internationally.

A. Tomiura: Your joke is very understandable, unlike other Jewish jokes. When I heard your final comment regarding cooperation between developed and developing countries for solving environmental problems, it seemed that the preliminary sessions of the coming COP (The Conference of the Parties to the United Nations Framework Convention on

Climate Change) had already started. As for knowledge and wisdom, because I have worked for long years in industry, what I meant is very simple; knowledge is to know something, while wisdom is to do something based on knowledge. In my experience in industry, an engineer who has a lot of knowledge but does not know how to utilize his knowledge is merely a walking dictionary, but an engineer who always makes an effort how to apply his knowledge, how to get the best out of his knowledge for problem solving, is a man of wisdom. I believe that technology which started from an individual's empirical knowledge is not always universal. In other words, it does not work well under different conditions and locations where this empirical knowledge was not acquired. To increase the certainty of this empirical knowledge, we introduced many scientific principles and laws by applying repetitive experiments, and obtained the experimental knowledge or theoretical knowledge. Some of them finally reached a level of scientifically manifested knowledge. In the meantime, recently, as a result of the progress of computer technologies, we could develop artificially programmed knowledge that can be applied to heterogeneous and nonlinear systems like the automatic operation of large-scale and complex facilities or financial engineering. The next problem may be how we can integrate social and cultural knowledge into technology, because every industry has to take into account not only natural science-based knowledge but also social and cultural science for solving environmental and energy concerns and other difficult issues. What I want to emphasize is that in industry we need plenty of knowledge, but what is important is how to convert knowledge into wisdom for problem solving. Frankly speaking, it seems to me that universities give a lot of knowledge to students, but little wisdom. That's my concern, and in this context, I share the same thoughts as Mr. Ichihara.

Well, we had a lot of discussion, and so far I think that we put them into a few categories, if I can summarize them. The first is, how can we develop personal awareness or how can people be well informed on issues of energy and the environment? The second is harmonization of the environment and energy with the economy. In connection with personal awareness, I personally want to know if religious customs or practices are effective for improvement of education on energy and the environment, supplementing school education, because many Japanese are Buddhist, and, whenever we discuss this kind of issue, differences of religion between Japan and other nations come up. Dr. Celata said that his family is Catholic. We also have a participant who is a mother. I would like to ask Mrs. Kalman a question first, and then Dr. Celata. Now, Mrs. Kalman, please.

R. Kalman: I would like to make two comments, first as a mother. I'm aware of the influence that the TV has on children. I think the suggestion of Professor Assael was a good one. I support it.

The second comment is that the government in Japan decided about 11 years ago on a new system for garbage disposal, and all the Japanese people followed the instructions precisely. I think the credit can go to the government as much as to the Japanese people, who just went along with it and contribute to preserving the environment. If we would try to do the same in Israel, for instance, I'm not sure it would be exactly as it is in Japan. However, I will try to make my small contribution as a librarian at the Ben-Gurion University. I am working also as a curator of exhibitions. In March through April I am going to show an exhibition on the culture of Japan. Besides all the other beautiful things that I will show, I will also show the Japanese efforts to preserve the environment.

A. Tomiura: How do you observe the relationship between religious customs and education in your country? I really don't know the Jewish practice of disciplining children. I am curious as to whether your religion is strongly connected with your school education.

R. Kalman: Yes. The education in the Jewish culture is very important; it doesn't cover environmental preservation and energy problems.

A. Tomiura: Thank you very much. Dr. Celata, please.

G. P. Celata: I am of the same opinion. My daughter goes to a Catholic school. And they get a good knowledge of religion but not about the environment. This is what I already said, that next Saturday I will be at the school talking with the students. It's on my own initiative.

A. Tomiura: When I was in the United States I asked the same question of a friend who is a university professor. My question was whether Sunday school was effective for supplementing school education. His answer was very simple: "I don't think so, because our family doesn't go to church so often." I don't know whether it is a current fashion in the United States or not, but Sunday school is no longer supplementing school education. Is that correct, Professor Sloan?

E. D. Sloan: It certainly is true. I think people in any kind of religious denomination in the United States find themselves in the gross minority of the population. But, at the same time, you should say that the Sunday school education is better than if it were absent entirely.

A. Tomiura: I think so. Because Sunday school actually supplemented school education in my time.

I would like to move to the next topic. I am very interested in how the public can be well informed. Who can give a comment on this issue? Professor Englezos, maybe?

P. Englezos: I just want to reiterate things that were said before, but in my own words. One is school. And I think it's much more effective to do it at school. The other is family. And it's very nice, but there is a problem with family, because both parents work or in some cases one parent is not there due to divorce or other reasons. So we have to recognize this constraint. It's not my role to criticize but I have to take that into account. So I cannot rely too much on the family, simply because the time might not be there to do that. I think the school is where we have to focus and make children aware of energy and environmental issues, using examples such as that presented by Professor Watanabe. We should follow a simple approach that touches people's lives and enables important concepts to stick in children's minds so that they will know what energy is and they can associate it very easily with their activities.

A. Tomiura: In connection with a simple approach for letting people know, Dr. Miller mentioned the steps. Can you elaborate on your idea?

W. M. Miller: Yes, I did. I think the point I was trying to make was that it is simply not sufficient for us to be telling the public. In my case, I really feel that we have to also focus our efforts on the public at large, the adult population. But we have to recognize that communication is a very, very difficult matter. Telling somebody something is not the same as their having wisdom and understanding it. And that's why I raised the issue of people not necessarily adopting the best behavior in cases like smoking or taking drugs. How many times do we have initiatives to tell children not to take drugs? And yet they ignore us. So taking steps, you have to take them one at a time. And I think the first step might be some research to fully understand how best to communicate information and wisdom to people. This is the problem, and maybe it is a psychological problem, but imparting wisdom to people is what we are trying to do. And in terms of improving the environment, we cannot allow ourselves to be burnt twice. We have to use the adage "Learn from other people's mistakes." And that's what the steps in communication are trying to achieve.

A. Tomiura: So in this case, what is the relationship between education and family discipline?

W. M. Miller: I think we are getting on difficult ground with things like this. I fully agree that the family is the starting point for trying to instill social understanding and social behavior. I think there is a tendency in some busy families in the United Kingdom to see the school as having the role of teaching social discipline to children. By the time a child starts school, at the age of 4, it is far too late to *start* instilling this.

Nonetheless, schoolteachers should expect respect from the children, and should continue the good work begun at home.

A. Tomiura: In other words discipline in the home is now diminishing.

W. M. Miller: Yes. I think there was sort of an implicit suggestion in your earlier questions regarding religion that somehow there might be an improved sense of social discipline in terms of religion. And I would like to just suggest that is a very dangerous road to go down.

A. Tomiura: In this case, can you rely on schoolteachers in elementary school?

W. M. Miller: For discipline?

A. Tomiura: Yes. Do you expect schoolteachers to give pupils good disciplining?

W. M. Miller: I think we're getting off the subject. But in terms of discipline, no, because in the United Kingdom there is concern among teachers about disciplining children. For example, the worst case is a male teacher trying to discipline a female student. All sorts of problems can arise in the United Kingdom, and have. So there is a tendency for teachers to back off on any one-to-one discipline.

A. Tomiura: The same is true in Japan. Discipline in the home is now diminishing, and schoolteachers are nervous about giving discipline to pupils. It is a serious problem. Frankly speaking, a family cannot rely on descipline in elementary school.

W. M. Miller: No. It should be a minor element that is building on what is learned at home.

A. Tomiura: However, many parents are weekend fathers and mothers. How can we find a solution for this contradiction? Do you have some comments on this matter?

M. J. Assael: I think the answer is not a clear-cut one, but a complicated one involving many factors. I listened very carefully to what Mr. Ichihara said. He talked about schools visiting industry. He furthermore stated that industry is already doing a lot about environmental issues, more than we seem to appreciate although some academics don't seem to do much. I believe Mr. Ichihara is influenced by the activities of TEPCO. Unfortunately not all industries behave like this; I can name many companies that consciously pollute the rivers, for example, or the air.

So, I believe the answer is a collective one. It means that industry should do what you proposed. Industry should actually cooperate with teachers and schools, coorganize events and visits and help in making pupils aware by showing them the problems. I also believe that families and parents should help. In Greece we are very happy to still have close family ties and because of this we spend a lot of time with our children. So, we can influence their ideas.

Hence, the family should play an important role. Of course industry should play an important role in the way Mr. Ichihara said. But I think academics should also play an important role. In this respect, I would disagree with Mr. Ohmura. Mr. Ohmura said that he wants more courses taught on environment issues. I think the environment is not one course, it is everything and everywhere. That means that it is the duty of the academics in their lectures to inform students and make them aware of the environmental concerns. Special courses on the environment are not enough.

Also, I disagree that you must have courses on ethics. I think "ethics" is not something that you can learn through books and courses, although we teach ethics in Greece. I think "ethics" can only be shown to you with examples throughout your school and university years.

So in conclusion, I believe that the answer to the question posed involves all the aforementioned factors. Mr. Ichihara is absolutely right. The academics should do a better job. We should always do a better job. But also, industry should try to cooperate more with teachers and schools, and parents should talk more with their children. So it's everything together, I think.

A. Tomiura: Yes. I agree that environmental or engineering ethics is an integral part of education. However, in connection with the issue that the public must be better informed, mass media give the public the information in the form of sensational stories with a flashing caption. This kind of information sometimes misleads the public. Do you have any comment on this issue? Professor Mori, please.

Y. H. Mori: Yes. That is a point on which I would like to make a brief comment. I agree with the panelists in stressing the importance of education or transferring knowledge concerning the energy and environment to the public. But the problem is "How accurately can such knowledge be transferred to children or to the public?" Last week I watched a television program in which some well-known persons were interviewed about their opinions on energy problems. In that program, one of the former prime ministers of Japan gave his opinion. He stressed the necessity of nuclear energy, but he also gave a negative view about fusion energy. At the same time, he said that hydrogen energy could probably be an alternative to nuclear energy. It seems that he has never considered the energy to produce hydrogen from seawater. He ought to have received a good elementary education about energy transformation. And such misleading statements can be found in many places. I'm afraid that some sort of knowledge may be transferred, sometimes even intensionally, in such a way as to mislead the people.

In another television program I watched several months ago, some professor from a Japanese university explained to a newscaster on his new research results about the formation of methanol from methane. But he never talked about how much energy was required for such a transformation process. Then, the newscaster expressed her very positive impression on the study of methanol formation. So it is very important, I believe, that concerning the energy and environmental issues, the knowledge should be transferred to the public in a proper and honest way, not to mislead the people.

A. Tomiura: Your comment leads us to the issue of the responsibility of academics to better inform the public. In this connection, Mr. Sakurai, do you have any comment?

H. Sakurai: Yes. Public awareness is very important. And the public must be informed of science and technology and their consequences. However, even scientists have difficulty in communicating to the public effectively. The communication to the public is more effectively done by the mass media, religious leaders, and politicians. They have better channels to communicate to the public. However, those three groups of people are the least science and technology literate. One of my Japanese professor friends stated a similar thing. I completely agree with the opinion. Japan has a very difficult problem of how to get those public opinion leaders brought up to a minimum acceptable level of science literacy first. Next will be technology literacy, which overlaps to some extent with science literacy, but technology is something of a different animal.

I understand that the National Academy of Engineering of the United States has started a committee to study what level of technology literacy is needed for leadership and for the general public. They just started that committee early this year. And I think one of the ways this could be done is if we can promote science and technology literacy in the schools first, particularly technology education in grade and middle schools. There is no curriculum in Japanese grade and secondary schools for industrial technology—none at all. It's unbelievable from outside. Japanese science education in primary and secondary schools is ruled by pure scientists, and they teach basic introductory science in grade schools and in secondary schools. And they do not teach any industrial technology. They teach technology as a part of household maintenance. And they teach a little craftsmanship too, but they never teach industrial technology in Japanese primary and secondary schools. And that is the major cause of the lack of science and technology literacy and the knowledge on the relationship between science and technology. We need a lot of improvement in our education system. Many of the professors in leading universities or the leaders in industry, like Dr. Tomiura, are on advisory committees of the Ministry of Education. So they must speak up in those places to include necessary science and technology in the curricula of the grade schools. It may take generations to educate the mass media, politicians, and religious leaders. But we must start from the base.

A. Tomiura: I recall Professor Feynman. He wrote several good books on popular science. He also gave a nice lecture once a month or so and tried to educate the public on science. I have very interesting statistics developed by a Japanese professor. He analyzed scientific literacy based on the data obtained from statistics in OECD countries. Let us suppose the horizontal axis indicates mathematics and science marks in international examination, and the vertical axis is the scientific literacy of the public. All countries other than Japan have good correlation between the examination marks and scientific literacy. While Japanese school students obtained high marks in mathematics and science, there is extremely deplorable scientific literacy among the public. This analysis may be good evidence to supplement Mr. Sakurai's comment.

Professor Sloan, in the United States do university professors still try to make such kind of effort as Professor Feynman did in the past? Dr. Celata told us that he goes to schools and kindergartens, and gives lectures to pupils and kids about energy development. How about the situation in the United States?

E. D. Sloan: I couldn't really pretend to speak for the United States. And I am certainly ignorant about a lot of this, so I would hesitate to give an answer. But I would say that Professor Feynman, while he was a wonderful physicist, and had served to educate much of the world about physical phenomena, he was one of the least spiritual professors in the United States, and one of the most amoral. So, while he was a wonderful educator and could see things like molecular orbitals in frisbees flying across lunchrooms, and that sort of thing, he had almost no regard for the spiritual component of life, in my estimation.

E. K. Longmire: I'll make an additional comment, which is that at my university, which is a state university, a lot of what is called outreach is done in terms of interacting with the public. And I think that a lot of the faculty are involved in interaction with students in schools, starting in the elementary schools, and including up to high-school level. And I think that's a frequent occurrence.

A. Tomiura: Is it a public service?

E. K. Longmire: Well, I guess you could say that a large part of our position is supposed to be a public service.

A. Tomiura: More directly, without an honorarium?

E. K. Longmire: No honorarium is expected or given.

P. Englezos: I would like to continue along this theme and talk a little about our university. It's a publicly funded university. In addition to initiatives such as those described earlier, for example, faculty visiting high schools and elementary schools to talk about the environmental issues, the university has started another new initiative, that is, to give access to its programs to less-privileged individuals from the city of Vancouver. So you have people who could never imagine going to listen to a university lecture coming to the university and listening to lectures. I believe that institutions should have an increasing community involvement as our University does.

A. Tomiura: It is a great story. I can give you an example of the contribution to schools by industry. TEPCO as well as Nippon Steel cooperated with municipal elementary schools and junior high schools, and performed scientific demonstrations. They were not lectures, which always bore pupils, but just a play utilizing various experimental facilities that we provide to schools. Kids are excited and they run around from one experiment to another, because they have extremely few occasions to have experiments in school.

Anyway, time is short. Professor Ohnishi, do you have any comment?

K. Ohnishi: I would like to add some comments on education on energy and environmental problems. In the education, the conclusions sometimes look very simple, that is, industry is wrong, and they should be responsible. As pointed out in every chapter, the energy and environmental problems are very complicated. A simple solution does not exist. The solutions must include technological aspects, economical balance, and political decisions. But it is clear that the most important key for solution is technology. In this meaning, scientific literacy of the public is important.

In addition to literacy, we should notice that the future evaluation of the solution for the energy and environmental problem will involve ethics, which include the responsibility to our society. The solution will need a long time, much effort, and patience. To continue this kind of activity, we should evaluate the right solution highly. For example, it is easy to emphasize new energy. But the alcohol engine has a lot of problems. Solar cells have also many problems. There must be so many ways to solve these kinds of problems. Unless we evaluate each solution correctly, we cannot attract very good students in this kind of problems. Clever students will escape such problems, because responsibility is so high. That is one reason why many students would like to avoid engineering education. That is a tragedy.

I would like to say that the education should include correct evaluation of the solution for energy and environmental problems.

A. Tomiura: You raised significant issues for engineering education. Although I would like to elaborate a little bit more on this problem, the time is very limited. The organizing committee members are a little nervous about moving on to the next program, and I believe that hungry makes angry. Before the participants become angry, I would like to invite a few final comments. Professor Maeda, do you have any comments?

M. Maeda: That may sound as a joke, but I have a proposal to ask you, the wise men of Greece, to create a god a specialized in solving the pollution problem.

Once I met a French guy on my flight from Tokyo to Paris, he was a cartoon buff. He believed Hayao Miyazaki is a god of cartoons. And his cartoon created a god of forest. He came to believe the god of forest would exist, and talked very much on the conservation of the environment. He said he was changed so much. The effectiveness of education must be important.

A. Tomiura: Very good closing remarks. It is very difficult to summarize the contents of this panel discussion. First, we agreed that education, which starts from elementary school, accompanied by disciplining in the home, to university, and the development of public awareness are extremely important and indispensable premises for solving environmental and energy concerns. Second, in connection with educational issues, the responsibility and the role of academia, academics, and industry are emphasized. Third, technological innovation can be indispensable for a better tomorrow in terms of energy and the environment. With regard to this question, it is pointed out that developed countries have a responsibility to contribute to improve energy and environmental concerns in developing countries, and international cooperation on this issue is crucial.

I was somewhat astonished by the facts that many foreign guests know very well Japanese electronic games such as Pokemon and Tamagotchi, although I heard some critical comments on these games. When I have a chance to meet top management people of game companies, I shall be sure to request them to provide excellent and educational programs for energy conservation and environment improvement.

That's my summary of this panel discussion. Thank you for your contribution.

Y. Ichihara: I would like to ask you to add one more item. And this is the importance of the research and invention of the methodology for communication or education on science and technology.

A. Tomiura: That's very important. Thank you, Mr. Ichihara. At this moment, I would like to close this panel discussion. Thank you again for your wonderful contribution to the panel. Thank you very much.

Contributors to the Panel Discussion

Assael, Marc J., Aristotle University, Thessaloniki, Greece (Panelist)
Bose, Bimal K., University of Tennessee, Knoxville, TN, U.S.A. (Panelist)
Celata, Gian Piero, Institute of Thermal-Fluid Dynamics, ENEA Casaccia, Rome, Italy
Englezos, Peter, University of British Columbia, Vancouver, Canada (Panelist)
Fujiwara, Akiko, Keio University (student), Yokohama, Japan
Ichihara, Yoshiaki, Tokyo Electric Power Co., Tokyo, Japan
Kalman, Haim, Ben-Gurion University of the Negev, Beer-Sheva, Israel (Panelist)
Kalman, Ruthie, Ben-Gurion University of the Negev, Beer-Sheva, Israel
Longmire, Ellen K., University of Minnesota, Minneapolis, MN, U.S.A.
Maeda, Masanobu, Keio University, Yokohama, Japan
Miller, William M., QuantiSci Ltd., Melton Mowbray, Leicestershire, U.K. (Panelist)
Mori, Yasuhiko H., Keio University, Yokohama, Japan
Mujumdar, Arun S., National University of Singapore, Singapore
Ohmura, Ryo, Keio University (student), Yokohama, Japan
Ohnishi, Kouhei, Keio University, Yokohama, Japan
Park, Jong-Keun, Seoul National University, Seoul, Korea
Sakurai, Hiroshi, The Engineering Academy of Japan, Tokyo, Japan
Sloan, E. Dendy, Jr., Colorado School of Mines, Golden, CO, U.S.A
Taylor, Alexander M.K.P., Imperial College of Science, Technology and Medicine, London, U.K.
Tomiura, Azusa, Nippon Steel Corp., Tokyo, Japan (Chair)
Watanabe, Edson H., COPPE/Federal University of Rio de Janeiro, Rio de Janeiro, Brazil

Key Word Index

MIX
Papier aus verantwortungsvollen Quellen
Paper from responsible sources
FSC® C105338

If you have any concerns about our products,
you can contact us on
ProductSafety@springernature.com

In case Publisher is established outside the EU,
the EU authorized representative is:
Springer Nature Customer Service Center GmbH
Europaplatz 3, 69115 Heidelberg, Germany

Printed by Libri Plureos GmbH
in Hamburg, Germany